Testing of Communicating Systems

IFIP – The International Federation for Information Processing

IFIP was founded in 1960 under the auspices of UNESCO, following the First World Computer Congress held in Paris the previous year. An umbrella organization for societies working in information processing, IFIP's aim is two-fold: to support information processing within its member countries and to encourage technology transfer to developing nations. As its mission statement clearly states,

> IFIP's mission is to be the leading, truly international, apolitical organization which encourages and assists in the development, exploitation and application of information technology for the benefit of all people.

IFIP is a non-profitmaking organization, run almost solely by 2500 volunteers. It operates through a number of technical committees, which organize events and publications. IFIP's events range from an international congress to local seminars, but the most important are:

- the IFIP World Computer Congress, held every second year;
- open conferences;
- working conferences.

The flagship event is the IFIP World Computer Congress, at which both invited and contributed papers are presented. Contributed papers are rigorously refereed and the rejection rate is high.

As with the Congress, participation in the open conferences is open to all and papers may be invited or submitted. Again, submitted papers are stringently refereed.

The working conferences are structured differently. They are usually run by a working group and attendance is small and by invitation only. Their purpose is to create an atmosphere conducive to innovation and development. Refereeing is less rigorous and papers are subjected to extensive group discussion.

Publications arising from IFIP events vary. The papers presented at the IFIP World Computer Congress and at open conferences are published as conference proceedings, while the results of the working conferences are often published as collections of selected and edited papers.

Any national society whose primary activity is in information may apply to become a full member of IFIP, although full membership is restricted to one society per country. Full members are entitled to vote at the annual General Assembly, National societies preferring a less committed involvement may apply for associate or corresponding membership. Associate members enjoy the same benefits as full members, but without voting rights. Corresponding members are not represented in IFIP bodies. Affiliated membership is open to non-national societies, and individual and honorary membership schemes are also offered.

Testing of Communicating Systems

Volume 10

IFIP TC6 10th International Workshop on Testing of Communicating Systems, 8–10 September 1997, Cheju Island, Korea

Edited by

Myungchul Kim

Sungwon Kang

and

Keesoo Hong

Korea Telecom
Seoul, Korea

SPRINGER-SCIENCE+BUSINESS MEDIA, B.V.

First edition 1997

Originally published by Chapman & Hall in 1997
MyCopy version of the original edition 1997

DOI 10.1007/978-0-387-35198-8

A catalogue record for this book is available from the British Library

∞ Printed on permanent acid-free text paper, manufactured in accordance with ANSI/NISO Z39.48-1992 and ANSI/NISO Z39.48-1984 (Permanence of Paper).
www.springer.com/mycopy

CONTENTS

Preface viii
Committee Members and Reviewers ix

PART ONE Past and Future of Protocol Testing

Invited Talk

1 Future directions for protocol testing, learning the lessons from the past
D. Rayner 3

PART TWO TTCN Extensions

2 PerfTTCN, a TTCN language extension for performance testing
I. Schieferdecker, B. Stepien and A. Rennoch 21

3 Real-time TTCN for testing real-time and multimedia systems
T. Walter and J. Grabowski 37

PART THREE Wireless Testing

4 Integration of test procedures and trials for DECT handsets
A. Alonistioti, P. Kostarakis, K. Dangakis, A.A. Alexandridis, A. Paschalis, N. Gaitanis, S. Xiroutsikos, E. Adilinis, A. Vlahakis and P. Katrivanos 57

5 Development of an ETSI standard for Phase-2 GSM/DCS mobile terminal conformance testing
S. Hu, J.-C. Wattelet and N. Wang 65

PART FOUR Data Part Test Generation

6 Automatic executable test case generation for extended finite state machine protocols
C. Bourhfir, R. Dssouli, E. Aboulhamid and N. Rico 75

7 A method to derive a single-EFSM from communicating multi-EFSM for data part testing
I. Hwang, T. Kim, M. Jang, J. Lee, H. Oh and M. Kim 91

PART FIVE Test Coverage and Testability

8 Basing test coverage on a formalization of test hypotheses
O. Charles and R. Groz 109

9 Design for testability: a step-wise approach to protocol testing
H. König, A. Ulrich and M. Heiner 125

PART SIX Theory and Practice of Protocol Testing

Invited Talk

10 Developments in testing transition systems
E. Brinksma, L. Heerink and J. Tretmans 143

11 Checking experiments with labeled transition systems for trace equivalence
Q.M. Tan, A. Petrenko and G. v. Bochmann 167

12 An approach to dynamic protocol testing
S. Yoo, M. Kim and D. Kang 183

13 Sensitivity analysis of the metric based test selection
J.A. Curgus, S.T. Vuong and J. Zhu 200

14 Analyzing performance bottlenecks in protocols based on finite state specifications
S. Zhang and S.T. Chanson 220

PART SEVEN Test Generation for Communicating State Machine

15 A conformance testing for communication protocols modeled as a set of DFSMs with common inputs
A. Fukada, T. Kaji, T. Higashino, K. Taniguchi and M. Mori 239

16 On test case generation from asynchronously communicating state machines
O. Henniger 255

17 Fault detection in embedded components
A. Petrenko and N. Yevtushenko 272

18 A pragmatic approach to generating test sequences for embedded systems
L.P. Lima Jr. and A.R. Cavalli 288

PART EIGHT Tools and Environments

19 The European initiative for the development of infrastructural tools: the INTOOL programme
P. Cousin 311

20 HARPO: testing tools development
E. Algaba, M. Monedero, E. Pérez and O. Valcárcel 318

21 Application of a TTCN based conformance test environment on the Internet email protocol
J. Bi and J. Wu 324

22 The INTOOL/CATG European project: development of an industrial tool in the field of computer aided test generation
E. Desécures, L. Boullier and B. Péquignot 330

PART NINE Applications of Protocol Testing

Invited Talk

23 Modeling and testing of protocol systems
D. Lee and D. Su 339

24 A pragmatic approach to test generation
E. Pérez, E. Algaba and M. Monedero 365

25 Towards abstract test methods for relay system testing
J. Bi and J. Wu 381

26 Applying SaMsTaG to the B-ISDN protocol SSCOP
J. Grabowski, R. Scheurer, Z.R. Dai and D. Hogrefe 397

27 Design of protocol monitor emulating behaviors of TCP/IP protocols
T. Kato, T. Ogishi, A. Idoue and K. Suzuki 416

28 A two-level approach to automated conformance testing of VHDL designs
J. Moonen, J. Romijn, O. Sies, J. Springintveld, L. Feijs and R. Koymans 432

Index of contributors 449

Keyword index 451

PREFACE

The IFIP International Workshop on Testing of Communicating Systems (IWTCS'97) is being held in Cheju Island, Korea, from 8 to 10 September, 1997. IWTCS'97, continuing the IFIP International Workshop on Protocol Test Systems (IWPTS), is the tenth of a series of the annual meetings sponsored by the IFIP Working Group 6.1. The nine previous workshops were held in Vancouver (Canada, 1988), Berlin (Germany, 1989), Mclean (USA, 1990), Leidschendam (The Netherlands, 1991), Montréal (Canada, 1992), Pau (France, 1993), Tokyo (Japan, 1994), Evry (France, 1995) and Darmstadt (Germany, 1996).

As in the years before, the workshop aims at bringing researchers and practitioners, promoting the exchange of views, and correlating the work of both sides. Forty seven papers have been submitted to IWTCS'97 and all of them have been reviewed by the members of the Program Committee and additional reviewers. Based on these reviews, the workshop consists of 19 regular and 6 short reviewed papers, 3 invited papers, all of which are reproduced in this volume, as well as a panel discussion and tool demonstrations.

IWTCS'97 was organized under the auspices of IFIP WG 6.1 by Korea Telecom. It was financially supported by Korea Telecom and Commission of the European Communities.

We would like to express our thanks to everyone who has contributed to the success of the conference. In particular, we are grateful to the authors for writing and presenting their papers, the reviewers for assessing and commenting on these papers, the members of the program committee, Samuel Chanson and Bernd Baumgarten, who shared their experience in organizing the workshop with us, and all the cooperative people participating in the local organization.

Myungchul Kim
Sungwon Kang
Keesoo Hong

Cheju Island, Korea, September 1997

CONFERENCE CO-CHAIRS

Myungchul Kim, Sungwon Kang, and Keesoo Hong
Korea Telecom Research & Development Group
Seocho Ku, Seoul, 137-792, Korea
{mckim, kangsw}@sava.kotel.co.kr

PROGRAM COMMITTEE

B. Baumgarten (GMD, Germany)
G.v. Bochmann (Univ. of Montréal, Canada)
E. Brinksma (Univ. of Twente, The Netherlands)
R. Castanet (Univ. of Bordeaux, France)
A.R. Cavalli (INT, France)
S.T. Chanson (Hong Kong Univ. of Science & Technology, Hong Kong)
B. Chin (ETRI, Korea)
Y. Choi (Seoul National Univ., Korea)
W. Chun (Chungnam National Univ., Korea)
A.T. Dahbura (Digital Equipment Corp., USA)
R. Dssouli (Univ. of Montréal, Canada)
J.-P. Favreau (NIST, USA)
R. Groz (CNET, France)
T. Higashino (Osaka Univ., Japan)
D. Hogrefe (Univ. of Luebeck, Germany)
G.J. Holzmann (Bell Laboratories, USA)
S. Kim (Pukyung National Univ., Korea)
H. König (Technical Univ. of Cottbus, Germany)
E. Kwast (KPN Research, The Netherlands)
J. Lee (Yonsei Univ., Korea)
M.T. Liu (Ohio State Univ., USA)
C.S. Lu (Telecom Lab., Taiwan)
J. de Meer (GMD, Germany)
T. Mizuno (Shizuoka Univ., Japan)
K. Parker (Telstra Research Labs., Australia)
A. Petrenko (CRIM, Canada)
O. Rafiq (Univ. of Pau, France)
B. Sarikaya (Univ. of Aizu, Japan)
K. Tarnay (KFKI, Hungary)
J. Tretmans (Univ. of Twente, The Netherlands)
H. Ural (Univ. of Ottawa, Canada)
S.T. Vuong (Univ. of British Columbia, Canada)
J. Wu (Tsinghua Univ., China)
N. Yevtushenko (Tomsk State Univ., Russia)

FURTHER REVIEWERS

O. Charles
G. Csopaki
P.R. D'argenio
S. Dibuz
A. En-Nouaary
S. Fischer
J. Grabowski
J. He
S. Heymer
M. Higuchi
D. Hristov
K. Karoui
B. Koch
L.P. Lima, Jr.
P. Maigron
A. Rennoch
M. Salem
F. Sato
M. Schmitt
J.-F. Suel
M. Tabourier
M. Toma
M. Törö
A. Ulrich
T. Vassiliou-Gioles
C.H. Wu
H. Xuan
Y. Yang
J. Zhu

ORGANIZATION COMMITTEE

Kwangjin Lee (Chair), Youngsoo Seo, and Mijeong Hong
Korea Telecom Research & Development Group
Seocho Ku, Seoul, 137-792, Korea
{sys01,lhdo}@sava.kotel.co.kr

Electronics and Telecommunications Research Institute (ETRI)
Hanwha Corporation - Telecom
Korean Institute of Communication Sciences (KICS)
Korea Information Science Society (KISS), SIG on Information Network
National Computerization Agency (NCA)
Open Systems Interconnection Association (OSIA)
Shinsegi Telecom
SK Telecom

PART ONE

Past and Future of Protocol Testing

1

Future directions for protocol testing, learning the lessons from the past

Dr D. Rayner
National Physical Laboratory
Teddington, Middlesex, UK, TW11 0LW, phone: +44 181 943 7040, fax: +44 181 977 7091, e-mail: Dave.Rayner@npl.co.uk

Abstract

A reflection on the history of all aspects of protocol testing reveals both successes and failures, although many observers take a purely negative view. Important lessons need to be learned if protocol testing in the future is to be regarded in a more positive light. There are now new signs of hope in some the most recent developments. Drawing upon the lessons from the past and the current encouraging developments, this paper sets out to recommend the directions that should be taken in the future if the subject is to recover from its slump in popularity and its good prospects are to be realised.

Keywords

Protocol testing, conformance, interoperability, cost-effective testing

Testing of Communicating Systems, M. Kim, S. Kang & K. Hong (Eds)
Published by Chapman & Hall © 1997 IFIP

1 INTRODUCTION

After over 17 years in the protocol testing business, there is a lot to reflect back upon. By learning the lessons from the history of our subject we may more confidently predict the way which ought to be followed in the future. This paper sets out to present some of the more important lessons from the past, to take stock of where we are, and to point the way forward, more out of pragmatism than idealism.

2 STANDARDISED CONFORMANCE TESTING

Lessons from the past

The concept of standardised protocol conformance testing emerged out of the strong collective desire for Open Systems Interconnection (OSI) in the early 1980s. It was widely assumed that there would be an enormous market for OSI products from many different vendors and that the only way confidence could be given in the interoperability of these products would be to give an assurance that they met all the requirements of the relevant protocol standards. This meant that conformance testing was considered necessary for the success of OSI and it was assumed that such testing would be required by public procurement agencies.

Given this motivation, the OSI conformance testing methodology and framework standard was developed as a five-part standard published in 1991 and 1992, with second editions and two extra parts being published three years later (ISO/IEC 9646, 1994 and 1995). This standard was a tremendous achievement, the result of a consensus painstakingly developed among representatives of testing laboratories, major suppliers, network operators, researchers, and consultants representing users. At the beginning in 1983 there was considerable mutual mistrust between the suppliers and testing laboratories, but gradually each came to understand the other's point of view and they began to work together constructively towards the common goal: an agreed methodology and framework for objective, standardised, protocol conformance testing. The result was the most comprehensive testing methodology standard ever produced in any area of information technology, published not only by ISO/IEC but also by ITU-T and CEN.

Unfortunately, this methodology was developed as an afterthought in the whole OSI enterprise, and its developers were seen as a separate group of testing specialists unconnected with the protocol specifiers. Hence, the testing specialists were forced to try to apply their testing methodology to protocols that had been developed with no thought for testability. Guidance given on how to produce more testable protocols was largely ignored. Furthermore, those who ended up developing the standardised test suites and related standards were separate from those who had developed each of the protocols. Thus, there was a lack of real understanding of the protocols which meant that the test designers were unable to make informed decisions to improve the cost-effectiveness of the test suites; instead they tended to apply the methodology in a rather mechanical unimaginative, even unintelligent way. This was compounded by the fact that there was no feedback loop - no possibility of the work of the test designers feeding back into improvements to the protocol designs. With hindsight, we can see that it was almost inevitable that the test suites would be too large, too expensive and insufficiently focused on real market needs - in short, not at all cost-effective.

Standardised conformance testing was thought to be applicable uniformly to all protocols which fitted within the OSI seven-layer architecture, apart from those protocols in the Physical layer which concern the signals relevant to a particular physical medium (e.g. electrical, optical or radio signals). Little thought was given to the question of when to standardise and when not to, nor to the question of when the coverage of the test suite needed to be comprehensive and when it could be rather more superficial or patchy. Furthermore, if it was suggested that the methodology should be applied to a non-OSI protocol, the tendency was to adapt the protocol to OSI rather than to adapt the methodology to non-OSI protocols.

Current situation

Although a lot of conformance test suites have been produced, relatively few in the voluntary sphere (outside telecoms) have been standardised internationally, and even fewer are being used or properly maintained. In contrast, in the telecoms field, especially in the regulatory sphere, standardised test suites are being produced, used and maintained, particularly by ETSI.

There is a move now towards more flexible application of the methodology. It is being applied with appropriately focused coverage to suppliers' needs for development testing. It is being adapted to apply to non-OSI protocols. It is being enhanced to meet the more stringent requirements of testing security protocols, using what is called "strict conformance testing" (see Barker, 1997).

The way forward

There is much to value and hold on to in the conformance testing methodology and framework, but it must be applied flexibly and with appropriate adaptation to meet the market needs in each case in the most cost-effective way. Sometimes there will still be a requirement for standardisation of test suites, e.g. to meet regulatory requirements or the needs of high-risk areas like security or safety-critical software, but often standardisation will be unnecessary. Coverage should be chosen to be appropriate to match the risks associated with a failure to conform to the requirements of the protocol standard. To achieve this protocol designers and/or product implementors must be involved in the test design process.

There should also be provision for feedback from the test design process into improved protocol design, but more of this in the next section.

3 PROTOCOL DESIGN AND TESTABILITY

Lessons from the past

A major contributory factor in the failure of OSI was the fact that most of the protocols were too complex, with too many options, too hard to implement, and too hard to test. Profiles were invented to try to overcome the complexity by making appropriate consistent selections of options from a stack of protocols. Hence, profiles were intended to solve the problem caused by the failure to design the protocols properly in the first place. Thus, this was tackling the symptom rather than the cause of the problem. Unfortunately, it made matters considerably worse, because there were too many profiles, each with still too many options, and the consequential protocol profile testing methodology (ISO/IEC 9646-6, 1994) was far too complicated and expensive to operate.

Since the early 1980s it has been said by some that there should be a formal description for each protocol standard. This could then be used to validate the protocol and would provide the basis for automated test generation. The reality was rather different. Firstly, there were too many different formal description techniques, whose advocates seemed to spend more time attacking the others than promoting the use of formal techniques. Secondly, the vast majority of people involved in protocol development could not understand the formal description techniques and would not trust the few who did understand them. Thirdly, there was inadequate tool support for the techniques. Fourthly, such formal descriptions as were developed were produced as an afterthought by academics, without adequate contact with the original protocol specifiers, usually without the full functionality of the protocol, and with no intention of maintaining the specification in line with the standard.

Current situation

It is now accepted that we must learn from the success of Internet, in which protocols are kept simple with very few options, and are implemented before being finalised, to demonstrate ease of implementation and understandability of the specification.

Profiles are now largely seen as irrelevant. Where a stack of protocols is to be tested together, it now seems more appropriate to use interoperability testing rather than profile conformance testing with all its complexity and cost.

Some significant progress has been made in the acceptance of formal specifications. Firstly, the OSI Distributed Transaction Processing protocol was standardised in 1992 and revised in 1996 (ISO/IEC 10026-3, 1996) including two annexes giving informative formal descriptions in LOTOS and Estelle. Although they were only informative these specifications were authoritative and accepted by the protocol defining group as being complete descriptions of the protocol. There was even clause by clause parallelism between the text and the LOTOS specification. Unfortunately, this protocol failed to be widely accepted by the market, and soon afterwards both LOTOS and Estelle fell largely into disuse in the protocol standardisation community in ISO/IEC JTC1.

Secondly, SDL (ITU-T Z.100, 1994, and ITU-T Z.105, 1994) has now become the dominant formal description technique within the protocol standardisation community, increasingly used in ETSI and ITU-T. A breakthrough came when an ETSI committee helped ITU-T develop the text and SDL specifications in parallel for INAP (Intelligent Network Application Protocol) CS-2, as specified in ITU-T Q.1224 (1997) and ITU-T Q.1228 (1997). This resulted in the formal specification being published by ITU-T as a normative annex to the standard, although in cases of conflict the text takes precedence over the SDL. Tools were used to perform protocol validation at various stages during the development. The result was faster, cheaper and better quality protocol development.

The way forward

What is needed is to build upon the INAP CS-2 experience, leading to the parallel development of text and SDL becoming the norm, with the SDL becoming the normative specification, with much of the text just having the status of informative commentary on the SDL. There will, however, probably always be a need for some normative text to express those requirements that are hard to express in SDL and to avoid the SDL having to become too detailed and complex. The SDL specification should then become the basis for validation, animation and automated test generation. It could also, if appropriate, be the basis for production of a trial or reference implementation, but experience shows that such an implementation is likely to require some non-trivial hand-coding and therefore even an implementation derived from the SDL is likely to need testing.

Moreover, the protocol design process should aim to minimise complexity, building in testability, focusing requirements on what is really needed to achieve interoperability, and including trial implementation before finalisation of the specification.

In order to achieve these objectives, it is important that future protocol design groups should include all the expertise necessary to do the job properly. This includes expertise in protocol design, the intended field of use, testability, testing methodology, formal specification using SDL, other supporting techniques (e.g. state tables, ASN.1, message sequence charts), and use of relevant software tools. It should be stressed that protocol designers can now invest with confidence in the training necessary to become knowledgeable in all the key techniques (testing methodology, TTCN, SDL, ASN.1, MSCs) because of their maturity and stability.

4 TEST SUITE DESIGN

4.1 Test purposes and test suite coverage

Lessons from the past

Perhaps the biggest problem with ISO/IEC 9646 has been its guidance on the development of test suite structures and test purposes. The importance of test purposes is not in doubt, for they provide a simple, easy to understand description of what a test case is to achieve. They provide the appropriate level at which to discuss test suite coverage. They also facilitate an understanding of the verdicts produced when the derived test cases are run. The problem is that if the guidance given in ISO/IEC 9646-2 (1994) is followed mechanically, without any consideration for which test purposes are likely to be most effective, then it is all too easy to produce a test suite which is much too large, with many test purposes that are frankly irrelevant to achieving interoperable products.

Current situation

Test purposes are now playing a vital role in automated test generation. It is now increasingly accepted that automated test generation should not simply go directly from a formal specification of the protocol to the output of a test suite. Instead automation can be used in two stages, either separately or in combination. The first stage is the production of the test purposes from the formal specification, performed with appropriate parameterisation and interaction with the test designer. The test designer can then review the test purposes and make adjustments as necessary, possibly altering the parameterisation and going through the process several times until an acceptable set of test purposes is produced. Test purposes output from this process could either be written in stylised natural language or in Message Sequence Charts (MSCs) as specified in ITU-T Z.120 (1993). They could be designed for standardisation or for development testing of a specific product.

The second stage is to use test purposes as the input to test case generation, ensuring that only the desired test cases are generated. For this process the test purposes need to be expressed in a suitable formalism; MSCs seem to be a natural and increasingly popular choice. The test purposes for this second stage may have been generated using an automated tool or may have been hand-written.

The way forward

Much more guidance needs to be produced on adopting flexible approaches to deciding on test suite coverage. There needs to be informed analysis of where the practical problems and threats to interoperability lie in implementing a particular protocol and what therefore most needs to be tested. It needs to be decided what the overall objective of the test suite is and the coverage needs to be appropriate to achieving that objective.

In some cases, a very small number of test cases will suffice, each perhaps with a high-level test purpose requiring a long sequence of state/event pairs to be exercised. In such cases, they are probably best identified by product developers getting together to share their knowledge of where the problems lie in implementation, just as they do to produce interoperability test suites for EuroSInet.

In other cases, a rather larger number of test cases may be needed, but perhaps skewed to focus on the most important features of the protocol. Only in a minority of cases should it be expected that large test suites with very even coverage of rather narrowly focused, atomic test purposes would be appropriate.

The development of coverage metrics could be useful, provided that they are used to inform the test designer rather than dictate what coverage must be used.

Automated test purpose generation will be useful in those cases where large test suites are still required, or where in development testing it is necessary to continually change the tests to focus on different problems in the implementation being tested; but for the development of very small test suites that are not continually changing such tools are probably unnecessary. Automated test case generation from test purposes should soon progress to the point where the only serious limitation to its use will come from the availability of authoritative SDL specifications, but before long even the lack of an SDL specification should not prove to be a limitation because practical automated test generation techniques will be applied to the extended finite state descriptions that are common in almost all protocol specifications. For input to such tools, we can expect that MSCs should become the dominant formalism for expressing test purposes.

4.2 Test cases

Lessons from the past

An abstract test case is a detailed implementation of a test purpose using a particular test method and expressed in a test specification language. The test method will be chosen from the set of test methods defined in ISO/IEC 9646-2 (1994) which has stood the test of time.

Abstract test cases are invariably written in TTCN (the Tree and Tabular Combined Notation, ISO/IEC 9646-3, 1992). Although TTCN is clearly the product of committee compromise, containing various arbitrary restrictions and inconsistencies, it has proved itself to be a widely applicable test specification language. There has been considerable investment in the development of TTCN tools and, although there were serious problems in the past with the quality of the tools, these problems have now been overcome.

The finalisation of the second edition of TTCN has been a long drawn-out process, caused primarily by the fact that all the members of the ISO/IEC editing group lost their funding for the work. This shows up a serious weakness in the voluntary standardisation process: if all the key people lose their support for the work, then any standards project can be brought to a halt. The risk of this happening is increased by the long timescales on which international standardisation operates.

Automated test generation has been criticised as being impractical in the past, not only for leading to inappropriate test suite coverage, but also for producing rather incomplete test cases which require a lot of manual editing.

Current situation

The second edition of TTCN is now technically complete and has undergone a process of validation by several European TTCN tool suppliers and others in the TTCN community in ETSI. A review copy (ISO/IEC 9646-3, 1997) was sent in early May to ISO/IEC JTC1/SC21 for national body and liaison organisation review. In parallel, the editor, Os Monkewich, is performing the final editorial work, including production of contents list and indexes. The main new features of the second edition are the use of ASN.1 94 (ISO/IEC 8824-1, 1994), concurrency, encoding operations, formalised test suite operations, modularity, and active defaults. Concurrency is useful not only for multi-party test cases, but also to provide structuring for multi-protocol or embedded test cases by using a different test component for each protocol used in the test case.

There is now a healthy competitive market in TTCN tools, especially in Europe, with most tools now supporting the main features of the second edition. The tools include editors, error checkers, compilers, interpreters, test case generators, test case validators, and behaviour simulators.

For some time, EWOS and ETSI have produced guidance documents related to TTCN, especially the TTCN style guide (ETG 25, 1994). ETSI is now involved in producing guides on the use of the second edition of TTCN and on the use of TTCN with ASN.1 94. Amongst this guidance there is expected to be guidance on relaxing the definition of a point of control and observation (PCO) to allow for communication with the test system if necessary, which is allowed by TTCN but not by ISO/IEC 9646-2 (1994).

The way forward

There now needs to be a period of consolidation based upon the complete second edition of TTCN. All the TTCN tools need to be brought into line with it. The ETSI guidance documents related to the second edition need to be completed and given widespread distribution. An effective maintenance process needs to be established for TTCN, given that there is nobody left in ISO/IEC JTC1/SC21 to maintain the standard after it is published.

It is clear that to be successful, automated test generation must produce good quality complete TTCN test cases, including declarations, constraints, preambles and postambles, as well as the test bodies.

5 TEST SYSTEMS

Lessons from the past

The requirements of ISO/IEC 9646-4 (1994) regarding test systems largely concern the information to be stored in the conformance log. These requirements have proved to be valuable in giving a common basis of objective information on which to base a test report. If statements in a test report are questioned, they can usually be substantiated by looking at the relevant part of the conformance log.

The main problem with test systems has been the difficulty and cost of mapping the TTCN abstract test cases into the appropriate executable test cases. The problem is especially bad in those test systems which use points of control and observation different from those used in the abstract test cases.

Current situation

There is now a trend towards support for TTCN in test systems, both in providing a TTCN compiler or interpreter and in providing results analysis directly related to the TTCN. The European Commission's INTOOL projects have recently delivered a set of interface specifications which when implemented will enable tools from different suppliers to be used in combination. Perhaps the most important of these is the GCI interface (generic TTCN compiler/interpreter interface) which enables a single TTCN compiler or interpreter to be used with multiple test systems.

Another recent advance is the publication by ETSI of the TSP1 (test synchronisation protocol 1) specification (ETR 303, 1997). TSP1 should be usable as a general mechanism for implementing test coordination procedures.

The way forward

There seems to be no need to modify the requirements on conformance logs, but online interpretation of the logs in terms of the TTCN should become the norm. Test system development should concentrate on implementing the GCI interface in order to allow a TTCN front-end to be provided to each test system. This will avoid the need to spend a lot of effort translating TTCN into executable test languages and in maintaining the alignment between abstract and executable test suites. TTCN compilers need to be further developed to support more fully the second edition of TTCN. Care should be taken to ensure that they make full use of PICS (protocol implementation conformance statement) and PIXIT (protocol implementation extra information for testing) information, to minimise the need for interaction with the test engineer during compilation.

Further work should be done to investigate the practicality of using ETSI's TSP1 (ETR 303, 1997), the interface specifications from the Open Testing Environment INTOOL project, and any other potentially useful APIs (application programming interfaces) to improve flexibility and cost-effectiveness in the use of test tools.

6 TESTING SERVICES

6.1 Conformance testing

Lessons from the past

The whole conformance assessment process specified in ISO/IEC 9646-5 (1994) has proved to be very robust. It is widely accepted as providing the definitive basis on which objective conformance testing services should be run, not least by accreditation bodies in their assessment of testing laboratories offering protocol testing services.

The interpretation of accreditation requirements in information and communication technologies (ICT) has needed internationally accepted guidance. This is because the traditional concepts of calibration and traceability which work well in areas involving physical measurement are not applicable to software and protocol testing in ICT. In their place the concept of validation of test tools had to be developed. Furthermore, guidance was also necessary to help determine which types of testing were sufficiently objective to be accreditable. A key idea in this was that the results should be repeatable at the one testing laboratory and reproducible by other testing laboratories.

Current situation

The interpretation of accreditation requirements in the ICT field has been agreed and published as ISO/IEC TR 13233 (1995). Its forerunner in Europe, ELA-G5 (1993), ought to be updated to align it to ISO/IEC TR 13233, but this is not a high priority at present for either the accreditation bodies or the European Commission.

For several years accreditation of protocol testing services was on the increase, coupled in Europe with the growth of agreement groups under ECITC (the European Committee for IT Testing and Certification). Now, however, because of the lack of a market demand for third party protocol conformance testing services in the voluntary sphere, the number of accredited laboratories has declined and ECITC agreement groups are disappearing. What are left are mostly testing laboratories operating in the telecommunications area, primarily in the regulatory sphere, plus a few offering testing services aimed at the US government procurement market.

The way forward

The problems of how to accredit protocol testing services are now solved, but in future such accreditation will be applied primarily in the telecommunications field. In addition, we can forecast a growth in security testing services, including the testing of security protocols. Given that the risks of non-conformity are much higher in the security field than in most other areas of ICT, and given that increased used of testing within security evaluation can reduce the costs and timescales to meet more closely the needs of industry, we can foresee the need for accredited "strict conformance testing" services. There is likely also to be a growth in safety-critical software testing, for similar reasons, but the protocol testing component of this is much less well developed. Thus, further research and development is needed in this area.

6.2 Interoperability testing

Lessons from the past

There have been two main approaches developed and applied to the provision of interoperability testing, but there is no standardised methodology. One approach was developed by SPAG (the Standards Promotion and Application Group), called PSI (process for system interoperability). In this approach, interoperability testing was an additional step that followed thorough conformance testing. The interoperability testing was based upon an SIS (system interoperability statement), which played a similar role to the PICS for conformance testing. It was conducted in a thorough and objective manner. The whole scheme was regarded by many as effectively an expensive certification scheme which was a long way from meeting market needs. As a result it never really took off, even though it was taken up both by JITC (the Joint Interoperability Testing Center) in the USA and by X/Open.

In contrast, there is the approach used by EuroSInet and other regional members of OSI^{one}. For years they paid lip service to the idea that suppliers who engaged in testing their products in the interoperability testing workshops should have first ensured that their products had passed conformance testing. In reality, however, suppliers simply brought pre-release products into a cooperative multi-vendor testing workshop in order to conduct tests which could never be performed in their own laboratories. The concept was simple and the cost was very low. Product developers got together to agree what scenarios they wanted to test. These were written down in much the same way as conformance test purposes. They would then get together for a week of cooperative testing, performing pairwise tests with a large matrix on the wall showing who had tested with who, and was currently testing with who. Originally physical interconnection might have been via wide area networks or local X.25, but more recently Ethernet local area networks (LANs) have been used instead. Two problems emerged: the suppliers were very reluctant to publish the results thereby reducing the visibility of the activity to users and procurers; and as the membership changed from large multi-national suppliers to small niche market players, EuroSinet got locked into only applying its testing approach to X.400 and X.500 testing. The problem of publication of results was overcome by EuroSInet publishing a sanitised version of the report of the workshop, showing who tested with who and what was tested, but withholding the detailed results as these only applied to pre-release versions of the products; any bugs found should be corrected before product release.

Other approaches were advocated, notably by the European Commission CTS-4 interoperability testing project. Also EWOS published guidance documents including a vocabulary, a classification scheme, and a survey of what was happening around the world. However, these all remained rather theoretical.

Current situation

With PSI effectively dead, the only practical interoperability testing scheme is that still being used by EuroSInet. EuroSInet requested EOTC (the European Organisation for Testing and Certification) support to apply the approach more widely, but in the absence of even a clear timetable for a response EuroSInet got together with the EWOS Expert Group on Conformance Testing (EGCT) to propose a way forward. EWOS set up a project team (number 34) on interoperability testing, including involvement from the chairman of EuroSInet and the chairman of OSI^{one}. This produced a comprehensive report that led to the development of two specific project proposals which were approved by the EWOS Technical Assembly in December 1996.

One was for the production of a new set of guidance documents on interoperability and cost-effective testing, including the development of new ideas on built-in interoperability and built-in testability. There would also be work on an interoperability testing framework, SIS proformas, interoperability test suites, and pilot trials of the new ideas in three areas relevant to the GII (global information infrastructure). This project did not, however, go ahead because of a lack of funding available to EWOS.

The other project proposal was for the establishment of interoperability and capability demonstration facilities. This was essentially the idea of applying the EuroSInet interoperability testing concept across the whole scope of EWOS activities, but extending it to use the same low cost multi-vendor set up to demonstrate the capabilities of products based on EWOS specifications to potential users and procurers. This project did at least go ahead to the extent of holding a first interoperability testing workshop in Brussels in May 1997, focused on X.500 testing, to demonstrate the practicality of the idea and to advocate its use across the scope of whatever organisation takes over from EWOS later this year.

The way forward

A cost-effective testing approach should be developed based on bringing together the best of both conformance testing and interoperability testing to provide a consistent approach to testing that can be applied right through the product life-cycle. It should be flexible enough to cater for the needs of development testing, independent conformance testing, multi-vendor interoperability testing, operational monitoring and diagnostic testing, and regression testing. The concepts of built-in interoperability and built-in testability should be developed to see whether they can make a practical contribution to cost-effective testing.

Built-in interoperability means that the protocol design enhances rather than diminishes the prospects for successful interoperability of different suppliers' products. This implies keeping the number of options to a minimum and ensuring that the requirements expressed are all necessary to achieve successful interoperability.

Built-in testability means including features within the protocol to provide a self-testing capability or to facilitate more control over the conduct of conformance or interoperability testing. The difficulty is to find an organisation to take the lead in such work, now that the future of EWOS activities is in doubt both with regard to the funding and the organisational stability.

Where simple multi-vendor low cost interoperability testing is appropriate, the EuroSInet approach should be adopted. According to need, any appropriate available physical network could be used in place of the Ethernet LAN.

7 RECOMMENDATIONS

To summarise, the following recommendations are made regarding the future direction of protocol testing:

a) Standardised conformance test suites are only necessary in regulatory areas and in areas of high risk, like security and safety-critical software.
b) Test suite coverage should be chosen to match the risks of not testing.
c) Protocol designers, test designers and product implementors all need to work together to improve the effectiveness of protocol specifications and test specifications.
d) Normative SDL specifications should be developed together with the text description of telecoms protocols, and the SDL should be the basis for validation, animation, reference implementation, and automated test generation.
e) The protocol design process should aim to minimise complexity, building in testability, focusing on the essential interoperability requirements, and getting feedback from trial implementation before finalisation.
f) Protocol design groups should include expertise on testing methodology, TTCN, SDL, ASN.1, and MSCs.
g) Test purposes should be expressed in MSCs.
h) Automated test purpose generation should be developed for use in those cases where large test suites or continually changing test suites are required.
i) Automated test case generation from test purposes should be developed to start from either SDL or an extended finite state description and to produce complete TTCN test cases.
j) All TTCN tools and test suite developments should be aligned to the second edition of TTCN.
k) Test systems should either directly support TTCN input or should support the GCI interface so that they can be used with TTCN compilers or interpreters.
l) The practicality and cost-effectiveness of using TSP1 and other APIs should be investigated.

m) ISO/IEC 9646-5 and ISO/IEC TR 13233 provide a sound basis for the accreditation of protocol testing laboratories.
n) Accredited testing will mainly be needed to meet regulatory requirements or the requirements of high risk areas, like security and safety-critical software.
o) The use of strict conformance testing should be encouraged in the security area, to reduce costs and timescales and thereby be more attractive to industry.
p) A cost-effective testing approach should be developed, based on the best of conformance and interoperability testing.
q) Flexible cost-effective testing should be applied consistently right across the product life-cycle.
r) The concepts of built-in interoperability and built-in testability need to be studied to determine their practicality and cost-effectiveness.
s) The EuroSInet approach to interoperability testing should be applied wherever low cost focused multi-vendor interoperability testing is needed.

8 REFERENCES

Barker, R.M., Hehir, S., Kelly, G., Lampard, R.P., Lovering, A., Parkin, G.I., Stevens, M.J. and Szczygiel, B.M. (1997) Strict conformance testing, NPL.

ELA-G5 (1993) Interpretation of accreditation requirements in ISO/IEC Guide 25 and EN 45001 - Guidance for information technology and telecommunications testing laboratories for software and communications testing services, WELAC and ECITC, Brussels.

ETG 25 (1994) The TTCN style guide, rev. 1, EWOS, Brussels.

ETR 303 (1997) Methods for testing and specification; Test synchronization; Architectural reference; Test synchronization protocol 1 specification, ETSI, Sophia Antipolis, France.

ISO/IEC 8824-1 (1995) Abstract syntax notation one (ASN.1) Part 1: Specification of basic notation. Also ITU-T X.680.

ISO/IEC 9646-1 (1994) OSI conformance testing methodology and framework Part 1: General concepts. Also ITU-T X.290.

ISO/IEC 9646-2 (1994) OSI conformance testing methodology and framework Part 2: Abstract test suite specification. Also ITU-T X.291.

ISO/IEC 9646-3 (1992) OSI conformance testing methodology and framework Part 3: The tree and tabular combined notation. Also ITU-T X.292.

ISO/IEC 9646-3 (1997) OSI conformance testing methodology and framework Part 3: The tree and tabular combined notation. Mock-up for SC21 review, version 9.6.

ISO/IEC 9646-4 (1994) OSI conformance testing methodology and framework Part 4: Test realization. Also ITU-T X.293.

ISO/IEC 9646-5 (1994) OSI conformance testing methodology and framework Part 5: Requirements and guidance for test laboratories and their clients for the conformance assessment process. Also ITU-T X.294.

ISO/IEC 9646-6 (1994) OSI conformance testing methodology and framework Part 6: Profile test specification. Also ITU-T X.295.

ISO/IEC 9646-7 (1995) OSI conformance testing methodology and framework Part 7: Implementation conformance statements. Also ITU-T X.296.

ISO/IEC 10026-3 (1996) Distributed transaction processing Part 3: Protocol specification.

ISO/IEC TR 13233 (1995) Interpretation of accreditation requirements in ISO/IEC Guide 25 - Accreditation of information technology and telecommunications testing laboratories for software and protocol testing services.

ITU-T Q.1224 (1997) Distributed function plane for intelligent networks capability set 2.

ITU-T Q.1228 (1997) Interface recommendation for intelligent networks capability set 2.

ITU-T Z.100 (1994) Specification and description language.

ITU-T Z.105 (1994) SDL combined with ASN.1.

ITU-T Z.120 (1993) Message sequence charts.

9 BIOGRAPHY

Dave Rayner joined NPL in 1975, became head of the protocol standards group in 1980, head of the open systems group in 1993, and head of information systems engineering in 1996. In 1983 he became ISO rapporteur for OSI conformance testing, leading the development of ISO/IEC 9646 until the present day. From 1993 to 1996 he was convenor of ISO/IEC JTC1/SC21/JWG9 and editor of ISO/IEC TR 13233. From 1989 to 1996 he was chairman of the ECITC/OTL group, and during this time he edited ELA-G5. In 1993 he became chairman of EWOS/EGCT, a post he still holds. He assesses protocol testing laboratories for accreditation for UKAS, SWEDAC and DANAK.

PART TWO

TTCN Extensions

2

PerfTTCN, a TTCN language extension for performance testing

I. Schieferdecker, B. Stepien, A. Rennoch
GMD FOKUS
Hardenbergplatz 2, D-10623 Berlin, Germany,
tel.: (+49 30) 254 99 200, fax: (+49 30) 254 99 202,
e-mail: {Schieferdecker, Stepien, Rennoch}@fokus.gmd.de

Abstract

This paper presents a new approach to test the performance of communication network components such as protocols, services, and applications under normal and overload situations. Performance testing identifies performance levels of the network components for ranges of parameter settings and assesses the measured performance. A performance test suite describes precisely the performance characteristics that have to be measured and procedures how to execute the measurements. In addition, the performance test configuration including the configuration of the network component, the configuration of the network, and the network load characteristics is described. PerfTTCN - an extension of TTCN - is a formalism to describe performance tests in an understandable, unambiguous and re-usable way with the benefit to make performance test results comparable. First results on the description and execution of performance tests will be presented.

Keywords

Performance Testing, Test Suite, TTCN, Quality of Service

Testing of Communicating Systems, M. Kim, S. Kang & K. Hong (Eds)
Published by Chapman & Hall

1 MOTIVATION

Non-functional aspects of today's telecommunication services (e.g. multimedia collaboration, teleteaching, etc.) and in particular Quality-of-Service (QoS) aspects became as important as the functional correctness of telecommunication systems. Different approaches for guaranteeing certain QoS levels to the end users were developed. They include approaches for QoS negotiation between the end users and service and network providers, QoS guarantees of transmission services, QoS monitoring and QoS management, for example in self-adapting applications.

This paper considers QoS in the area of testing. Testing is a general method to check whether a network component meets certain requirements. Network components are considered to be communication protocols, telecommunication services, or end user applications. The requirements on network components are often described in a specification. The tested network component is also called implementation under test (IUT). Testing is either oriented at the conformance of an IUT with respect to the specification of the network component, the interoperability between the IUT and other network components, the quality of service of the IUT, or at its robustness.

QoS testing checks the service quality of the IUT against the QoS requirements of the network component. A specific class of QoS is that of performance-oriented QoS. Performance-oriented QoS requirements include requirements on delays (e.g. for response times), throughputs (e.g. for bulk data transfer), and on rates (e.g. for data loss). We concentrate exclusively on performance-oriented QoS, other classes of QoS are not considered. Subsequently, we use the term performance instead of QoS and refer therefore to performance testing.

One of the well-established methods in testing is that of conformance testing. It is used to check that an implementation meets its functional requirements, i.e. that the IUT is functionally correct. Since conformance testing is aimed at checking only the functional behavior of network components, it lacks in concepts of time and performance. Timers are the only means to impose time periods in the test execution. Timers are used to distinguish between network components that are too slow, too fast or do not react at all. In conformance testing, the correctness of the temporal ordering and exchanged protocol data units (PDUs) or of abstract service primitives (ASPs) have been the main target.

Performance testing is an extension to conformance testing to check also QoS requirements. Performance tests make use of performance measurements. Traditionally, performance measurements in a network consist of sending time stamped packets through a network and of recording delays and throughput. Once measurement samples have been collected, a number of statistics are computed and displayed. However, these statistics are sometimes meaningless since the actual conditions in which these measurements have been performed are unknown.

Different strategies can be used to study performance aspects in a communication network. One consists in attempting to analyze real traffic load in a network and to correlate it with the test results. The other method consists of creating artificial traffic load and of correlating it directly to the behavior that was observed during the performance test. The first method enables one to study the performance of network components under real traffic conditions and to confront unexpected behaviors. The second method allows us to execute more precise measurements, since the conditions of an experiment are fully known and controllable and correlations with observed performance are less fuzzy than with real traffic. Both methods are actually useful and complementary. A testing cycle should involve both methods: new behaviors are explored with real traffic load and their understanding is further refined with the help of the second method by attempting to reproduce them artificially and to test them. The presented approach to performance testing attempts to address both methods.

This paper presents a new approach to describe performance tests for network components and to test their performance under normal and overload situations. Certain performance levels of an IUT can be identified by means of repeating performance tests with varying parameter settings. On the basis of a thorough analysis, the measured performance can be assessed.

A performance test suite describes precisely the performance characteristics that have to be measured and procedures how to execute the measurements. In addition, a performance test has to describe the configuration of the IUT, the configuration of the network, and the characteristics of the artificial load. The exact description of a test experiment is a prerequisite to make test results repeatable and comparable. The description of the performance test configuration is an integral part of a performance test suite.

The objectives of performance testing can be realized with a variety of existing languages and tools. However, there is only one standardized, well known and widely used notation for the description of conformance tests: TTCN - the tabular and tree combined notation (ISO/IEC 1991, 1996 and Knightson, 1993). In addition, a number of TTCN tools are available. We decided to base our work on TTCN due to its wide acceptance. We define an extension of the TTCN language to handle concepts of performance testing. Only a limited number of additional declarations and functionalities are needed for the definition of performance tests. PerfTTCN - an extension of TTCN with notions of time, traffic loads, performance characteristics and measurements - is a formalism to describe performance tests in an understandable, unambiguous and re-usable way with the benefit to make the test results comparable.

The proposal introduces also a new concept of time in TTCN. The current standard of TTCN considers time exclusively in timers, where the execution of a test can be branched out to an alternative path if a given timer expires. New proposals by Walter and Grabowski (1997) introduce means to impose timing deadlines during the test execution by means of local and global timing constraints. In contrast to that, a performance test gathers measurement samples of occurrence times of selected test events and computes various performance characteristics on the basis of several samples. The computed performance characteristics are then used to check performance constraints, which are based on the QoS criteria for the network component.

Although the approach is quite general, one of its primary goals was the study of the performance of ATM network components. Therefore, the approach is in line with the ATM Forum performance testing specification (ATM Forum, 1997) that defines performance metrics and measurement procedures for the performance at the ATM cell level and the frame level (for layers above the ATM layer).

In this paper, we first discuss the objectives, main concepts, and architectures for performance tests, next we present the language features of PerfTTCN to describe the new concepts, and finally we present some results of experiments on an example handling queries to an HTTP server using a modified test generator of some well known TTCN design tool.

2 INTRODUCTION TO PERFORMANCE TESTING

2.1 Objectives of performance testing

The main objective of performance testing is to test the performance of a network component under normal and overload situations. The normal and overload situations are generated by artificial traffic load on the network component. The traffic load follows traffic patterns of a well-defined traffic model. For performance testing, the conformance of an IUT is assumed. However, since overload may degrade the functional behavior of the IUT to be faulty, care has to be taken to recognize erroneous functional behavior in the process of performance testing.

Another goal of performance testing is to identify performance levels of the IUT for ranges of parameter settings. Several performance tests will be executed with different parameter settings. The testing results are then interpolated in order to adjust that range of parameter value, where the IUT shows a certain performance level.

Finally, if performance-oriented QoS requirements for an IUT are given, performance testing should result in an assessment of the measured performance, whether the network component meets the performance-oriented QoS requirements or not.

The main advantage of the presented method is to describe performance tests unambiguously and to make test results comparable.This is in contrast with informal methods where test measurement results are provided only with a vague description of the measurement configuration, so that it is difficult to re-demonstrate and to compare the results precisely. The presented notation PerfTTCN for performance tests has a well-defined syntax. The operational semantics for PerfTTCN is under development. Once given, it will reduce the possibilities of misinterpretations in setting up a performance test, in executing performance measurements, and in evaluating performance characteristics.

2.2 Concepts of performance testing

This section discusses the basic concepts of the performance test approach. The concepts are separated with respect to the test configuration, measurements and analysis, and test behavior.

2.2.1 Test components

A *performance test* consists of several distributed foreground and background test components. They are coordinated by a main tester, which serves as the control component.

A *foreground test component* realizes the communication with the IUT. It influences the IUT directly by sending and receiving PDUs or ASPs to and respectively from the IUT. That form of discrete interaction of the foreground tester with the IUT is conceptually the same interaction of tester and IUT that is used in conformance testing. The discrete interaction brings the IUT into specific states, from which the performance measurements are executed. Once the IUT is in a state that is under consideration for performance testing, the foreground tester uses a form of continuous interaction with the IUT. It sends a continuous stream of data packets to the IUT in order to emulate the foreground load for the IUT. The foreground load is also called foreground traffic.

A *background test component* generates continuous streams of data to cause load for the network or the network component under test. A background tester does not directly communicate with the IUT. It only implicitly influences the IUT as it brings the IUT into normal or overload situations. The background traffic is described by means of traffic models. Foreground and background tester may use load generator to generate traffic patterns.

Traffic models describe traffic patterns for continuous streams of data packets with varying interarrival times and varying packet length. An often used model for the description of traffic patterns is that of Markov Modulated Poison Processes (MMPP). We selected this model for traffic description due to its generosity and efficiency. For example, audio and video streams of a number of telecommunication applications as well as pure data streams of file transfer or mailing systems have been described as MMPPs (Onvural, 1994). For the generation of MMPPs traffic patterns, efficient random number generator and an efficient finite state machine logic are needed only. Nonetheless, the performance testing approach is open to other kinds of traffic models.

Points of control and observation (PCOs) are the access points for the foreground and background test components to the interface of the IUT. They offer means to exchange PDUs or ASPs with the IUT and to monitor the occurrence of test events (i.e. to collect the time stamps of test events). A specific application of PCOs is their use for monitoring purposes only. Monitoring is needed to observe for example the artificial load of the background test components, the load of real network components that are not controlled by the performance test or to observe the test events of the foreground test component.

Coordination points (CPs) are used to exchange information between the test components and to coordinate their behavior. In general, the main tester has access via a CP to each of the foreground and background test components.

To sum up, a *performance test* uses an ensemble of foreground and background tester with well-defined traffic models. The test components are controlled by the main tester via coordination points. The performance test accesses the IUT via points of control and observation. A performance test suite defines the conditions under which a performance test is executed. Performance characteristics and measurements define what has to be measured and how. Only a complete performance test suite defines a performance test unambiguously, makes performance test experiments reusable and performance test results comparable.

2.2.2 Performance test configurations

In analogy to conformance testing, different *types of performance test configurations* can be identified. They depend on the characteristics of the network component under test. We distinguish between performance testing the implementation (either in hardware, software, or both) of

- an end-user telecommunication application,
- an end-to-end telecommunication service, or
- a communication protocol.

Of course, additional test configurations for other network components can be defined. The test configurations for these three types of performance tests are given in Figure 1, 2, and 3. The notion System Under Test (SUT) comprises all IUT and network components. For simplification, we omit the inclusion of the main tester in the figures.

The three test configurations differ only in the use of foreground tester. The use of background tester that generate artificial load to the network, and the use of monitors, that measure the actual real load in the network are the same in each of

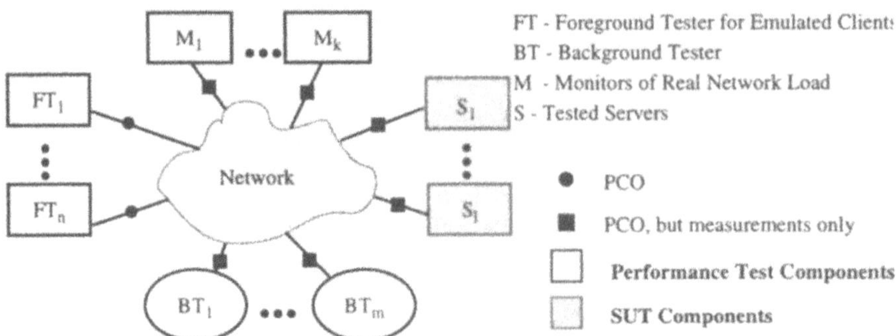

Figure 1 Performance test configuration for a server.

these test configurations.

In the case of performance testing a server in Figure 1, foreground tester emulate the clients. The test configuration for an end-to-end service in Figure 2 includes foreground tester at both ends of the end-to-end service, which emulate the service user. Performance testing of a communication protocol (Figure 3) includes foreground tester at the upper service access point to the protocol under test and at the lower service access point. This test configuration corresponds to the distributed test method in conformance testing (please refer to ISO/IEC, 1991 for other test methods). The service access points are reflected by points of control and observation.

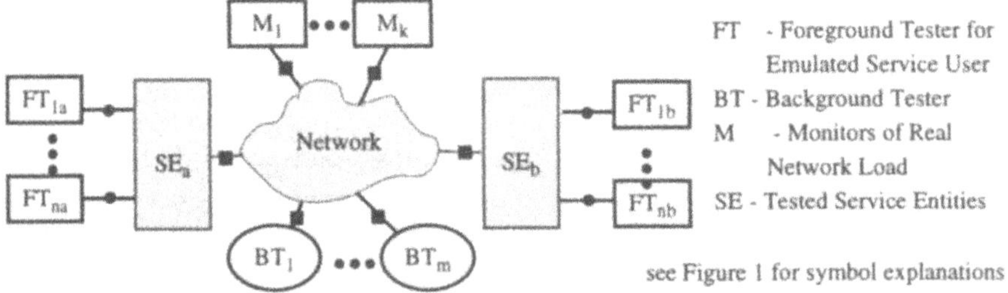

Figure 2 Performance test configuration for an end-to-end service

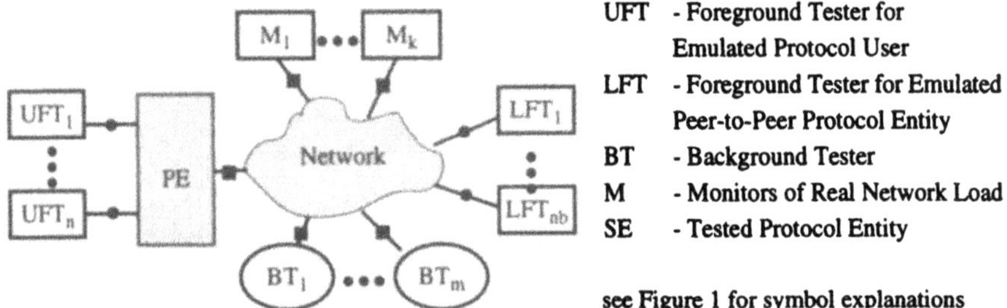

Figure 3 Performance test configuration for a protocol.

2.2.3 Measurements and Analysis

A *measurement* is based on the collection of time stamps of events. A measurement can be executed by monitoring components that are sensitive to specific test events only. The format of a test event that belongs to a measurement is described by constraints, so that the monitor can collect time stamps whenever an event at a certain PCO matches that format. The constraints used here are the same that are used in conformance testing. A measurement is started once and continues until it is cancelled by a test component or reaches its time duration. Currently, we investigate the need to control measurements explicitly in the dynamic behavior of a performance test.

Based on the measurements, more elaborated *performance characteristics* such as mean, standard deviation, maximum and minimum as well as the distribution functions can be evaluated. Their evaluation is based on predefined *metrics*, which have a well-defined semantics.

Performance characteristics can be evaluated either off-line or on-line. An off-line analysis is executed after the performance test finished and all samples have been collected.

On-line analysis is executed during the performance test and is needed to make use of *performance constraints*. Performance constraints allow us to define requirements on the observed performance characteristics. They can control the execution of a performance test and may even lead to the assignment of final test verdicts and to a premature end of performance tests. For example, if the measured response delay of a server exceeds a critical upper bound, a fail verdict can be assigned immediately and the performance test can finish.

2.2.4 Performance Test Behavior

A performance test suite has to offer features to start and cancel background and foreground test components, to start and cancel measurements, to interact with the IUT and to generate a controlled load to the IUT, as well as to access recent measurements via performance constraints.

At the end of each performance test, a final test verdict such as pass or fail has to be assigned. However, a verdict of a performance test should not only evaluate the observed behavior and performance of the tested network component to be correct or incorrect (i.e. by assigning pass or fail, respectively), but also return the measured performance characteristics that are of importance for the analysis of the test results.

3 PERFTTCN - A PERFORMANCE EXTENSION OF TTCN

This section presents the new language constructs of PerfTTCN for the declaration of traffic models and background traffic, for the declaration of performance measurements, performance characteristics and performance constraints, for the control of test components and measurements, for the use of performance constraints, and for the assignment of verdicts.

3.1 Traffic Models and Background traffic

The location of background test components and the orientation of the background traffic are defined in the test component configuration table (see also Table 1).

For each background test component, PCOs identify the location of the source of the background traffic (left side) and of the destination of the background traffic (right side). Lists of PCOs for the source or destination can be used to declare multipoint-to-multipoint background traffic.

The coordination points of a background test component are used to control its behavior, e.g. to start or to stop the traffic generation. The main tester sends in its dynamic behavior coordination messages to the background test components. The traffic patterns that are generated by a background test component are defined in a

Table 1 Integration of Background Test Components

Test Component Configuration Declaration			
Configuration name: CONFIG_2			
Components Used	**PCOs Used**	**CPs Used**	**Comments**
MTC	PCO_1	CP1,	
PTC1	PCO_2	MCP2, CP1	
Background Test Components			
Identifier	**PCOs Used**	**CPs Used**	**Comments**
traffic1	(PCO_B1) -> (PCO_B2)	BCP1, BCP2	Point to Point
traffic2	(PCO_B1) -> (PCO_B4)	BCP1, BCP2	Point to Point

traffic stream declaration table (see next section).

Specific, implementation dependent details of the location of a background test component, e.g. the connection information such as the VPI/VCI for an ATM connection, are subject to the protocol extra information for testing (PIXIT) document of the performance test suite.

3.2 Traffic models

The purpose of the background traffic is to create load on the communication network that traverses the communication links of the system under test. The background traffic is a continuous, uninterrupted, and predictable stream of packets following a well-defined traffic pattern.

The traffic pattern defines the data packet lengths and interarrival times of the data packets. Traffic patterns can simulate the traffic that is associated with different kinds of applications.

Table 2 MMPP Traffic Model Declaration

Traffic Model Declaration		
Name: on_off **Type**: MMPP **Comments**:		
Length	S1	10
Length	S2	1000
Rate	S1	2
Rate	S2	10
Transition	S1, S2	3
Transition	S2, S1	5

The traffic patterns are defined in traffic model declaration tables (see Table 2 and 3). The declaration selects the stochastic model and sets the corresponding parameters. Each model type has a varying number of parameters and different types of parameters. Therefore, PerfTTCN supports different tables for each type of traffic model. Each traffic model has a name so that it can be referenced later in the traffic stream declaration. Tables 2 and 3 illustrate the table format of an MMPP and CBR model, respectively.

Table 3 CBR Traffic Model Declaration

Traffic Model Declaration	
Name: const1 **Type**: CBR **Comments**:	
PCR	10 MBit/s

The background traffic stream declarations (see also Table 4) relate a traffic stream to a background test component. A traffic stream uses as many instances of a traffic model as necessary to produce significant load. A traffic stream is identified by a name that can be used in the dynamic behavior part to start the corresponding background traffic.

Table 4 Background Traffic Stream Declaration

Background Traffic Stream Declaration			
Traffic Name	**Background Test Component**	**Model Name**	**Nr. of Instances**
Load1	traffic1	on_off	6
Load2	traffic1	const1	2
Load3	traffic2	const1	8

3.3 Measurements and Analysis

The introduction of performance measurements into the testing methodology leads to new tables for the declaration of measurements, performance characteristics and performance constraints as well as to additional operations in the dynamic behavior description of test cases and test steps.

A measurement declaration (Table 5) consists of a metric and is combined with one or two test events that define the critical events of the measurement. For example, for a delay measurement the events define the start and end event. The events are observed only at specific PCOs. The direction of the event is also indicated: "!" means sending and "?" means receiving at that specific PCO (as seen from the test components).

A measurement uses standard metrics such as counter, delay, jitter, frequency, or throughput with predefined semantics. For example, DELAY_FILO is the delay between first bit send and last bit arrived*. User defined metrics (implemented by means of test suite operations) can also be used.

Table 5 Declaration of measurements

Measurement Declaration						
Name	**Metric**	**Unit**	**event 1**	**constr. 1**	**event 2**	**constr. 2**
response_delay	DELAY_FILO	ms	PCO_1 !Request	s_req_spc	PCO_1 ?Response	r_resp_spc

Measurements can be most effectively evaluated with the use of statistical indicators such as means, frequency distributions, maximum, minimums, etc. For that purpose, PerfTTCN offer the concept of performance characteristics. A performance characteristics is declared in a performance characteristics declaration table (see also Table 6). It refers to a single measurement. In order to be statistically significant, a performance characteristics should be calculated only if the measurement has been repeated several times. Therefore, it is possible to define a sample size or a time duration of the measurement for the calculation of a performance characteristic.

Table 6 Declaration of performance characteristics

Performance Characteristics Declaration				
Name	**Calculation**	**Measurement**	**Sample size**	**Duration**
res_delay_mean	MEAN	response_delay	20	
res_delay_max	MAX	response_delay		1 min

---

*. In general, four different semantics can be given to a delay measurement: FILO = first bit in, last bit out, FIFO = first bit in, first bit out, LIFO = last bit in, first bit out, and LILO = last bit in, last bit out.

3.4 Performance constraints and verdicts

Performance constraints are used for the on-line analysis of observed performance characteristics. For example, if performance falls below some set limits, the verdict should be set to fail. In contrast to constraints in TTCN, a performance constraint evaluation is based on repeated measurement of test events rather than the matching of a single event.

Therefore, we distinguish between functional constraints based on PDU and ASP value matching (that are the traditional constraints in TTCN) and performance constraints. The performance constraint declaration (Table 7) consists of a name and a logical expressions. The expression may use performance characteristics with individual thresholds. More than one performance characteristic can be used in a performance constraint. For example p_resp in Table 7 uses the performance characteristics res_delay_mean and res_delay_max.

Table 7 Declaration of performance constraints

Performance Constraint Declaration		
Name	**Constraint Value Expression**	**Comments**
p_resp	(res_delay_mean < 5) AND (res_delay_max < 10)	
n_p_resp	NOT (p_resp)	

Functional constraints are specified for each event line in the dynamic behavior of a test components. However, performance constraints apply only to the lines where measurements are performed.

3.5 Performance Test Behavior

The behavior of a performance test is specified in the dynamic part of the performance test suite. The main tester is defined in the test cases, while the other test components are specified by test steps.

Test components are created with the START construct. Either they execute their complete behavior or are cancelled explicitly via coordination messages. The control of performance measurements is specified similar to the control of timers, i.e. a measurement can be started and cancelled with START and CANCEL, respectively.

Performance constraints are indicated in the constraint reference column. However, performance constraints are evaluated differently from functional constraints. That is caused by the sample size required for statistical significance and/or the type of metrics used, where more than one observation is required to compute the metric such as the computation of a mean value. Whenever the sample size to evaluate the constraint has not yet been reached, the performance constraint is implicitly evaluated to "true". As soon as the sample size is reached through repeated sampling, the performance constraint is evaluated. If it evaluates to "false", the related event is consequently not accepted. Both, a functional and a performance constraint can be used at the same behavior line. Please note, that performance constraints can be used in qualifiers, too.

Table 8 provides an example of a test case behavior which includes a background test traffic identified by 'Load2', i.e. according to Table 3 it is a constant bit rate. After the background traffic has started (line 1) a series of 'Requests' occurs at PCO_1 (line 2).

The test system awaits from the SUT a 'Response' primitive (line 3 or 5). Due to the response_delay declaration of Table 5 delay measurements occur to determine the time between 'Request' and 'Response'. There are two possibilities to accept

'Response', which are distinguished by the different performance constraints 'p_resp' (line 3) and 'n_p_resp' (line 5). The resulting preliminary test verdict 'pass' or 'inconclusive' depends on these performance constraints.

The test cases finishes when the timer T_response_delay timeouts (line 7). In that case a final verdict is assigned. The reception of an event other than 'Response' terminates the test case (line 8) and measurements, timer, and background traffic are stopped. It is planned to return the measured performance characteristics in combination with the test verdicts in order to support an in-depth result analysis after a performance test finished.

Table 8 The behavior description of a performance test

Test Case Dynamic Behavior					
Test Case Name: www_Get **Group:** **Purpose:** **Configuration:** CONFIG_2 **Default:** **Comments:**					
Nr	**Label**	**Behavior Description**	**Constr. Ref**	**Verdicts**	**Comments**
1		BCP1 ! Start(Load2)			start backgr. traffic Load2
2	top	PCO_1 ! Request START response_delay START T_response_delay	s_req		start measurements
3		PCO_1 ? Response	p_resp	(pass)	acceptable performance
4		GOTO top			
5		PCO_1 ? Response	n_p_resp	(inconc)	unacceptable perf.
6		GOTO top			
7		? T_response_delay CANCEL response_delay			measurement terminates
8		BCP1 ! Stop(Load2)		R	stop background traffic
9		PCO_1 ? OTHERWISE CANCEL response_delay CANCEL T_response_delay		(fail)	unexpected event, stop measurements
10		BCP1 ! Stop(Load2)		R	stop background traffic
Detailed Comments:					

3.6 Comparison with TTCN

Concurrent TTCN has been designed as a test description language for conformance tests, only. It uses discrete test events such as sending and receiving of protocol data units and abstract service primitives. The conformance test suite and the implementation under test interact with each other by sending test events to and receiving test events from the opposite side. A test continues until the tester assigns a test verdict saying that the observed behavior of the implementation conforms (pass) or does not conform (fail) to the specification. In the case that the observed behavior can neither be assessed to be conformant or non-conformant, the inconclusive verdict is assigned. The basis for the development of a conformance test suite is the functional protocol specification only.

The development of a performance test suite is based on a QoS requirement specification that is combined with the functional specification of the implementation under test. The QoS requirements may include requirements on delays, throughputs, and rates of certain test events. A performance test uses not only discrete test events (those are used to bring the IUT in a controlled way into a well-defined state), but uses also a bulk data transfer from the tester to the IUT. Bulk data transfer is realized by continuous streams of test events and emulates different load situations for the IUT. A performance test assigns not only pass, fail or inconclusive, but also assigns the measured performance characteristics that are the basis for an in-depth analysis of the test results.

The new concepts of PerfTTCN have been introduced in Section 3. The existence of a mapping from PerfTTCN to ConcurrentTTCN would allow us to model performance tests on a level of abstraction that has been specifically defined for performance tests, and would enable us to re-use existing tools for Concurrent TTCN for the execution of performance tests. However, it turned out that some of the new concepts (in particular, traffic models, background tester, measurements, performance constraints) with their semantics can only hardly be represented in Concurrent TTCN. Predefined test suite operations with a given semantics seem to be an easy possibility to include the new concepts. Further study is needed in that area.

4 PERFORMANCE TEST EXAMPLES

Two studies were performed to show the feasibility of PerfTTCN. Performance tests for a SMTP and a HTTP server has been implemented. The experiments were implemented using the Generic Code Interface of the TTCN compiler of ITEX 3.1. (Telelogic, 1996) and a distributed traffic generator VEGA (Kanzow, 1994). VEGA uses MMPPs as traffic models and is a traffic generator software that allows us to generate traffic between a potentially large number of computer pairs using TCP/UDP over IP communication protocols. It is also capable of using ATM adaptation layers for data transmission such as these provided by FORE Systems on the SBA200 ATM adaptors cards. The traffic generated by VEGA follows the traffic pattern of the MMPP models.

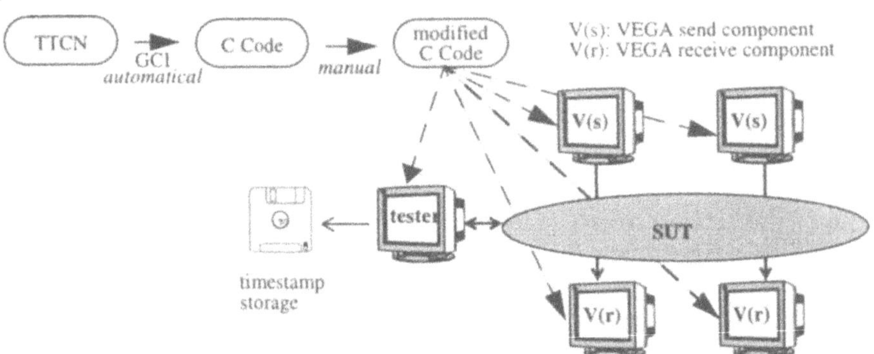

Figure 4 Technical Approach of the experiment.

The C-code for the executable performance tests was first automatically derived from TTCN by ITEX GCI and than manually extended

- to instantiate sender/receiver pairs for background traffic,

- to evaluate inter-arrival times for foreground data packets, and
- to locally measure delays.

Figure 4 illustrates the technical approach of executing performance tests: the derivation of the executable test suite and the performance test configuration. The figure presents also a foreground tester and several send/receive components of VEGA.

The performance tests for SMTP and HTTP server use the concepts of performance test configuration of the network, of the end system, and of the background traffic only. Other concepts such as measurements of real network load, performance constraints and verdicts will be implemented in the next version.

4.1 A performance test for an HTTP server

This example of a performance tests consists of connecting to a Web server using the HTTP protocol and of sending a request to obtain the index.html URL. If the query is correct, a result PDU containing the text of this URL should be received. If the URL is not found either because the queried site does not have a URL of that name or if the name was incorrect, an error PDU reply can be received. Otherwise, unexpected replies can be received. In that case, a fail verdict is assigned.

The SendGet PDU defines an HTTP request. The constraint SGETC defines the *GET /index.html HTTP/1.0* request. A ReceiveResult PDU carries the reply to the request. The constraint RRESULTC of the ReceiveResult PDU matches on "?" to the returned body of the URL: *HTTP/1.0 200 OK.*

The original purely functional test case in TTCN has been extended to perform a measurement of the response time of a Web server to an HTTP Get operation (see also Table 10). The measurement "MeasGet" has been declared to measure the delay between the two events SendGet and ReceiveResult as shown in Table 9.

Table 9 HTTP measurement declaration

Measurement Declaration						
Name	**Metric**	**unit**	**event 1**	**constr. 1**	**event 2**	**constr. 2**
MeasGet	DELAY	ms	SendGet	SGETC	ReceiveResult	RRESULTC

The repeated sampling of the measurement has been implemented using a classical TTCN loop construct to make this operation more visible in this example. The sampling size has been set to 10. The location of the operations due to "MeasGet" measurements are revealed in the comments column in the dynamic behavior of Table 10. It consists in associating a start measurement with the SendGet event and an end measurement with the ReceiveResult event as declared in Table 9. The delay between these two measurements will give us the response time to our request, which includes both network transmission delays and server processing delays.

The main program of the HTTP performance test is shown in Figure 5. The GCI TTCN code of the performance test case is initiated in Line 2. Line 3 instantiates a measurement entity to collect time stamps. The co-working between TTCN GCI and VEGA is initiated by vegaTtcnBridge (Line 4). Models for background traffic are declared and defined on Line 5-7. Background traffic components are declared on Line 8-9. Finally, lines 10-12 define and start the background traffic streams consisting of a background traffic component, a traffic model, and a number of instances. Line 13 starts the performance test case that controls the execution of the test and accesses the measurement entity. The test cases finishes with reporting the measured delays (Line 14). An example of the statistics with and without network load is shown in Figure 6.

This experiment has been performed on an ATM network using Sun workstations and TCP/IP over ATM layers protocols. The graph on the left of Figure 6 shows delay measurement under no traffic load conditions while the graph to the right shows results achieved with six different kinds of CBR and three different kinds of Poisson traffic flows between two pairs of machines communicating over the same segment as the HTTP client machines[†].

Table 10 Performance test case for the HTTP example

Test Case Dynamic Behavior					
Test Case: www_Get **Group:** **Purpose:** **Configuration:** **Default:** **Comments:**					
Nr	**Label**	**Behaviour Description**	**Constr. Ref**	**Verdicts**	**Comments**
1	Top	N ! Connect (NumTimes := 0)	CONNECTC		
2		N ! SendGet START T_Receive START MeasGet	SGETC		start measurement, begin delay sample
3		N ? ReceiveResult (NumTimes := NumTimes+1) CANCEL T_Receive	RRESULTC	(P)	acceptable response, end delay sample
4		[NumTimes < 10] GOTO Top			
5		[NumTimes >= 10] CANCEL MeasGet		R	measurement terminates
6		N ? ReceiveError CANCEL T_Receive CANCEL MeasGet	RERRORC	I	incorrect response
7		? T_Receive CANCEL MeasGet		F	no response
8		N ? OTHERWISE CANCEL T_Receive		F	unexpected response
Detailed Comments:					

5 CONCLUSIONS

The importance of Quality-of-Service aspects in multimedia communication environments and the lack of conformance testing to check performance oriented QoS requirements lead us to the development of a performance testing framework. The paper presents a first approach to extend Concurrent TTCN with performance features.

The main emphasis of our work is the identification and definition of basic concepts for performance testing, the re-usable formulation of performance tests and the development of a performance test run time environment. Thus, the concrete

---

†. Due to lack of space, we have no included the complete performance test suite into the paper. However, it is available on request.

```
1    int main( char* argc, int argv ) { ...
2     GciInit( ); CreatePCOsAndTimers();
3     WWWResponseEnt = new MeasurementEntity("GetWWW"); ...
4     vegaTtcnBridgeInit(argc,argv);
5     backgroungtraffic = new backGroundTraffic();
6     aModel = new vegaModel("cbr_slow", "cbr",10, 0.1, 0);
7     backgroungtraffic->addAModel(aModel); ...
8     aBackGroundDataflow=new BackGroundDataflow("traffic_1",
          "kirk","clyde","udp");
9     backgroungtraffic->addABGDataflow(aBackGroundDataflow); ...
10    aBackGroundTrafficLoad=
          new BackGroundTrafficLoad("traffic_1", "cbr_slow", 3);
11    backgroungtraffic->
          addABGBackGroundTrafficLoad(aBackGroundTrafficLoad);
12    backgroungtraffic->SetupBGTraffic(); ...
13    GciStartTestCase("www_GET"); ...
14    WWWResponseEnt->printStatistics(); ... }
```

Figure 5 Performance test configuration for the HTTP performance test.

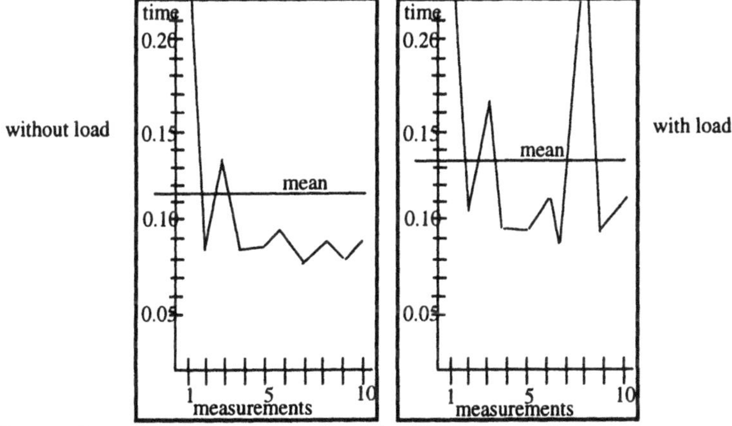

Figure 6 Performance test result of the HTTP example.

syntax of PerfTTCN is a minor concern, but also the basis for ongoing work.

An initial feasibility study of the approach on performance testing has been conducted using the SMTP and the HTTP protocols as examples. The usability of this approach has been demonstrated on a more complex example: A performance test suite to test the end-to-end performance of ATM Adaptation Layer 5 Common Part (AAL5-CP) has been defined only recently (Schieferdecker, Li, Rennoch, 1997).

In parallel, we are further exploring the possibility of re-using existing TTCN tools in a performance test execution environment. Therefore, we are working on a set of test suite operations (reflecting the new performance concepts) and on a mapping from PerfTTCN to TTCN by using these special test suite operations. The definition of the operational semantics of PerfTTCN is currently under work.

6 REFERENCES

Combitech Group Telelogic AB (1996): ITEX 3.1 User Manual.

ISO/IEC (1991): ISO/IEC 9646-1 „Information Technology - Open Systems Interconnection - Conformance testing methodology and framework - Part 1: General Concepts".

ISO/IEC (1996): ISO/IEC 9646-1 „Information Technology - Open Systems Interconnection - Conformance testing methodology and framework - Part 3: The tree and tabular combined notation".

Jain, R. (1996) *The Art of Computer Systems Performance Analysis*, John Wiley & Sons, Inc. Publisher.

Kanzow, P. (1994) Konzepte für Generatoren zur Erzeugung von Verkehrslasten bei ATM-Netzen. MSc-thesis, Technical University Berlin (in german only).

Knightson, K. G. (1993) *OSI Protocol Conformance Testing, IS9646 explained*, McGraw-Hill.

Onvural, R. O. (1994) *Asynchronous Transfer Mode Networks: Performance Issues.* Artech House Inc.

Schieferdecker, I. and Li, M. and Rennoch, A. (1997) An AAL5 Performance Test Suite in PerfTTCN, Proceedings of FBT'97, Berlin, Germany

The ATM Forum Technical Committee (1997): Introduction to ATM Forum Performance Benchmarking Specifications, btd-test-tm-perf.00.01.ps.

Walter, T. and Grabowski, J. (1997) A Proposal for a Real-Time Extension of TTCN, Proceedings of KIVS'97, Braunschweig, Germany.

7 BIOGRAPHY

Ina Schieferdecker studied mathematical computer science at the Humboldt University in Berlin and received her Ph.D. from the Technical University in Berlin in 1994. She attended the postgraduate course on open communication systems at the Technical University in Berlin. Since 1993, she is a researcher at GMD FOKUS - the Research Institute for Open Communication Systems - and a lecturer at Technical University Berlin since 1995. She is working on testing methods for network components, and executes research on formal methods, performance-enhanced specifications and performance analysis.

Bernard Stepien holds a Master degree from the University of Montpellier in France. Subsequently, he carried out research in Transportation Science with the Montreal Transportation Commission and worked as an economist for Bell Canada. He has been a private consultant in computer applications since 1975. He has been active in research on Formal Description techniques with the University of Ottawa since 1985. Currently he is involved in various aspects of communication protocols software with the Canadian Government (Department of Industry, Atomic Energy Control Board), Bell Canada and Nortel.

Axel Rennoch studied mathematics at the Free University of Berlin. His research interests include the application of Formal Description Techniques for testing methodologies and Quality of Service considerations. Currently he is employed as a scientist at the GMD - Research Institute for Open Communication System in Berlin.

3

Real-time TTCN for testing real-time and multimedia systems

Thomas Walter
Swiss Federal Institute of Technology Zurich, Computer Engineering and Networks Laboratory (TIK)
CH-8092 Zurich, Switzerland, Tel: +41-1-632 7007, Fax: +41-1-632 1035, e-mail: walter@tik.ee.ethz.ch

Jens Grabowski
Medical University of Lübeck, Institute for Telematics
Ratzeburger Allee 160, D-23538 Lübeck, Germany, Tel: +49-451-500 3723, Fax: +49-451-500 3722, e-mail: jens@itm.mu-luebeck.de

Abstract

In this paper we define real-time TTCN and apply it to several applications. In real-time TTCN, statements are annotated with time labels that specify their earliest and latest execution times. The syntactical extensions of TTCN are the definition of a table for the specification of time names and time units, and two new columns in the dynamic behaviour description tables for the annotation of statements with time labels. We define an operational semantics for real-time TTCN by mapping real-time TTCN to timed transition systems. Alternatively, we introduce a refined TTCN snapshot semantics that takes time annotations into account.

Keywords

Conformance testing, TTCN, test cases, real-time testing

Testing of Communicating Systems, M. Kim, S. Kang & K. Hong (Eds)
Published by Chapman & Hall © 1997 IFIP

1 INTRODUCTION

Testing, or to be precise *conformance testing*, is the generally applied process in validating communication software. A conformance testing methodology and framework (ISO9646-1 1994) have been established within the standardization bodies of ISO and ITU. An essential part of this methodology is a notation, called TTCN (Tree and Tabular Combined Notation) (ISO9646-3 1996), for the definition of conformance test cases. TTCN has been designed for testing systems for which in general timing between communicating entities has not been an issue. Test cases are specified as sequences of test events which are input and output events of abstract service primitives (ASP) or protocol data units (PDU). The relative ordering of test events is defined in a *test case behaviour description*.

The situation is changing now. We can identify two main new kinds of distributed systems: firstly, real-time systems which stem from the use of computers for controlling physical devices and processes. For these systems, real-time communication is essential for their correct behaviour. Secondly, multimedia systems which involve the transmission of several continuous streams (of bits) and their timely reproduction (e.g., synchronization of audio and video). However, as pointed out in, e.g., (Ates *et al.* 1996), TTCN is not an appropriate test notation for testing real-time and multimedia systems: Firstly, test events in TTCN are for message-based systems and not for stream-based systems. Secondly, in TTCN real-time can only be approximated. In this paper we define a real-time extension of TTCN as a contribution for solving the second problem.

Our extension of TTCN to *real-time TTCN* is on a syntactical and a semantical level. The syntactical extension is that we allow an annotation of test events with an earliest execution time (EET) and a latest execution time (LET). Informally, a test event may be executed if it has been enabled for at least EET units and it must be executed if it has been enabled for LET units. For the definition of an operational semantics of real-time TTCN we use timed transition systems (Henzinger *et al.* 1991).

A number of techniques for the specification of real-time constraints have been proposed which are besides others: time Petri Nets (Berthmieu *et al.* 1991, Merlin *et al.* 1976) and extensions of LOTOS (Bowman *et al.* 1994, Hogrefe *et al.* 1992, Léonard *et al.* 1994, Quemada *et al.* 1987), SDL (Hogrefe *et al.* 1992, Leue 1995) and ESTELLE (Fischer 1996). As in the cited literature, our approach allows the timing of actions relative to the occurrence of previous actions. The difference between the cited approaches and ours is that real-time TTCN is a hybrid method used for the specification of *properties* of test systems and *requirements* on implementations under test (IUT).

Section 2 gives a brief introduction to TTCN. Section 3 explains real-time TTCN. The applicability of our approach is shown in Section 4. Section 5 concludes the paper with an assessment of our work and the discussion of open issues.

2 TTCN - TREE AND TABULAR COMBINED NOTATION

TTCN is a notation for the description of test cases to be used in conformance testing. For the purpose of this paper we restrict our attention to TTCN concepts related to the description of the dynamic test case behaviour. Further details on TTCN can be found in: (Baumgarten and Giessler 1994, Baumgarten and Gattung 1996, ISO9646-3 1996, Kristoffersen *et al.* 1996, Linn 1989, Probert *et al.* 1992, Sarikaya 1989).

2.1 Abstract testing methods and TTCN

A test case specifies which outputs from an IUT can be observed and which inputs to an IUT can be controlled. Inputs and outputs are either *abstract service primitives* (ASPs) or *protocol data units* (PDUs). In general, several concurrently running distributed *test components* (TC) participate in the execution of a test case. TCs are interconnected by *coordination points* (CPs) through which they asynchronously exchange *coordination messages* (CMs). TCs and IUT logically communicate by exchanging PDUs which are embedded in ASPs exchanged at *points of control and observation* (PCOs), which are interfaces above and below the IUT. Since in most cases the lower boundary of an IUT does not provide adequate PCO interfaces, TCs and IUT communicate by using services of an underlying service provider.

2.2 Test case dynamic behaviour descriptions

The behaviour description of a TC consists of *statements* and *verdict assignments*. A verdict assignment is a statement of either PASS, FAIL or INCONCLUSIVE, concerning the conformance of an IUT with respect to the sequence of events which has been performed. TTCN statements are *test events* (SEND, IMPLICIT SEND, RECEIVE, OTHERWISE, TIMEOUT and DONE), *constructs* (CREATE, ATTACH, ACTIVATE, RETURN, GOTO and REPEAT) and *pseudo events* (qualifiers, timer operations and assignments).

Statements can be grouped into *statement sequences* and *sets of alternatives*. In the graphical form of TTCN, sequences of statements are represented one after the other on separate lines and being *indented* from left to right. The statements on lines 1 - 6 in Figure 1 are a statement sequence. Statements on the same level of indentation and with the same predecessor are alternatives. In Figure 2 the statements on lines 4 and 6 are a set of alternatives: they are on the same level of indentation and have the statement on line 3 as their common predecessor.

Test Case Dynamic Behaviour					
Nr	Label	Behaviour Description	CRef	V	Comments
1		CP ? CM	connected		RECEIVE
2		(NumOfSends := 0)			Assignment
3		REPEAT SendData UNTIL [NumOfSends > MAX]			Construct
4		START Timer			Timer Operation
5		?TIMEOUT timer			TIMEOUT
6		L ! N-DATA request	data		SEND

Figure 1 TTCN Behaviour Description - Sequence of Statements.

Test Case Dynamic Behaviour					
Nr	Label	Behaviour Description	CRef	V	Comments
1		[TRUE]			Qualifier
2	L1	(NumOfSends := NumOfSends + 1)			
3		+SendData			ATTACH
4		[NOT NumOfSends > MAX]			Alternative 1
5		-> L1			GOTO
6		[NumOfSends > MAX]			Alternative 2

Figure 2 TTCN Behaviour Description - Set of Alternatives.

2.3 Test component execution

A TC starts execution of a behaviour description with the first *level of indentation* (line 1 in Figure 1), and proceeds towards the last level of indentation (line 6 in Figure 1). Only one alternative out of a set of alternatives at the current level of indentation is executed, and test case execution proceeds with the next level of indentation relative to the executed alternative. For example, in Figure 2 the statements on line 4 and line 6 are alternatives. If the statement on line 4 is executed, processing continues with the statement on line 5. Execution of a behaviour description stops if the last level of indentation has been visited, a test verdict has been assigned, or a test case error has occurred.

Before a set of alternatives is evaluated, a *snapshot* is taken (ISO9646-3 1996), i.e., the state of the TC and the state of all PCOs, CPs and expired timer lists related to the TC are updated and frozen until the set of alternatives has been evaluated. This guarantees that evaluation of a set of alternatives is an *atomic* and *deterministic action*.

Alternatives are evaluated in sequence, and the first alternative which is *evaluated successfully* (i.e., all conditions of that alternative are fulfilled (ISO9646-3 1996)) is executed. Execution then proceeds with the set of alternatives on the next level of indentation. If no alternative can be evaluated successfully, a new snapshot is taken and evaluation of the set of alternatives is started again.

3 REAL-TIME TTCN

In real-time TTCN, statements are annotated with time labels for earliest and latest execution times. Execution of a real-time TTCN statement is instantaneous. The syntactical extensions of TTCN (Section 3.2) are the definition of a table for the specification of time names and time units and the addition of two columns for the annotation of TTCN statements in the behaviour description tables. We define an operational semantics for real-time TTCN (Section 3.3). For this we define a mapping of real-time TTCN to timed transition systems (Henzinger *et al.* 1991) which are introduced in Section 3.1. Applying timed transition systems has been motivated by our experiences with the definition of an operational semantics for TTCN (Walter *et al.* 1992, Walter and Plattner 1992). To emphasize the similarities of TTCN and real-time TTCN we also propose a refined snapshot semantics which takes time annotations into account and which is compliant with the timed transition system based semantics. In the following section we quote the main definitions of (Henzinger *et al.* 1991).

3.1 Timed transition systems

A *transition system* (Keller 1976) consists of a set V of variables, a set Σ of states, a subset $\Theta \subseteq \Sigma$ of initial states and a finite set $\mathcal{T}$ of transitions which also includes the idle transition t_I. Every transition $t \in \mathcal{T}$ is a binary relation over states; i.e., it defines for every state $s \in \Sigma$ a possibly empty set $t(s) \subseteq \Sigma$ of so-called t-successors. A transition t is said to be *enabled* on state s if and only if $t(s) \neq \emptyset$. For the idle transition t_I we have that $t_I = \{(s, s) \mid s \in \Sigma\}$.

An infinite sequence $\sigma = s_0 s_1 \ldots$ is a *computation* of the underlying transition system if $s_0 \in \Theta$ is an initial state, and for all $i \geq 0$ there exists a $t \in \mathcal{T}$ such that $s_{i+1} \in t(s_i)$, denoted $s_i \stackrel{t}{\longrightarrow} s_{i+1}$, i.e., transition t is *taken* at position i of computation σ.

The extension of transition systems to timed transition systems is that we assume the existence of a real-valued global clock and that a system performs actions which either advance time or change a state (Henzinger *et al.* 1991). Actions are executed instantaneously, i.e., they have no duration.

A *timed transition system* consists of an underlying transition system and, for each transition $t \in \mathcal{T}$, an earliest execution time $EET_t \in \mathbb{N}$ and a latest execution time $LET_t \in \mathbb{N} \cup \{\infty\}$ is defined.* We assume that $EET_t \leq LET_t$ and, wherever they are not explicitly defined, we presume the default values are zero for EET_t and ∞ for LET_t. EET_t and LET_t define timing constraints which ensure that transitions cannot be performed neither to early (EET_t) nor too late (LET_t).

A *timed state sequence* $\rho = (\sigma, T)$ consists of an infinite sequence σ of states and

*In principle, time labels may not only be natural numbers. For an in-depth discussion of alternative domains for time labels, the reader is referred to (Alur *et al.* 1996).

an infinite sequence T of times $T_i \in \mathbb{R}$ and T satisfies the following two conditions:

- *Monotonicity:* $\forall i \geq 0$ either $T_{i+1} = T_i$ or $T_{i+1} > T_i \wedge s_{i+1} = s_i$.
- *Progress:* $\forall t \in \mathbb{R}\, \exists\, i \geq 0$ such that $T_i \geq t$.

Monotonicity implies that time never decreases but possibly increases by any amount between two neighbouring states which are identical. If time increases this is called a *time step*. The transition being performed in a time step is the idle transition which is always enabled (see above). The progress condition states that time never converges, i.e., since $\mathbb{R}$ has no maximal element every timed state sequence has infinitely many time steps. Summarizing, in timed state sequences state activities are interleaved with time activities. Throughout state activities time does not change, and throughout time steps the state does not change.

A timed state sequence $\rho = (\sigma, T)$ is a *computation* of a timed transition system if and only if state sequence σ is a computation of the underlying transition system and for every transition $t \in \mathcal{T}$ the following requirements are satisfied:

- for every transition $t \in \mathcal{T}$ and position $j \geq 0$ if t is taken at j then there exists a position $i, i \leq j$ such that $T_i + EET_t \leq T_j$ and t is enabled on $s_i, s_{i+1}, \ldots, s_{j-1}$ and is not taken at any of the positions $i, i+1, \ldots, j-1$, i.e., a transition must be continuously enabled for at least EET_t time units before the transition can be taken.
- for every transition $t \in \mathcal{T}$ and position $i \geq 0$, if t is enabled at position i, there exists a position $j, i \leq j$, such that $T_i + LET_t \geq T_j$ and either t is not enabled at j or t is taken at j, i.e., a transition must be taken if the transition has been continuously enabled for LET_t time units.

A finite timed state sequence is made infinite by adding an infinite sequence of idle transitions or time activities.

3.2 Syntax of real-time TTCN

In real-time TTCN, timing information is added in the declarations and the dynamic part of a test suite.

As shown in Figure 3 the specification of time names, time values and units is done in an Execution Time Declarations table. Apart from the headings the table looks much like the TTCN Timer Declarations table. Time names are declared in the Time Name column. Their values and the corresponding time units are specified on the same line in the Value and Unit columns. The declaration of time values and time units is optional.

Execution Time Declarations			
Time Name	**Value**	**Unit**	**Comments**
EET	1	s	EET value
LET	1	min	LET value
WFN	5	ms	Wait For Nothing
NoDur		min	No specified value

Figure 3 Execution Time Declarations Table.

EET and LET* are predefined time names with default values zero and infinity. Default time values can be overwritten (Figure 3).

Besides the static declarations of time values in an Execution Time Declarations table, changing these values within a behaviour description table can be done by means of assignments (Figure 4). However, evaluation of time labels should alway result in EET and LET values for which $0 \leq EET \leq LET$ holds. As indicated in Figure 4 we add a Time and a Time Options column to Test Case Dynamic Behaviour tables (and similar for Default Dynamic Behaviour and Test Step Dynamic Behaviour tables). An entry in the Time column specifies EET and LET for the corresponding TTCN statement. Entries may be constants (e.g., line 1 in Figure 4), time names (e.g., the use of NoDur on line 3), and expressions (e.g., line 6).

In general, EET and LET values are interpreted relative to the enabling time of alternatives at a level of indentation, i.e., the time when the level of indentation is visited the first time. However, some applications may require to define EET and LET values relative to the execution of an earlier test event, i.e, not restricted just to the previous one. In support of this requirement, a label in the Label column may not only be used in a GOTO but can also be used in the Time column, so that EET and LET values are computed relative to the execution time of the alternative identified by the label: In Figure 4 on line 6 the time labels (L1 + WFN, L1 + LET) are referring to the execution time of the alternative in line 1 (for which label L1 is defined).

Entries in the Time Options column are combinations of symbols M and N. Similar to using labels in expressions, time option N allows to express time values relative to the alternative's own enabling time even though some TTCN statements being executed in between two successive visits of the same level of indentation. Thus, the amount of time needed to execute the sequence of TTCN statements in between two successive visits is compensated: If time option N is defined, then execution of this alternative is not pre-emptive with respect to the timing of all alternatives at the same level of indentation.

In some executions of a test case, a RECEIVE or OTHERWISE event may be evaluated successfully before it has been enabled for EET units. If it is intended to define EET as a mandatory lower bound when an alternative may be evaluated successfully, then time option M has to be specified. Informally, if time option M is specified and the corresponding alternative can be successfully evaluated before it has been enabled for EET units, then this results in a FAIL verdict.

*We use different font types for distinguishing between syntax, EET and LET, and semantics, EET and LET.

Test Case Dynamic Behaviour							
Nr	Label	Time	Time Options	Behaviour Description	C	V	Comments
1	L1	2, 4	M	A ? DATA und			Time label Mandatory EET
2				(NoDur := 3)			Time assignment
3		2, NoDur		A ! DATA ack			
4				(LET := 50)			LET update (ms)
5				A ? Data ind			
6		L1 + WFN, L1 + LET	M, N	B ? Alarm			Mandatory EET not pre-emptive

Figure 4 Adding EET and LET values to behaviour lines.

3.3 Operational semantics of real-time TTCN

The operational semantics of real-time TTCN is defined in two steps:

1. We define the semantics of a TC using timed transition systems. An execution of a TC is given by a computation of the timed transition system associated with that TC. As time domain we use the real numbers $\mathbb{R}$ which are an *abstract* time domain in contrast to the *concrete* time domain of TTCN which counts time in discrete time units. Progress of time however is, however, a continuous process adequately modelled by $\mathbb{R}$.
2. The semantics of a test system is determined by composing the semantics of individual TC (for details see (Walter *et al.* 1997)).

Given a TC we associate with it the following timed transition system: A state $s \in \Sigma$ of a TC is given by a mapping of *variables* to *values*. The set of variables V includes constants, parameters and variables defined for the TC in the test suite and, additionally, a variable for each timer. Furthermore, we introduce a *control variable* π which indicates the location of control in the behaviour description of the TC. π is updated when a new level of indentation is visited. We let PCOs and CPs be pairs of variables so that each holds a queue of ASPs, PDUs or CMs sent and received, respectively.

In the initial state of a TC all variables have assigned their initial values (if specified) or being undefined. All PCO and CP variables have assigned an empty queue and all timer variables have assigned the value stop. The control variable π has been initialized to the first level of indentation. If the TC is not running, i.e., the TC has not been created yet, then all variables are undefined.

The set $\mathcal{T}$ of transitions contains a transition for every TTCN statement in the TC behaviour description and the idle transition t_I. Furthermore, we have a transition t_E which models all activities performed by the environment, e.g., the updating of a PCO, CP or timer variables. Execution of t_E changes the state of the TC because shared PCO, CP or timer variables are updated.

In the following we assume that the current level of indentation has been expanded as defined in Annex B of (ISO9646-3 1996). After expansion its general form is $A_1[eexp_1, lexp_1], \ldots, A_n[eexp_n, lexp_n]$, where A_i denotes an alternative and $eexp_i$, $lexp_i$ are expressions for determining EET and LET values of alternative A_i. The evaluation of expressions $eexp_i$ and $lexp_i$ depends on whether $eexp_i$ and $lexp_i$ make use of a label Ln. If so, absolute time references are converted into time references relative to the enabling time of the current set of alternatives.

Let eval be a function from time expressions to time values for EET or LET. Let enablingTime(A_i) be a function that returns the time when alternative A_i has been enabled. Let executionTime(Ln) be a function that returns the execution time of an alternative at the level of indentation identified by label Ln. Function NOW returns the current global time. Notice that for all alternatives A_i in a set of alternatives, enablingTime(A_i) is the same. Since only one alternative of a set of alternatives is executed, executionTime(Ln) returns the execution time of the executed alternative. For the evaluation of time expressions the following rules apply:

1. If $eexp_i$ and $lexp_i$ do not involve any operator Ln then EET = eval($eexp_i$) and LET = eval($lexp_i$). It is required that $0 \leq EET \leq LET$ holds; otherwise test case execution should terminate with a test case error indication.
2. If $eexp_i$ and $lexp_i$ involve any operator Ln then, firstly, executionTime(Ln) is substituted for Ln in $eexp_i$ and $lexp_i$ resulting in expressions $eexp_i'$ or $lexp_i'$, and secondly, EET = eval($eexp_i'$) – NOW and LET = eval($lexp_i'$) – NOW. It is required that $0 \leq EET \leq LET$ holds; otherwise test case execution should terminate with a test case error indication.

We say that alternative A_i is *potentially enabled* if A_i is in the current set of alternatives. A_i is *enabled* if A_i is evaluated successfully (Section 2.3), A_i is *executable* if A_i is enabled and A_i has been potentially enabled for at least EET_i and at most LET_i time units.

We make the evaluation of a TC explicit by defining the following refined snapshot semantics (cf. Section 2.3).

1. The TC is put into its initial state.
2. A snapshot is taken, i.e., PCO, CP and timer variables are updated and frozen.

   (a) If the level of indentation is reached from a preceding alternative (i.e., not by a GOTO or RETURN) then all alternatives are marked *potentially enabled* and the global time is taken and stored. The stored time is accessible by function enablingTime(A_i).
   (b) If the level of indentation is reached by executing a GOTO or RETURN and enabling-Time(A_i) has been frozen (see Step 5. below) then all alternatives are marked *potentially enabled* but enablingTime(A_i) is not updated.
   (c) If the level of indentation is reached by executing a GOTO or RETURN but enablingTime(A_i) has not been frozen previously then all alternatives are mar-

ked *potentially enabled* and the global time is taken and stored. The stored time is accessible by function enablingTime(A_i).

(d) Otherwise, it is a new iteration of Steps 2. - 5.

EET and LET are computed as described above.
If for an A_i enablingTime(A_i) + LET_i < NOW then test case execution stops (FAIL verdict).

3. All alternatives which can be evaluated successfully are marked *enabled.* If no alternative in the set of alternatives can be evaluated successfully then processing continues with Step 2.
If for an enabled alternative, say A_i, time option M is set and if enablingTime(A_i)+ EET_i > NOW then test case execution stops with a FAIL verdict.
4. An enabled alternative A_i is marked *executable* provided that enablingTime(A_i)+ EET_i $\leq$ NOW $\leq$ enablingTime(A_i) + LET_i and if there is another enabled alternative A_j with enablingTime(A_j) + EET_j $\leq$ NOW $\leq$ enablingTime(A_j) + LET_j, then $i < j$, i.e., the i-th alternative precedes the j-th alternative in the set of alternatives.
If no alternative can be marked executable then processing continues with Step 2.
5. The alternative A_i marked executable in Step 4. is executed. If a label Ln is specified then the alternative's execution time is stored and which can be accessed by function executionTime(Ln). If time option N is specified for the executed alternative, enablingTime(A_i) is frozen for later use. Control variable π is assigned the next level of indentation.
Test case execution terminates if the last level of indentation has been reached or a final test verdict has been assigned; otherwise, evaluation continues with Step 2.

Remarks: If any potentially enabled alternative cannot be evaluated successfully before latest execution time then a specified real-time constraint has not been met and test case execution stops. Conversely, if an alternative can be evaluated successfully before it has been potentially enabled for EET (Step 3.) then a defined real-time constraints is violated, too, and test case terminates with an error indication. In Step 4., the selection of alternatives for execution from the set of enabled alternatives follows the same rules as in TTCN (ISO9646-3 1996). If a TC stops (Step 5.) then the finite timed state sequence is extended to an infinite sequence by adding an infinite sequence of idle transitions. Every iteration of Steps 2. - 5. is assumed to be *atomic*.

In terms of the definitions given in Section 3.1, a computation of a TC is a timed state sequence $\rho = (\sigma, T)$. By substituting *potentially enabled* for *enabled* and *executed* for *taken*, the refined snapshot semantics can be stated formally as:

1. If alternative A is executed at position j of ρ then there exists positions i and l, $i \leq l \leq j$, such that $T_i + EET \leq T_j$ and enablingTime(A) = T_i and alternative A is evaluated successfully on all states $s_l, s_{l+1}, \ldots, s_{j-1}$ and is not executed at any position $l, l+1, \ldots, j-1$; i.e., alternative A is potentially enabled for at least

Test Case Dynamic Behaviour							
Nr	Label	Time	Time Options	Behaviour Description	CRef	V	C
1	L1	2, 4	M	PCO1 ? N-DATA ind	info		
2				...			
3				-> L1			
4		0, INFINITY		PCO2 ? N-ABORT ind	abort		
5				...			

Figure 5 Partial Real-Time TTCN Behaviour Description.

EET time units before it is executed provided it can be evaluated successfully after having been potentially enabled, and

2. for position $i \geq 0$, if enablingTime(A) $= T_i$ then for position j, $i \leq j$, $T_i + LET \geq T_j$ and alternative A is not evaluated successfully on any state $s_l, \ldots, s_j$ or A is executed at j provided no other alternative A' exists for which these conditions hold and which precedes A in the set of alternatives; i.e., the first alternative evaluated successfully is executed at latest LET units after being potentially enabled.

Example 1 In ISDN (Integrated Digital Services Network) systems (Halsall 1994, Tanenbaum 1989), the B channels are used by applications for data exchange whereas the D channel is used for the management of connections between users or application processes. We consider a scenario where an ISDN connection between test system and IUT has been established and where PCO1 and PCO2 are the respective B and D channel interfaces. At the B channels we expect to receive user data every $EET_1 = 2$ to $LET_1 = 4$ time units. At any time the ISDN connection may be aborted on the D channel.

We consider the partial real-time TTCN behaviour description given in Figure 5. The first alternative may be evaluated successfully and may be executed only in the interval $EET_1 = 2$ and $LET_1 = 4$ because time option M is set on line 1. Let us assume that at T' with enablingTime(A_1) $+ EET_1 \leq T' \leq$ enablingTime(A_1) $+ LET_1$, an N-DATA indication is received. The first alternative may be executed at T'' with enablingTime(A_1) $+ EET_1 \leq T' \leq T'' \leq$ enablingTime(A_1) $+ LET_1$ (Step 4.) because no other alternative is executable (no N-ABORT indication has been received yet). A corresponding computation might be:

$$\ldots \longrightarrow (s, \text{enablingTime}(A_1)) \xrightarrow{t_I}$$
$$(s, T') \xrightarrow{t_E} (s', T') \xrightarrow{t_I} (s', T'') \xrightarrow{t_1} (s'', T'') \longrightarrow \ldots$$

The reception of an N-DATA indication at time T' is a state activity, $(s, T') \xrightarrow{t_E} (s', T')$, because a PCO variable is updated by the environment performing transition t_E. Transitions t_I are time activities, and transition t_1 is the transition that is derived from TTCN statement line 1.

Suppose that an N-DATA indication and an N-ABORT indication have been received from the environment at some $T''' : T' \leq T''' \leq T''$. Then, although both alternatives are executable, the first alternative is executed because of the ordering of alternatives in the set of alternatives (Step 4.). If an N-DATA indication is received at $T <$ enablingTime$(A_i) + EET_1$ then test case execution stops with a FAIL verdict (Step 3.).

If no N-DATA indication and no N-ABORT indication have been received before LET_1 time units after the alternatives have been potentially enabled, test case execution stops with a FAIL verdict (Step 2.).

3.4 Discussion of the proposal

If we assume that no time values are defined (in this case EET and LET are zero and infinity, respectively), execution of a test case results in the same sequence of state-transitions as in TTCN. Therefore, our definition of real-time TTCN is compatible to TTCN (ISO9646-3 1996, Baumgarten and Gattung 1996).

Real-time TTCN combines property and requirement oriented specification styles. Time labels for TTCN statements, in general, define real-time constraints for the test system. A test system should be implemented so that it can comply with all properties defined. Time labels for RECEIVE and OTHERWISE events, which imply a communication with the IUT, define requirements on the IUT and the underlying service provider. As well as the test system, the underlying service provider is assumed to be "sufficiently reliable for control and observation to take place remotely" (ISO9646-1 1994). For real-time TTCN, the underlying service provider should also be sufficiently fast with respect to the timing of activities. Therefore, if a timing constraint of a RECEIVE or OTHERWISE event is violated, this clearly is an indication that the IUT is faulty and the test run should end with a FAIL verdict assignment.

In Figure 6, a test case in TTCN is given for the one in Example 1. The timing constraints on the reception of N-DATA indications are expressed using timers T1 and T2. The alternatives coded on lines 2 and 8 in combination check that an N-DATA indication should not be received before EET (= timer T1); otherwise, test case execution results in a FAIL verdict (line 8). The TIMEOUT event on line 6 controls the latest execution time and if timer T2 expires then this gives a FAIL verdict.

Let us assume that test case execution is at the third level of indentation (lines 3, 5 and 6) and that TIMEOUT of timer T2 precedes reception of an N-DATA indication. Furthermore, let us assume that the system executing the test case is heavily loaded and therefore evaluation of a set of alternatives lasts too long, so that both events are included in the same snapshot. The late arrival of an N-DATA indication gets undetected because of the ordering of alternatives on line 3, 5 and 6. A fast system will take a snapshot which includes the TIMEOUT only whereas a slow system will take a snapshot which includes an N-DATA indication and a TIMEOUT. For the slow system, the RECEIVE succeeds over the TIMEOUT event. Unfortunately, the behaviour description does not comply with the requirement stated in (ISO9646-

Test Case Dynamic Behaviour					
Nr	**Label**	**Behaviour Description**	**CRef**	**V**	**C**
1	L1	START T1(EET)			
2		?TIMEOUT T1 START T2(LET-EET)			
3		PCO1 ? N-DATA indication	data		
4		-> L1			
5		PCO2 ? N-ABORT indication STOP T2	abort	INCONC	
6		?TIMEOUT T2		FAIL	
7		PCO2 ? N-ABORT indication STOP T1	abort	INCONC	
8		PCO1 ? OTHERWISE STOP T1		FAIL	

Figure 6 TTCN Behaviour Description for Example 1.

1 1994) "that the relative speed of the systems executing the test case should not have an impact on the test result" and thus is not valid.

In conclusion, real-time TTCN is more powerful than TTCN. The advantage of real-time TTCN is that all requirements on the behaviour of test systems and IUT are made explicit. The timing constraints that are to be met and thus the result of a test case is determined by the observed behaviour only.

4 APPLICATION OF REAL-TIME TTCN

In this section we continue the discussion of real-time TTCN by elaborating on an example taken from high speed networking.

In ATM (Asynchronous Transfer Mode) networks (Black 1995, Prycker 1995), network traffic control is performed to protect network and users to achieve predefined network performance objectives. During connection set up a *traffic contract specification* is negotiated and agreed between users and network. A contract specification consists of the connection traffic descriptor, given in peak cell rate and cell delay variation tolerance; the requested quality-of-service class, given in terms of required cell loss ratio, cell transfer delay and cell delay variation; and the definition of a compliant connection.

A connection is termed *compliant* as long as the number of non-conforming cells does not exceed a threshold value negotiated and agreed in the traffic contract. If the number of non-conforming cells exceeds the threshold then the network may abort the connection. The procedure that determines conforming and non-conforming cells is known as the *generic cell rate algorithm* (GCRA(T,τ)) (Figure 7). The variant we discuss is referred to as *virtual scheduling* and works as follows (Prycker 1995): The algorithm calculates the theoretically predicted arrival times (TAT) of cells assuming equally spaced cells when the source is active. The spacing between cells is determined by the minimum interarrival time T between cells which computes to $T = 1/R_p$ with R_p the peak cell rate (per seconds) negotiated for the connection. If the actual arrival time of a cell t_a is after TAT $- \tau$, τ the cell delay variation tolerance caused, for instance, by physical layer overhead, then the cell is a conforming cell; otherwise, the cell is arriving too early and thus is being considered as a non-conforming cell. Traffic control subsumes all functions necessary to control, monitor

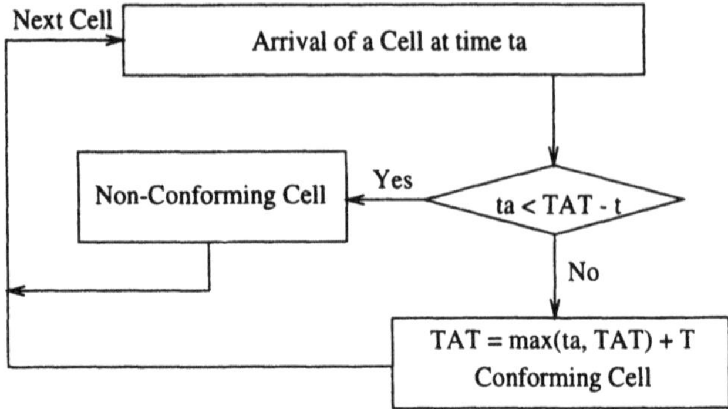

Figure 7 Generic Cell Rate Algorithm - Virtual Scheduling.

and regulate traffic at the user-network-interface (UNI). The correctly timed delivery of ATM cells at the UNI is important for a connection to be compliant.

A possible test purpose derivable from the informal definition of traffic contract specification and GCRA may be as follows: "It is to be tested that the amount of traffic (in terms of ATM cells) generated at the UNI is compliant to the traffic contract specification".

For the following discussion we assume a testing scenario as depicted in Figure 8. The IUT, i.e., the user's end-system, is connected to an ATM switch which in this scenario is the test system. Several ATM sources may generate a continuous stream of ATM cells which is, by the virtual shaper, transformed into a cell stream compliant with the traffic contract. Via the physical connection of end-system and ATM switch ATM cells are transferred. It is the test system that checks compliance of the received cell stream to the traffic contract.

The definition of a test case assumes that a connection has already been established so that a traffic contract specification is available. From the traffic contract, parameters R_p, T and τ can be extracted which are assigned to test case variables. The threshold value (for determining when a connection is to be aborted) is provided as a test suite parameter. For simplicity we let $\tau = 0$.

The definition of the dynamic test case behaviour (Figure 9) is based on the observation that according to the GCRA, except for the first cell, at most every $T (= EET)$ time units an ATM cell is expected from the IUT. Since we do not expect an ATM cell to arrive before T time units, time option M is defined. If an ATM cell arrives before T time units then the test case is terminated with a FAIL verdict.

This test case implies a threshold value of zero. If we allow for a number of non-conforming cells (NCC) greater than zero then the test case definition changes as shown in Figure 10. The difference compared to the previously discussed test case is that whenever an ATM cell arrives before T time units then counter NCC is incremented and is checked against the defined threshold. Time option N on line 2 instructs the system not to pre-empt the time constraint of the current set of alternatives. If control returns to level L2 from line 5 the enabling time is not updated.

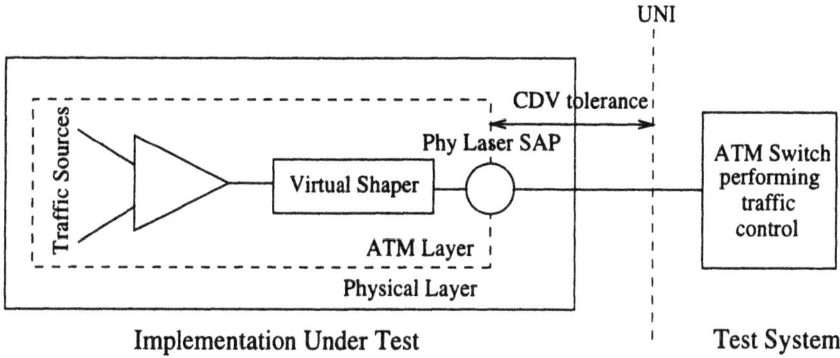

Figure 8 Generic Cell Rate Algorithm - Testing Scenario.

Test Case Dynamic Behaviour							
Nr	**Label**	**Time**	**Time Options**	**Behaviour Description**	**CRef**	**V**	**C**
1		0, INFINITY		UNI ? ATM-Cell	?		First cell to initialize GCRA
2	L2	T, INFINITY	M	UNI ? ATM-Cell	?		
3				-> L2			

Figure 9 Real-Time TTCN Behaviour Description for GCRA - Threshold = 0.

We have shown the use of time labels and time options. Without time options (in a previous paper, (Walter *et al.* 1997), we have used time labels only) the specification of both test cases would have been more complex. For the first test case it would have been necessary to introduce a second alternative similar to line 2 of Figure 10 instead of using time option M. For the second test case without time option N calculations of absolute and relative time values would have be necessary in order to adjust EET. Nonetheless, without real-time features, testing GCRA would have been impossible.

Test Case Dynamic Behaviour							
Nr	**Label**	**Time**	**Time Options**	**Behaviour Description**	**CRef**	**V**	**C**
1		0, INFINITY		UNI ? ATM-Cell (NCC := 0)	?		
2	L2	0, T	N	UNI ? ATM-Cell (NCC := NCC + 1)	?		
3				[NCC > Threshold]		FAIL	
4				[NCC <= Threshold]			
5				-> L2			
6		T, INFINITY		UNI ? ATM-Cell	?		
7				-> L2			

Figure 10 Real-Time TTCN Behaviour Description for GCRA - Threshold > 0.

5 CONCLUSIONS AND OUTLOOK

We have defined syntax and semantics of real-time TTCN. On a syntactical level TTCN statements can be annotated by time labels. Time labels are interpreted as earliest and latest execution times of TTCN statements relative to the enabling time of the TTCN statement. The operational semantics of real-time TTCN is based on timed transition systems (Henzinger *et al.* 1991). We have described the interpretation of real-time TTCN in timed transition systems. The applicability of real-time TTCN has been shown by an example: We have defined test cases for the generic cell rate algorithm employed in ATM networks for traffic control (Black 1995, Prycker 1995).

The motivation for our work has been given by the demand for a test language that can express real-time constraints. The increasing distribution of multimedia applications and real-time systems impose requirements on the expressive power of a test language that are not met by TTCN. Particularly, real-time constraints can not be expressed. However, for the mentioned new applications correctness of an implementation also with respect to real-time behaviour is essential and, thus, should also be tested.

In our approach a TTCN statement is annotated by time labels. The advantages of this approach are twofold: Firstly, only a few syntactical changes are necessary. Secondly, TTCN and real-time TTCN are compatible: If we assume that zero and infinity are earliest and latest execution times, a computation of a real-time TTCN test case is the same as in TTCN. A possible extension of our approach is to allow the use of time labels at a more detailed level, e.g., the annotation of test events, assignments and timer operations (an extension of (Walter *et al.* 1992, Walter and Plattner 1992)). Our future work will focus on these aspects.

Acknowledgements. The authors are indebted to Stefan Heymer for proofreading and for his detailed comments on earlier drafts of this paper. We are also grateful to the anonymous reviewers providing detailed comments and valuable suggestions which have improved contents and presentation of this paper.

6 REFERENCES

R. Alur, T. Feder, T. Henzinger. *The Benefits of Relaxing Punctuality*. Journal of the ACM, Vol. 43, 1996.

A. Ates, B. Sarikaya. *Test sequence generation and timed testing*. In Computer Networks and ISDN Systems, Vol. 29, 1996.

B. Baumgarten, A. Giessler. *OSI conformance testing methodology and TTCN*. Elsevier, 1994.

B. Baumgarten, G. Gattung. *A Proposal for the Operational Semantics of Concurrent TTCN*. Arbeitspapiere der GMD 975, 1996.

H. Bowman, L. Blair, G. Blair, A. Chetwynd. *A Formal Description Technique Supporting Expression of Quality of Service and Media Synchronization*. In Multimedia Transport and Teleservices. LNCS 882, 1994.

B. Berthomieu, M. Diaz. *Modeling and Verification of Time Dependent Systems Using Time Petri Nets*. In IEEE Transactions on Software Engineering, Vol. 17, No. 3, March 1991.

U. Black. *ATM - foundation for broadband networks*. Prentice-Hall, 1995.

A. Danthine, Y. Baguette, G. Leduc, Léonard. *The OSI 95 Connection-Mode Transport Service - The Enhanced QoS*. In High Performance Networking, IFIP, 1992.

S. Fischer. *Implementation of multimedia systems based on a real-time extension of Estelle*. In Formal Description Techniques IX Theory, application and tools, 1996.

F. Halsall, *Data Communications, Computer Networks and Open Systems*, Addison-Wesley, 1994.

D. Hogrefe, S. Leue. *Specifying Real-Time Requirements for Communication Protocols*. Technical Report IAM 92-015, University of Berne, 1992.

T. Henzinger, Z. Manna, A. Pnueli. *Timed Transition Systems*. In Real-Time: Theory in Practice. LNCS 600, 1991.

ISO/IEC. *Information Technology - OSI - Conformance Testing Methodology and Framework - Part 1: General Concepts*. ISO/IEC IS 9646-1, 1994.

ISO/IEC. *Information Technology - OSI - Conformance Testing Methodology and Framework - Part 2: Abstract Test Suite Specification*. ISO/IEC IS 9646-2, 1994.

ISO/IEC. *Information Technology - OSI - Conformance Testing Methodology and Framework - Part 3: The Tree and Tabular Combined Notation (TTCN)*. ISO/IEC IS 9646-3, 1996.

R. Keller. *Formal Verification of Parallel Programs*. In Communications of the ACM, Vol. 19, No. 7, 1976.

F. Kristoffersen, T. Walter. *TTCN: Towards a formal semantics and validation of test suites*. In Computer Network and ISDN Systems, Vol. 29, 1996.

L. Léonard, G. Leduc. *An Enhanced Version of Timed LOTOS and its Application to a Case Study*. In Formal Description Techniques VI, North-Holland, 1994.

S. Leue. *Specifying Real-Time Requirements for SDL Specifications - A Temporal Logic Based Approach*. In Protocol Specification, Testing and Verification XV, 1995.

R. Linn. *Conformance Evaluation Methodology and Protocol Testing*. In IEEE Journal on Selected Areas in Communications, Vol. 7, No. 7, 1989.

P. Merlin, D. Faber. *Recoverability of Communication Protocols*. In IEEE Transactions on Communication, Vol. 24, No. 9, September 1976.

M. Prycker. *Asynchronous transfer mode: solutions for broadband ISDN*, 3rd Edition, Prentice Hall, 1995.

R. Probert, O. Monkewich. *TTCN: The International Notation for Specifying Tests of Communications Systems*. In Computer Networks and ISDN Systems, Vol. 23, 1992.

J. Quemada, A. Fernandez. *Introduction of Quantitative Relative Time into LOTOS*. In Protocol Specification, Testing and Verification VII, North-Holland, 1987.

B. Sarikaya. *Conformance Testing: Architectures and Test Sequences.* In Computer Networks and ISDN Systems, Vol. 17, 1989.
I. Sommerville. *Software engineering.* 3rd Edition, Addison-Wesley, 1989.
A. Tanenbaum, *Computer Networks*, Prentice-Hall, 1989.
T. Walter, J. Ellsberger, F. Kristoffersen, P.v.d. Merkhof. *A Common Semantics Representation for SDL and TTCN.* In Protocol Specification, Testing and Verification XII, North-Holland, 1992.
T. Walter, B. Plattner. *An Operational Semantics for Concurrent TTCN.* In Proceedings Protocol Test Systems V, North-Holland, 1992.
T. Walter, J. Grabowski. *A Proposal for a Real-Time Extension for TTCN.* In Kommunikation in Verteilten Systemen '97, Informatik aktuell, Springer Verlag, 1997.

7 BIOGRAPHY

Thomas Walter received his Diploma in Informatics and his Doctorate degree in electrical engineering in 1987 and 1993, respectively. Since 1986 he is with the Computer Engineering and Networks Laboratory (TIK) of the Swiss Federal Institute of Technology (ETHZ). In the years 1991 and 1992 he led a project at the ETSI and in 1994 he participated in an EWOS project; both projects were in conformance testing. His current research interests include formal methos for specification and validation of real-time systems. Besides this he is also active in setting up an infrastructure for teleteaching at ETHZ.

Jens Grabowski studied Computer Science at the University of Hamburg, Germany, where he graduated with a diploma degree in 1990. From 1990 to 1995 he was research scientist at the University of Berne, Switzerland, where he received his Ph.D. degree in 1994. Since October 1995 Jens Grabowski is researcher and lecturer at the Institute for Telematics at the University of Lübeck, Germany. His research activities are directed towards network analysis, formal methods in protocol specification and conformance testing.

PART THREE

Wireless Testing

Integration of test procedures and trials for DECT handsets

A. Alonistioti, P. Kostarakis, K. Dangakis, A. A. Alexandridis, A.Paschalis, N. Gaitanis, S. Xiroutsikos, E. Adilinis, A. Vlahakis, P. Katrivanos
N.C.S.R. "Demokritos", Institute of Informatics and Telecommunications,153 10 Ag. Paraskevi, Athens, Greece.
TEL: +30 1 65 20 847, FAX: +30 1 65 32 175
e-mail: kdang@iit.nrcps.ariadne-t.gr

Abstract

In this paper we elaborate on a generic integrated reference modelling of trials execution procedure and test procedures for the Digital European Cordless Telecommunications (DECT) physical layer conformance testing. This model provides for a generic reference approach to trials execution steps and test case selection, based on the specific physical entities that have to be tested. Trials are presented, in the form of elementary test procedures pertinent to a specific physical entity that has to be tested (i.e., time, frequency, power) and auxiliary test procedures for synchronisation, calibration, etc.

Keywords

DECT, Conformance Testing, Test Procedures, Integration, Modelling

Testing of Communicating Systems, M. Kim, S. Kang & K. Hong (Eds)
Published by Chapman & Hall © 1997 IFIP

1 INTRODUCTION

For the deployment of the CTS-3/DECT test laboratories aiming to provide a conformance test service on DECT equipment in Europe, a test system and corresponding test suite have been developed. The framework for launching the DECT test facilities adheres to the ISO/IEC 9646 standard testing methodology (ISO/IEC 9646, 1993).

In this paper, the integration of trials execution procedure with respect to test procedures is introduced and a generic model for the mapping of elementary test procedures to trials execution steps is presented. This integration aims to advance the level of understanding of DECT test execution aspects in the communications industry and provide a framework addressed to formal/informal reference modelling development methodology for test procedures and trials execution mapping.

2 TEST SYSTEM CONFIGURATION

Conformance testing of the DECT physical layer involves the measuring of a physical entity (e.g. a modulated RF signal or a bit pattern) of the physical medium, that is, of the air interface. Thus, the DECT physical layer conformance testing requires an Equipment Under Test (EUT) and a Lower Test Unit (LTU) for controlling and observing the EUT through the air interface. All relevant procedures are described in (ISO/IEC 9646, 1993), (Alexandridis, 1995), (CEC/CTS3/DECT, 1994).

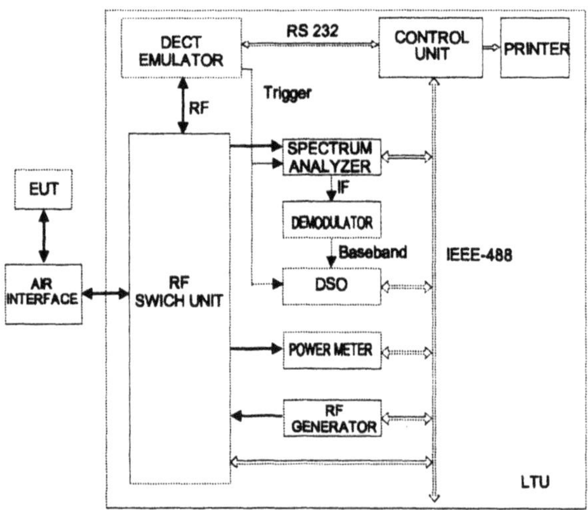

Figure 1 Test system configuration.

The configuration of LTU, that we have developed for performing the DECT physical layer conformance test suite is shown in Figure 1 as a part of the whole test system, based on (Papavramidis, 1992), (CEC/CTS3/DECT 1992). The EUT in our case is a DECT handset.

3 INTEGRATION OF TEST PROCEDURES INTO TRIALS

The test case execution tool is the "heart" of the integrated DECT test system. The test case execution tool is implemented in the form of test procedures. Every test procedure is considered as a primitive software module related with the physical entity that has to be tested according to the specific test case. Such entities are time (e.g., measurements of jitter), frequency (e.g., measurements of accuracy of RF carriers) and power (e.g., measurements of spurious emissions). In that sense, the following three basic categories of test procedures are defined and described in (Alexandridis, 1995): *time test procedures, frequency test procedures, power test procedures* In addition to the elementary test procedures, there is also another category of *auxiliary test procedures.* They include software modules for the initialization and setting of the test system's instruments and units, the selection of the appropriate RF signal path through the RF switch unit, the control and programming of the DECT Emulator. Every executable test case consists of one or more of the elementary test procedures that are accompanied by other auxiliary test procedures. This way the test cases can form groups with respect to the category of the test procedures used in each test case. Each group is symbolised by the initial letters of the test procedures used for a complete test execution (T: time test procedures, F: frequency test procedures, P: power test procedures and A: auxiliary test procedures).The corresponding trials execution can be mapped on the respective group test procedures. Four groups can be identified. The 4 groups are characterised by the test procedures they comprise, namely, 1) the TPA group, 2) the TFA group, 3) the TA group and 4) the PA group.

All the DECT test cases currently covered by the test case execution tool in association with the 4 test groups previously defined are given below:

TPA : Transmitter Release/Attack Time, Transmitter Minimum/Maximum Power, Maintenance of Transmission after Packet End, Transmitter idle power output, Peak Power per Transceiver.

TFA : RF Carrier Modulation, Accuracy and Stability of RF Carriers.

TA : Timing Accuracy and Stability (Jitter from slot-to-slot), Reference Timer Accuracy and Stability of a RFP.

PA : Emissions due to Modulation, Emissions due to Transmitter Transients, Spurious Emissions when allocated a transmit channel.

These test procedures are based on those described abstractly in (prTBR6, 1992). The trials execution procedures for each group are illustrated in Figures 2 and 3. The main basic steps for complete test execution and the values returned by each step are also indicated.

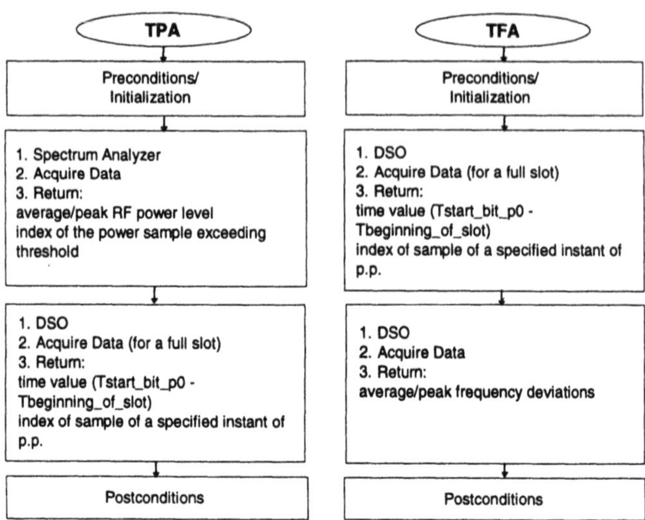

Figure 2 Trials execution procedures for the TFA and TPA groups.

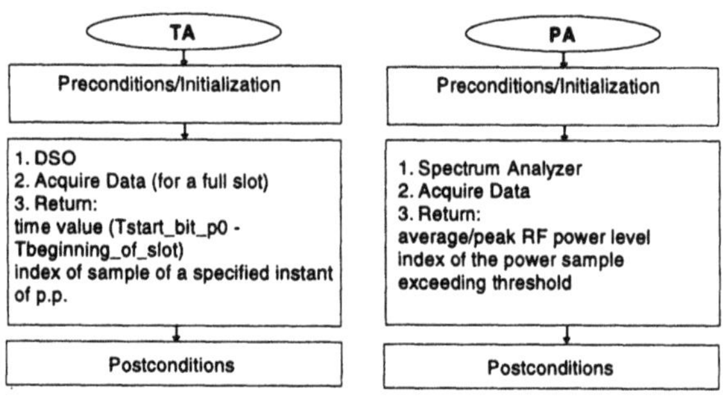

Figure 3 Trials execution procedures for the TA and PA groups.

Integration of the test procedures and trials provide an overview reference modelling of the manner in which trials are executed with respect to general, elementary test procedures. This aims at providing a common view with the communications industry of the test execution procedures to be followed during the DECT test service provisioning. Another target is to facilitate test selection with focus on the intended entities to be tested on the handset. The mapping of test procedures to trials execution procedures for the TPA and TFA test case groups are introduced in Figure 4.

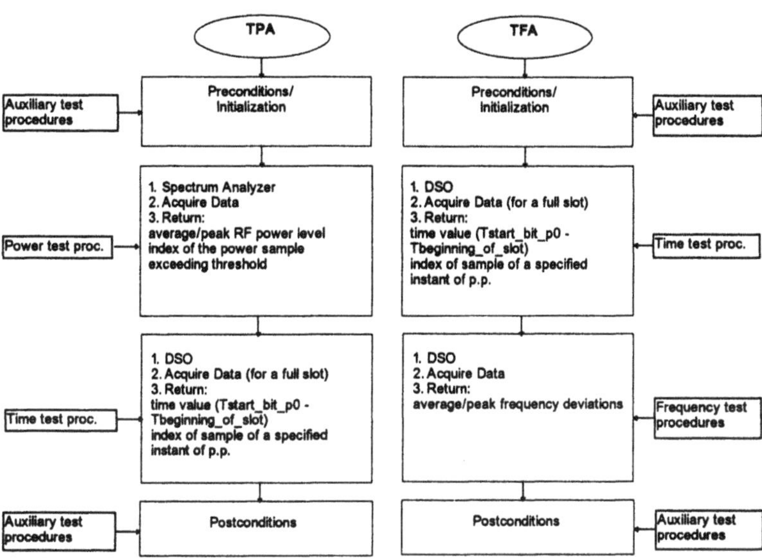

Figure 4 Trials execution procedures for the TFA and TPA groups and mapping of test procedures to trials execution procedures.

The mapping for the TA and PA test case groups is performed in a similar way. The mapping of test procedures and trials execution procedure provides a generic, integrated reference model for the trials execution steps with respect to test procedures and test case groups involved. The integrated model is illustrated in Figure 5. This provides for a generic evolutionary trials execution procedure and for efficient overview of test selection.

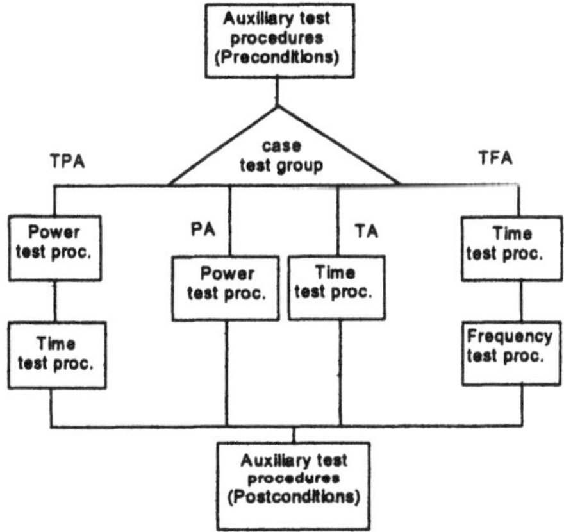

Figure 5 Integration of trials execution procedures and test procedures groups.

4 TEST TRIALS

Test suites, covering all the necessary test cases for DECT physical layer conformance testing, have been carried out on an equipment (DECT - handset) that was provided by a certain manufacturer. The use of the generic overview of trials execution procedures with respect to test procedures in the implementation of executable test cases is clarified below by describing, step-by-step, the execution of a specific test case.

The test case belongs to the TFA group and refers to measurement of accuracy and stability of RF carriers, that is measurement of the DECT portable part carrier frequency relative to an absolute frequency reference, set to the nominal frequency of the corresponding DECT RF channel. The steps during the execution of this test case are the following:

Using auxiliary test procedures, the DSO is initialised and programmed to start data acquisition at the beginning of a specific slot carrying the physical packet transmitted by the EUT. First, the frequency calibration module of the frequency test procedures generates the appropriate look-up table that maps the voltage levels at the demodulator output to frequency deviations from the nominal carrier frequency. The RF Switch Unit is programmed to select the appropriate RF signal path. The DECT Emulator (DECT Fixed Part) is given the command to establish a communication with the EUT (DECT Portable Part) in a specified channel and slot. As a result, the EUT is placed in a test mode whereby it performs the loopback function.

Using time test procedures, the DSO starts digitising the received physical packet for a full-slot duration and the acquired data are recorded. Then, the time, T_0, of the start of bit p0 (first bit of the physical packet) is returned. This time reference is used to specify a time window containing only the bits of the loopback field of a packet transmitted by the EUT during the first 1 sec after its transition from a non-transmitting mode to a transmit mode.

Using frequency test procedures the captured data contained in the predefined time window are processed and the EUT carrier frequency is calculated as the average of the measured frequencies corresponding to the voltage levels of the acquired samples.

Using auxiliary test procedures, the postconditions are set and this measurement is repeated following the same procedures as above but in this case data acquisition starts after allowing the EUT to be in an active locked state for more than 1 sec. This is accomplished by a time procedure delaying properly the start of the physical packet data acquisition.

For the specific EUT that was used, the measured carrier frequency deviations for the two cases mentioned above were equal to 8.985 kHz and 7.122 kHz respectively. Since these values were within ± 100 kHz and ± 50 kHz of the nominal DECT carrier frequency, that are the limits set by the standard (ISO/IEC 9646, 1993) for the two cases respectively, it is concluded that the EUT passed the test.

5 CONCLUSIONS

For acquiring a compact view of the DECT test service offered by the CTS-3/DECT test service laboratory, the test system configuration and the trials execution procedures were referenced. The elementary test procedures for DECT test cases execution were identified and an integrated reference model for the trials execution procedures modelling with respect to elementary test procedures was introduced in this paper. The model is evolutionary in the sense that it can be extended to cover also other test cases' execution that would cover future user needs. It is also generic, as it maps the test execution steps to elementary test procedures, providing a uniform way of execution. Furthermore, due to the methodology followed for its construction it provides an overview tool for test case selection. This model increases readability and understanding of test case execution procedures and serves as a base for common ground of understanding between the test service laboratory and the industrial customer for the trials provision.

6 REFERENCES

Alexandridis, A.A., Paschalis, A., Dangakis, K., Kostarakis, P., Lazarakis, F. and Gaitanis, N. (1995) Test Procedures for DECT Physical Layer Conformance Testing. International Conference on Telecommunications, ITC'95, Bali, Indonesia, 106-109.

CEC/CTS3/DECT project Deliverable 4 (1992) *List of Equipment and Facilities.*

CEC/CTS3/DECT project Deliverable 11 (1994) *Demonstration of Services.*

ISO/IEC 9646 (1993) *IT - OSI - Conformance Testing Methodology and Framework.*

Papavramidis, A., Paschalis, A., Myrtue, O., Dangakis, K., Tombras, G. and Kostarakis, P. (1992) A Conformance Test System for DECT Physical Layer. Proc. of the 3rd IEEE International Symposium on Personal, Indoor and Mobile Radio Communications, Boston, Massachusetts, 357-361.

7. BIOGRAPHY

Athanassia A. ALONISTIOTI is research assistant at the I.I.T. of the N.C.S.R. "DEMOKRITOS". Her current interests include third generation mobile communications (UMTS), SDL, formal methods and testing, modelling, simulation and design of object oriented platforms and protocols for advanced communications systems.

Panos D. KOSTARAKIS is research Director at the I.I.T. of the N.C.S.R. "DEMOKRITOS". He is actively involved in testing and type approval and he is Lloyd's registered auditor for ISO 9000 and member of the Total Quality forum.

Kostas P. DANGAKIS is researcher at the I.I.T. of the N.C.S.R. "DEMOKRITOS", and now head of the Mobile Communications Laboratory. His current interests include Mobile Communications and, specifically, Digital modulation and data transmission techniques, Spread Spectrum Systems and CDMA techniques.

Antonis A. ALEXANDRIDIS is researcher at the I.I.T. of the N.C.S.R. "DEMOKRITOS", in Mobile Communications Lab. His current interests include Mobile Communications, Propagation Models, Digital Modulation Techniques, and, specifically, Spread Spectrum Systems and CDMA techniques.

Antonis M. PASCHALIS is researcher at the I.I.T. of the N.C.S.R. "DEMOKRITOS". His current interests include research in the area of Testing, Design for Testability, Built-In Self Test, Automatic Test Generation, Self-Checking and Conformance Testing.

Nikos GAITANIS is R&D scientist at the I.I.T. of the N.C.S.R. "DEMOKRITOS", in the Hardware Technology Lab. His current interests include Conformance Testing, Formal Methods and Test case derivation.

Spyros K. XIROUTSIKOS is R&D scientist at the I.I.T. of the N.C.S.R. "DEMOKRITOS", in the Hardware Technology Lab. His current interests include Design and Development of control and communication systems.

Eleftherios S. ADILINIS is a technician at the I.I.T. of the N.C.S.R. "DEMOKRITOS", in Mobile Communications Lab. His current interests include Design and Development of advanced electronic circuits and software.

Antonis B. Vlahakis is R&D scientist at I.I.T. of the N.C.S.R. "DEMOKRITOS", in Mobile Communications Lab . His current interests are in the area of Communications, Power Electronics, Biomedical Electronics.

Panagotis L. Katrivanos is R&D scientist at I.I.T. of the N.C.S.R. "DEMOKRITOS", in Mobile Communications Lab . His current interests are in the area of Mobile Communications, Propagation aspects, channel characterization, simulation and modelling.

5

Development of an ETSI Standard for Phase-2 GSM/DCS Mobile Terminal Conformance Testing

*Hu., Wattelet. and Wang.**
ETSI
Route des Lucioles, 06921 Sophia Antipolis CEDEX, France
+33 4 9294 4369, +33 4 9365 3851, shicheng.hu@etsi.fr
jean-claude.wattelet@etsi.fr nwang@cs.ecnu.edu.cn

Abstract

The European Telecommunications Standards Institute (ETSI) is responsible for the production and publication of telecommunications standards in Europe and the marketing of these standards world-wide. The ETSI Special Mobile Group (SMG) is responsible for defining and specifying the GSM (Global System for Mobile communications) and UMTS (Universal Mobile Telecommunications System) standards. One of its Sub-Technical Committees, SMG7, is in charge of the specifications of mobile terminal testing standards. This experiences paper describes one of the major results from SMG7: a first Abstract Test Suite (ATS) developed as an European Telecommunication Standard (ETS) for mobile terminal conformance testing.

Keywords

ATS, DCS, ETSI, GSM, mobile, SMG, signalling test, standard, TTCN, validation

---

* Prof. Wang's current address: East China Normal University, Department of Computer Science
3663 N.Zhongshan Rd., 200062 Shanghai, China

Testing of Communicating Systems, M. Kim, S. Kang & K. Hong (Eds)
Published by Chapman & Hall © 1997 IFIP

1 GENERAL

The Phase-2† GSM mobile terminal conformance testing standards are specified in an ETSI standard, ETS 300 607 (GSM 11.10), in multiple parts. The testing standard covers the conformance requirements for mobile terminals at the radio interface, the reference point Um, for the frequency bands 900 for GSM and/or 1800 Mhz for DCS. The standard includes radio, speech, Subscriber Identity Module (SIM) testing, the Data Link (L2) and the Network Layer (L3) testing, and different kinds of service-related L3 protocol testing. More than 600 test cases are described and specified in English prose in the standard ETS 300 607-1.

No TTCN specification is available for the Phase-1 testing. In order to use full advantages of the testing methodology described in the ISO 9646, to reduce the costs of signalling (Network protocol) tests and the cost of the tester itself, and to avoid possible ambiguity of test descriptions, and therefore to increase the quality and applicability of the testing standard, it was decided to convert the existing L3 signalling test descriptions in the ETS 300 607-1 into an ATS in TTCN.

The development of the ATS for Phase-2 GSM/DCS mobile terminal conformance testing was started in September 1994. The design of the ATS was undertaken by ETSI project teams which were funded under European Commission Mandates‡. The ATS was published by ETSI in September 1996 as ETS 300 607-3. During this period, four ETSI TBRs (Technical Basis for Regulation) TBR19, TBR20, TBR31 and TBR32 were specified by SMG for the technical requirements to be met by mobile terminals capable of connecting to a public GSM/DCS network. A subset of the ATS has been selected as a part of mobile regulatory (approval) testing in Europe. In parallel to the ATS production, a TTCN stand-alone tester was developed by a UK-based company, Anite Systems, under the contract of the GSM Memorandum of Understanding association. Currently, the ETSI GSM ATS is used by 6 European test houses and more than 20 world-wide mobile manufacturers and GSM operators.

2 ABSTRACT TEST SUITE

The ATS was developed manually. The Test Suite Structure (TSS) and Test Purposes (TP) of the ATS are identical to those in the prose test specifications. The one-to-one mapping of TSS&TP ensures that an implementation of the ATS in a simpler test system (stand-alone TTCN tester) can have comparable test

---

† The development of GSM standards and the deployment of the network in terms of facilities available to users were undertaken in two mutually compatible phases: Phase-1 and Phase-2. The Phase-2+ evolves to introduce new technology and features on the Phase-2 basis.

‡ The projects were funded under the EC Mandates: BC-T237, BC-T247, BC-T-316 and co-funded by ETSI and EC under the EC mandate BC-T-342.

results with outcomes from an implementation of the prose test specification on a complex test platform (system simulator).

The GSM L3 protocol at the radio interface consists of three sub-layers for peer-to-peer communications: Radio Resource (RR) management functions, Mobility Management (MM) functions and Connect Management functions (CM). The CM sub-layer is composed of Call Control (CC), Short Message Service (SMS) support and Supplementary Services (SS) support. For each entity a corresponding test group tests all elementary procedures belonging to the entity. In addition, a test group called Structured Procedures underlines the testing of the relationships and interworking of procedures between different entities. Invalid and inopportune tests are sensible for radio protocols because of higher error rates. The BIBO group tests the mobile error handling behaviours. The Initial and Idle Mode test groups are similar to the basic interconnection test in the ISO 9646 conventions. An EGSM test group testing the mobile using the extended GSM frequency band. Finally, the General test group is contributed to the basic bearer or tele- service-related general signalling tests.

More than 700 essential conformance requirements have been identified for the mobile signalling testing. In order to reduce the number of tests most test cases have combined TPs containing 2-3 primary TPs on average. Combined TPs are mostly related to the same elementary procedure. Either they share initial test conditions, or they are described in a consecutive manner. The resulting ATS is more compact and has a total of 324 test cases without changing the test coverage. The ATS has 4.3 MB codes in MP form, under which 3% codes for Overview, 16% for Declarations, 26% for Constraints and 55% for Dynamic parts.

3 TEST METHOD AND TEST MODEL

The whole ATS is based on the Distributed test method in an SPyT context. The PCO consists of an L2 SAPI0 and an SAPI3. Different SAPI values indicate the respective data links. All test events are specified in terms of L2 Primitives and L3 PDUs: RR, MM, CC, SMS or SS message units. Because of having sub-layer model, messages are often chaining embedded. An SMS command as application layer PDU is embedded in a short message (SM) transfer PDU which is in turn embedded in an SM relay protocol (RP). The RP PDU is again embedded in an SM control PDU which is then embedded in an L3 SMS message. An another example is that different kinds of SS components in ASN.1 are embedded in a Facility information element of a CC or SS message in the TTCN definitions.

Since Dm control channels are distributed on several types of channels / sub-channels in several possible cells, a parameter referred to as logical channel is introduced to all L2 ASP type definitions to ensure correct distributions of the L3 messages on various types of channel. In order to check whether a message from the mobile under test is received in the correct time, another parameter for received TDMA (Time Division Multiple Access) frame number is defined as a time stamp in all DL_INDICATION Primitives type definitions, indicating the received first frame number of a received message by the tester.

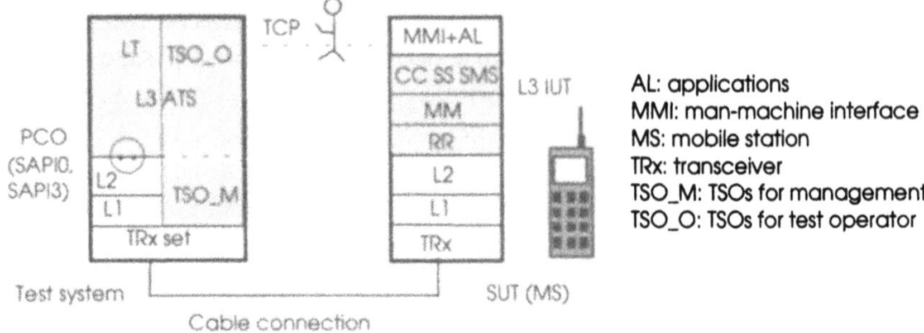

Figure 1 Distributed test method.

Besides the PCO, the ATS needs an additional management interface to communicate with the L2 or L1 of the tester in order to simulate the required GSM network behaviours for testing. The management interface was the most difficult design in the ATS. Effort was made to keep this interface independent of real implementations. Through this interface the ATS is able:

- to configure, deactivate or reactivate a radio channel within a cell;
- to stop a working cell or to generate lower layer failures;
- to manipulate a sending message on a required frame number;
- to prepare for frequency hopping, handover or to down-load a ciphering key;
- to read an L1 header;
- to pre-filter out periodically received measurement reports.

The management interface is specified by defining Test Suite Operations (TSOs). It was a design intention, not to define an additional PCO or a specific set of Primitives for building the management interface, but to keep a clear separation between the well-standardised interface represented by the key word PCO and an interface being standardised in the future version of ISO 9646-3. Nearly 30 TSO_M functions have been specified in the current ATS for this purpose.

To achieve the synchronisation between the (lower) tester and test operator (as upper tester) and management of information exchanges during testing nearly 70 test co-ordination procedures (TCP) are defined as TSOs (TSO_O).

4 PRACTICAL AND TECHNICAL CHALLENGES

It is generally expected within ETSI that an ETSI test suite can be compiled with little human intervention. In order to reach this goal, a series of questions had to be answered in the ATS design phase, for instance:

- how to manage channels;
- how to handle L3 periodic and synchronous signalling;
- how to initialise the tester and the mobile under test;
- how to fill the notation gaps between TTCN and ASN.1;
- how to handle upward compatibility.

4.1 Channel Management

Each logical channel needs to be mapped onto a physical channel which has many radio parameters, such as the frequency, time slots, timing advance, etc. to characterise the physical channel. One or more physical channels are mapped onto a radio transceiver. It was a design trade-off as to whether or not the ATS sees transceivers. To avoid over-standardisation, each test case needs to know only which logical channel is currently in use, and what is the physical mapping, and to leave transceivers out of the specification scope.

Furthermore, the stand-alone tester simulates the GSM network functionality at the radio interface. The channel management for each test case has been specified to a certain extent to ensure that test cases can run reasonably.

4.2 Handling synchronous signalling

The GSM L3 protocol contains several synchronous signalling. At the downlink (sending) direction System Information (SI) messages are periodically and synchronously broadcast on the Broadcasting Control Channel (BCCH) and on all Slow Associated Control Channels (SACCHs). At the uplink (receiving) direction the tester periodically receives synchronous measurement reports on the SACCHs. However, test events are based on request/ acknowledgement. To extract test events from a synchronous signalling background additional test semantics are defined.

An additional SI buffer is needed in the tester which is capable of storing all SI messages being sent. When an SI message is sent via the PCO, the SI will not be passed to the TTCN out-buffer, but is down-loaded to the SI buffer. As soon as the corresponding BCCH is configured or a SACCH channel is allocated and initialised, the tester controls the SI buffer sending out the stored SI messages periodically and correctly according to the TDMA timing.

The tester is in the position to prevent the measurement reports from entering the TTCN in-buffer and to provide a control way via the management interface.

4.3 Test initial conditions and preambles

The initial conditions for mobile testing vary to a large extent. It is necessary at the beginning of each test case to bring the mobile under test and the tester itself into pre-defined initial states irrespective of the tester or mobile current state or the SIM contents. This ensures test case independency from each other and is essential when interrupting test execution in the middle of a test, or changing a test execution sequence without changing any of the test results.

Moreover, 25 bearer and tele-services are defined in the Phase-2. It is also necessary before a call establishment to allocate a suitable traffic channel and to prepare an appropriate bearer capability for the service selected for the test.

To cope with these requirements and to avoid specifying a large number of test preambles and constraints for different initial test conditions several general, but highly parameterised test preambles, are designed in the ATS. All initial conditions are presented in terms of a set of independent parameters.

4.4 Bridging gaps between TTCN and ASN.1

The SS definitions in ASN.1 came from the GSM Mobile Application Part (MAP) based on the CCITT X.208 (1988) but in addition Ellipsis Notation ("...") is used wherever future MAP protocol extensions are foreseen. All SS specifications in the MAP protocol make use of the Remote Operation Services (ROS) which are served for the exchange of specific SS PDUs' invoking operations and reporting of results or errors. However, the current TTCN does not support:

- the ASN.1 ROS macro definitions in X208 (1988);
- the ASN.1 Ellipsis Notation.

To bridge the notation gaps between ASN.1 and TTCN, all GSM SS used have been redefined in a form which is acceptable by the TTCN and provides the ASN.1 data structures identical to the ROS expansions. The Ellipsis Notation is not tested at all in the ATS. The mobile under test is not allowed to send any protocol extension in the SS protocol.

4.5 Upward compatibility

The GSM is one of the fastest moving standardisation areas. When a new type of the Phase-2 mobile emerges on the market it can already encompass several new Phase-2+ features on top of the Phase-2 implementation. Thus, the ATS must:

- not fail the mobiles having implemented the new GSM features,
- be easily adapted if adding new features to the existing test cases.

To achieve the upward compatibility nearly 100 ICS and 150 IXIT parameters are defined in the ATS. The corresponding ICS/IXIT questions need to be answered by mobile manufacturer. The provided ICS/IXIT parameter values control not only test case selection, but also the selection of an executing path within a test case or a test step, or assigning appropriate values to a constraint according to the features and characteristics of the mobile. For instance, by using a few ICS/IXIT parameters the GSM and DCS mobile tests share the same ATS. Furthermore, the ATS can be also adapted to the PCS testing for mobiles working at the frequency band 1900 Mhz.

5 VALIDATION AND MAINTENANCE

The ATS has experienced more than one year of validation and maintenance. By exercising executive test cases at the TTCN stand-alone tester the ATS was intensively validated against 10 reference mobile terminals from 4 different sources. The validation results were audited by a third party. Discovered problems were reported to ETSI. A project team was responsible for the ATS maintenance. Based on the problem reports more than 250 change requests (CRs) were produced; 5 versions and 25 revisions of the ATS were delivered. To ensure a minimum quality of each Delivery the ATS was analysed by using 4 different TTCN tools for the syntax and static semantic cross checking. Under all the CRs, 63% were caused by fixing bugs, 22% clarified the informally specified TSO_M interface, 15% followed the changes in the GSM standards.

6 CONCLUSIONS

ETSI has developed a complex GSM ATS for GSM/DCS Phase-2 mobile conformance testing within a short period of time. This paper has highlighted the practical and technical challenges of ATS development. Despite the complexity of the GSM standards and various conformance requirements much effort has been made to ensure that the ATS is compilable, feasible, easily adaptable, upward compatible and largely reusable.

The ATS has been validated and is widely used both for mobile regulatory testing and in-house testing. The application of a TTCN specification to mobile terminal testing is now well accepted by the GSM society. However, using TTCN does not necessarily guarantee the correctness of the GSM ATS itself. The validated ATS and the continuous maintenance have added considerable values to the quality of mobiles.

New technology and features in the Phase 2+ GSM standards need to be introduced to the market as fast as possible. Requests have been received to develop new TTCN test specifications for the Phase-2+ mobile testing. The introduction of the ISO 9646-3 edition 2 Mock-Up to ETSI and its implementation in new TTCN tools are essential to meeting the challenge of producing high-quality GSM Phase-2+ mobile conformance test specifications.

7 REFERENCES

ETS 300 599 (GSM 09.02): "Digital cellular telecommunications system (Phase 2); Mobile Application Part (MAP) specification".

ETS 300 607-1 (GSM 11.10-1): "Digital cellular telecommunications system (Phase 2); Mobile Station (MS) conformance specification; Part 1: Conformance specification".

ETS 300 607-2 (GSM 11.10-2): "Digital cellular telecommunications system (Phase 2); Mobile Station (MS) conformance specification; Part 2: PICS Proforma".

ETS 300 607-3 (GSM 11.10-3): "Digital cellular telecommunications system (Phase 2); Mobile Station (MS) conformance specification; Part 3: ATS".

TBR 19: "European digital cellular telecommunications system; Attachment requirements for Global System for Mobile communications (GSM) mobile stations; Access".

TBR 20: "European digital cellular telecommunications system; Attachment requirements for Global System for Mobile communications (GSM) mobile stations; Telephony".

TBR 31: "European digital cellular telecommunications system; Attachment requirements for mobile stations; in the DCS 1800 band and additional GSM 900 band; Access".

TBR 32: "European digital cellular telecommunications system; Attachment requirements for mobile stations; in the DCS 1800 band and additional GSM 900 band; Telephony".

CCITT Recommendation X.208 (1988): "Specification of Abstract Syntax Notation One (ASN.1)".

8 BIOGRAPHY

Shicheng Hu

Is one of the permanent nucleus at the ETSI PEX competence centre for specification, validation and testing, supports SMG, was the project leader for GSM/DCS ATS, managed as a senior engineer several European testing projects at SIGOS , Germany, worked as a security specialist for Philips.

Jean-Claude Wattelet

Is a specialist for test and TBR specifications of the ETSI GSM and DECT protocols, maintained GSM/DCS ATS, worked as an engineer for CAP SESA.

Neng Wang

Is a Professor at the East China Normal University for computer networking, communications and protocol testing, developed GSM/DCS ATS, directed a numerous research projects, was a visiting professor at the University of Erlangen, Germany and the University of Illinois, USA.

PART FOUR

Data Part Test Generation

6

Automatic executable test case generation for extended finite state machine protocols

C. Bourhfir[1], R. Dssouli[1,2], E. Aboulhamid[1], N. Rico[2]

[1] DIRO, Pavillon André Aisenstadt, C.P. 6128, succursale Centre-Ville, Montréal, Québec, H3C-3J7, Canada.

E-mail: {bourhfir, dssouli, aboulham}@iro.umontreal.ca.

[2] Nortel, 16 Place du commerce, Verdun, H3E-1H6

Abstract

This paper presents a method for automatic executable test case and test sequence generation which combines both control and data flow testing techniques. Compared to published methods, we use an early executability verification mechanism to reduce significantly the number of discarded paths. A heuristic which uses cycle analysis is used to handle the executability problem. This heuristic can be applied even in the presence of unbounded loops in the specification. Later, the generated paths are completed by postambles and their executability is re-verified. The final executable paths are evaluated symbolically and used for conformance testing purposes.

Keywords

EFSM, Conformance testing, Control flow testing, Data flow testing, Executability, Cycle Analysis, Symbolic evaluation, Test case generation.

Testing of Communicating Systems, M. Kim, S. Kang & K. Hong (Eds)
Published by Chapman & Hall © 1997 IFIP

1 INTRODUCTION

In spite of using a formal description technique for specifying a system, it is still possible that two implementations derived from the same specification are not compatible. This can result from incorrect implementation of some aspects of the system. This means that there is a need for testing each implementation for conformance to its specification standard. Testing is carried out by using test sequences generated from the specification.

With EFSMs, the traditional methods for testing FSMs such as transition tours, UIOs, distinguishing sequences (DS), or W-Method are no longer adequate. The extended data portion which represents the data manipulation has to be tested also to determine the behaviors of the implementation. Quite a number of methods have been proposed in the literature for test case generation from EFSM specifications using data flow testing techniques (Sarikaya, 1986) (Ural, 1991) (Huang, 1995). However, they have focused on data flow testing only and control flow has been ignored or considered separately, and they do not consider the executability problem. As to control flow test, applying the FSM-based test generation methods to EFSM-based protocols may result in non-executable test sequences. The main reason is the existence of non-satisfied predicates and conditional statements. To handle this problem, data flow testing has to be used.

The generation of test cases in the field of communication protocols, combining both control and data flow techniques, has been well studied. In (Chanson, 1993), the authors presented a method for automatic test case and test data generation, but many executable test cases were not generated. This method uses symbolic evaluation to determine how many times an influencing self loop should be executed. An influencing transition is a transition which changes one or more variables that affect the control flow, and a self loop is a transition which starts and ends at the same state. The variables are called influencing variables. (Ural, 1991) does not guarantee the executability of the generated test cases because it does not consider the predicates associated with each transition. Also control flow testing is not covered. (Huang, 1995) generates executable test cases for EFSM-based protocols using data flow analysis and control flow is not tested. To handle the executability problem, this method uses a breadth-first search to expand the specification graph, according to the inputs read and to the initial configuration. It is a kind of reachability analysis. Hence, it has the same disadvantage, i.e. state explosion.

In this paper, we present a method which alleviates some of the existing problems. This method is different from (Huang, 1995) because it combines control and data flow testing instead of using only data flow testing. Unlike (Chanson, 1993) which verifies the executability after all the paths are generated and which considers only the self loops to solve the executability, our method verifies the executability during path generation which prevents from generating paths which will be discarded later. To make the non-executable paths executable, Cycle Analysis is performed in order to find the shortest cycle to be inserted in a path so that it becomes executable. A cycle is one or many transitions $t_1,t_2,..,t_k$ such that the ending state of t_k is the same as the starting state of t_1. Our method can also generate test cases for specifications with unbounded loops.

In the next section, concepts such as the FSM and EFSM models, conformance testing, data flow and control flow testing are described. Section 3 presents the general algorithm for executable test case and test sequence generation. In sections 4 and 5, the algorithm for executable definition-uses paths (or du-paths) generation is presented. This latter checks the executability during the du-path generation and uses cycle analysis to make the non-executable paths executable. Finally, in the last sections, we will compare the results obtained by our tool to those of another method and conclude the paper.

2 PRELIMINARIES

2.1 The FSM and EFSM models

Formalized methods for the specification and verification of systems are developed for simplifying the problems of design, validation and implementation. Two basically different approaches have been used for this purpose: modeling by FSMs, and specifications using high-level modeling languages.

The FSM model falls short in two important aspects: the ability to model the manipulation of variables conveniently and the ability to model the transfer of arbitrary values. For this reason, an FSM becomes cumbersome for simple problems (state explosion) because the number of states grows rapidly. This type of problems can be alleviated when EFSMs are used.

An EFSM is formally represented as a 6-tuple $<S, s_0, I, O, T, V>$ where

1. S is a non empty set of states,
2. s_0 is the initial state,
3. I is a nonempty set of input interactions,
4. O is a nonempty set of output interactions,
5. T is a nonempty set of transitions,
6. V is the set variables.

Each element of T is a 5-tuple t=(initial_state, final_state, input, predicate, block). Here *initial_state* and *final_state* are the states in S representing the starting state and the tail state of t, respectively. *input* is either an input interaction from I or empty. *predicate* is a predicate expressed in terms of the variables in V, the parameters of the input interaction and some constants. *block* is a set of assignment and output statements.

We assume that the EFSM representation of the specification is deterministic and that the initial state is always reachable from any state. In order to simplify the determination of the control and data flow graphs of a formal specification, it is convenient to transform the specification into an equivalent form containing only the so-called "*Normal Form Transitions*" (NFT). A method for generating a normal form specification from an ESTELLE specification is given in (Sarikaya, 1986).

2.2 Conformance testing

There are two approaches for checking conformance between an implementation and a specification. One approach is verification and the other is conformance testing. While verification techniques are applicable if the internal structure of the implementation is known, conformance testing aims to establish whether an implementation under test (IUT) conforms to its specification. If the implementation is given as a *black box*, only its observable behavior can be tested against the observable behavior of the specification. During a conformance test, signals are sent to (inputs) and received from (outputs) the implementation. The signals from the implementation are compared with the expected signals of the specification. The inputs and the expected outputs are described in a so-called test suite. A test suite is structured into a set of test cases. The execution of a test case results in a test verdict. From the test verdicts a conclusion about the conformance relation is drawn.

In recent years, several approaches have been developed for conformance test generation; these techniques are based upon traditional finite automata theory and usually assume a finite-state machine (FSM).

2.3 Fault models and control flow testing

The large number and complexity of physical and software failures dictate that a practical approach to testing should avoid working directly with those physical and software failures. One method for detecting the presence or absence of failures is by using a fault model to describe the effects of failures at some higher level of abstraction (logic, register transfer, functional blocks, etc.) (Bochmann, 1991).

The purpose of control flow testing is to ensure that the IUT behaves as specified by the FSM representation of the system and the fault model used to test it is the FSM model. The most common types of errors it tries to find are *transition (or operation) errors* which are errors in the output function and *transfer errors* (errors in the next state function) in the IUT.

Many methods for control flow testing exist. They usually assume that the system to be tested is specified as an FSM (transition tours, DS, W, etc.). Many attempts were made to generalize these methods to EFSM testing (Ramalingom, 1995) (Chanson, 1993). For control flow testing, we choose the UIO sequence for state identification since the input portion is normally different for each state and the UIO sequence for a state distinguishes it from all other states.

2.4 Data flow analysis

This technique originated from attempts in checking the effects of test data objects in software engineering. It is usually based on a data flow graph which is a directed graph with the nodes representing the functional units of a program and the edges representing the flow of data objects. The functional unit could be a statement, a transition, a procedure or a program. Data flow analyzes the data part of the EFSM in order to find data dependencies among the transitions. It usually uses a data-flow graph where the vertices represent transitions and the edges represent data and control dependencies. The objective is to test the dependencies between each definition of a variable and its subsequent use(s).

Definitions

A transition T has an assignment-use or **A-Use** of variable x if x appears at the left hand side of an assignment statement in T. When a variable x appears in the input list of T, T is said to have an input-use or **I-Use** of variable x. If a variable x appears in the predicate expression of T, T has a predicate-use or **P-Use** of variable x. T is said to have a computational-use or **C-Use** of variable x if x occurs in an output primitive or an assignment statement (at the right hand side). A variable x has a definition (referred to as **def**) if x has an A-Use or I-Use.

We now define some sets needed in the construction of the path selection criteria: def(i) is the set of variables for which node i contains a definition, C-Use(i) is the set of variables for which node i contains a C-use and P-Use(i,j) is the set of variables for which edge (i,j) contains a P-use. A path $(t_1,t_2...t_k,t_n)$ is **a def-clear-path** with respect to (w.r.t) a variable x if $t_2,..,t_k$ do not contain definitions of x.

A path $(t_1,...,t_k)$ is **a du-path** w.r.t a variable x if $x \in def(t_1)$ and either $x \in C-Use(t_k)$ or $x \in P-Use(t_k)$, and $(t_1,...,t_k)$ is a def-clear path w.r.t x from t_1 to t_k.

When selecting a criterion, there is, of course, a trade-off. The stronger the selected criterion, the more closely the program is scrutinized in an attempt to detect program faults. However, a weaker criterion can be fulfilled, in general, using fewer test cases. As the strongest criterion *all-paths* can be very costly, we will use the second strongest criterion *all-du-paths* (see (Weyuker, 1985) for all the criteria). P satisfies the *all-du-paths* criterion if for every node i and every $x \in def(i)$, P includes every *du-path* w.r.t x.

The main difference between the "all definition-use" or "all du" criterion and a fault model such as FSM fault model is the following: in the case of the "all du", the objective is to satisfy the criterion by generating test cases that exercise the paths corresponding to it. Exercising the paths does not guarantee the detection of existing faults because of variable values that should be selected. If the right values are selected then certain "du" criteria are comparable to fault models.

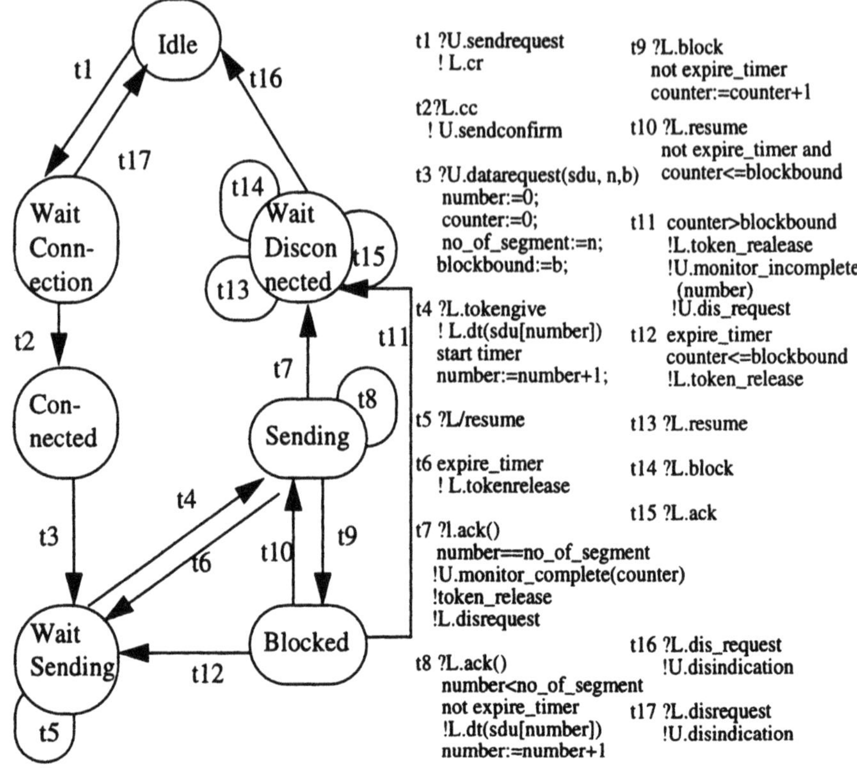

Figure 1. Example of an EFSM specified protocol (same as in (Huang, 1995)).

For transition t_3 in figure 1, I-Use(t3)={sdu, n, b}, A-Use(t3)={number, no_of_segement, blockbound, counter}, C-Use(t3)={n, b} and P-Use(t3)=$\varnothing$.

3 TEST CASE GENERATION

3.1 Choosing the values for the input parameters

The choice of the values of the input parameters has a sure impact on the test cases. These values may influence the number of times a cycle should be repeated. The user may specify valid and invalid values for each input parameter and our tool will choose randomly a value within the valid domain. If no value is specified, then if the input parameter influences the control flow, the user will be asked to enter a value for that input parameter.

3.2 Test case and test sequence generation

As we mentioned earlier, our method combines both control and data flow testing techniques to generate complete test cases (a complete test case is a test case which starts and ends at the initial state). Also, it verifies the executability during the du-path generation. The following algorithm illustrates the process of generating automatically executable test cases.

Algorithm *EFTG (Extended Fsm Test Generation)*

```
Begin
  Read an EFSM specification
  Generate the dataflow graph G form the EFSM specification
  Choose a value for each input parameter influencing the control flow
  Executable-Du-Path-Generation(G)
  Remove the paths that are included in others
  Add state identification to each executable du-path
  Add a postamble to each du-path to form a complete path
  For each complete path
      Re-check its executability
      If the path is not executable
          Try to make it executable
      EndIf
      If the path is still not executable Discard it
      EndIf
  EndFor
  For each uncovered transition T
      Add a path which covers it (for control flow testing)
  EndFor
  For each executable path
      Generate its input/output sequence using symbolic evaluation
  EndFor
End;
```

Procedure *Executable-Du-Path-Generation(flowgraph G)*

```
Begin
  Generate the set of A-Uses, I-Uses, C-Uses and P-Uses for each transition in G
  Generate the shortest executable preamble for each transition
  For each transition T in G
      For each variable v which has an A-Use in T
          For each transition U which has a P-Use or a C-Use of v
              Find-All-Paths(T,U)
          EndFor
      EndFor
  EndFor
End.
```

Table 1 presents the shortest executable preambles for the transitions in the EFSM in figure 1 (both input parameters n and b are equal to 2).

Table 1. Executable preambles for the EFSM's transitions in figure 1

Trans	*Executable Preamble*	*Trans*	*Executable Preamble*
t2	t1, t2	t10	t1, t2, t3, t4, t9, t10
t3	t1, t2, t3	t11	t1, t2, t3, t4, t9, t10, t9, t10, t9, t11
t4	t1, t2, t3, t4	t12	t1, t2, t3, t4, t9, t12
t5	t1, t2, t3, t5	t13	t1, t2, t3, t4, t8, t7, t13
t6	t1, t2, t3, t4, t6	t14	t1, t2, t3, t4, t8, t7, t14
t7	t1, t2, t3, t4, t8, t7	t15	t1, t2, t3, t4, t8, t7, t15
t8	t1, t2, t3, t4, t8	t16	t1, t2, t3, t4, t8, t7, t16
t9	t1, t2, t3, t4, t9	t17	t1, t17

The reason we start by finding the shortest executable preamble for each transition is as follow: Suppose we want to find all executable du-paths between t_3 and t_7. Since t_3 needs a preamble, then any path from t_3 to t_7 cannot be made executable unless an executable (or feasible) preamble is attached to it.

When finding the preambles and postambles, we try to find the shortest path which does not contain any predicate. If we fail to find such a path, then we choose the shortest path and try eventually to make it executable.

4 EXECUTABLE DU-PATH GENERATION

In (Chanson, 1993), after adding preambles and postambles to the du-paths, their executability is verified. However, many paths remain non-executable and are discarded because the predicates associated with some transitions are not satisfied. To overcome this problem, we verify the executability of each path during its generation. Below is the algorithm which finds all the paths between two transitions.

```
Procedure Find-all paths(T1, T2, var)
Begin
  If a preamble, a postamble or a cycle is to be generated
     Preamble:=T1
  Else
     Preamble:= the shortest executable preamble from the first transition to T1
  EndIf
  Generate-All-Paths(T1,T2,first-transition, var, preamble)
End;
```

The following algorithm is the algorithm used to find all executable preambles and all executable du-paths between transition T1 and transition T2 with respect to the variable var defined in T1.

```
Procedure Generate-All-Paths(T1, T2, T, var, Preamble)
Begin
  If (T is an immediate successor of T1) (e.g. t3 is an immediate successor of t2)
    If (T=T2 or (T follows T1 and T2 follows T in G)) (e.g. t4 follows t2)
      If we are building a new path
        Previous:= the last generated du-path (without its preamble)
        If (T1 is present in the previous path)
          Common:= the sequence of transitions in the previous path before
          T1
        EndIf
      EndIf
      If we are building a new path
        Add Preamble to Path, Add var in the list of test purposes for Path
      EndIf
      If Common is not empty
        Add Common to Path
      EndIf
      If (T = T2)
        Add T to Path, Make-Executable(Path)
      Else
        If T is not present in Path (but may be present in Preamble) and T does
          not have an A-use of var
          Add T to Path
          Generate-All-Paths(T, T2, first-transition, var, Preamble)
        EndIf
      EndIf
    EndIf
  T:= next transition in the graph
  If (T is not Null) Generate-All-Paths(T1, T2, T, var, Preamble)
  Else
    If (Path is not empty)
      If (the last transition in Path is not an immediate precedent of T2)
        Take off the last transition in Path
      Else
        If (Path is or will be identical to another path after adding T2)
          Discard Path
        EndIf
      EndIf
    EndIf
  EndIf
End.
```

The algorithm used to find the postambles and the cycles is also similar, except that it does not call the procedure Make-Executable(Path).

Suppose P1=(t_1,t_2,..t_{k-1},t_k). Make-Executable(P1) finds the non-executable transition t_k in P1 if it exists. Then it finds if another executable du-path P2=(t_1,t_2,..,t_{k-1},...,t_k) exists. If such path exists, P1 is discarded. If not, the procedure Handle-Executability(P1) is called (see next section). This verification enables to save time generating the same path or an equivalent path (the same du-path with different cycles in it) more than once. Handle-Executability(Path) starts by verifying if each transition in Path is executable or not. In each transition, each predicate is interpreted symbolically until it contains only constants and input parameters and the algorithm can determine if the transition is executable or not (especially for simple predicates). However, for some specifications with unbounded loops, Handle-Executability may not be able to make a non-executable path executable.

Table 2 shows all the du-paths (with the preamble (t_1, t_2, t_3, t_4)) form t_9 to t_{10} w.r.t the variable counter and the reason why some paths were discarded. All the paths that were discarded because the predicate became (3=2) cannot be made executable, because the influencing transition (t_4 or t_8) appears more than it should be.

Table 2. All du-paths form t9 to t7 w.r.t. counter

Du-Path	*Discarded*	*Reason path is discarded*
1,2,3,4,9,10,6,4,7	no	-
1,2,3,4,9,10,6,4,8,7	yes	predicate in t7 become (3=2)
1,2,3,4,9,10,6,5,4,7	no	-
1,2,3,4,9,10,6,5,4,8,7	yes	predicate in t7 become (3=2)
1,2,3,4,9,10,7	yes	will be equivalent to the first path after solving the executability
1,2,3,4,9,10,8,6,4,7	yes	predicate in t7 become (3=2)
1,2,3,4,9,10,8,6,5,4,7	yes	predicate in t7 become (3=2)
1,2,3,4,9,10,8,7	no	-
1,2,3,4,9,12,4,7	no	-
1,2,3,4,9,12,4,8,7	yes	predicate in t7 become (3=2)
1,2,3,4,9,12,5,4,7	no	-
1,2,3,4,9,12,5,4,8,7	yes	predicate in t7 become (3=2)

In the next section, we will show what cycle analysis is and how it can be used to make the non-executable paths executable.

5 HANDLING THE EXECUTABILITY OF THE TEST CASES

The executability problem is in general undecidable. However, in most cases, it can be solved. (Ramalingom, 1995) deals essentially with the executability of the preambles and postambles, and not with the executability of the du-paths covering the data flow. (Huang, 1995) overcame this problem by executing the EFSM. This method does not cover the control flow and may not deal with large EFSMs. (Chanson, 1993) used static loop analysis and symbolic evaluation techniques to determine how many times the self loop should be repeated so that test cases become executable. This method is not appropriate for specifications where the influencing variable is not updated inside a self loop, such as the EFSM in figure 1, and cannot be used if the number of loop iterations is not known. For these reasons, the following heuristic was developed in order to find the appropriate cycle to be inserted in a non-executable path to make it executable.

```
Procedure Handle_Executability(path P)
Begin
  Cycle:= not null
  Process(P)
  If P is still not executable Remove it
  EndIf
End;

Procedure Process(path P)
Begin
  T:= first transition in path P
  While (T is not null)
     If (T is not executable)
          Cycle:= Extract-Cycle(P,T)
     EndIf
     If (Cycle is not empty)
          Trial:=0
          While T is not executable and Trial<Max_trial Do
             Let Precedent be the transition before T in the path P
             Insert Cycle in the path P after Precedent
             Interpret and evaluate the path P starting at the first transition
             of Cycle to see if the predicates are satisfied or not
             Trial:= Trial+1
          EndWhile
     Else
          Exit
     EndIf
     T:= next transition in P
  EndWhile
End.
```

We would like to mention that our tool makes a difference between two kinds of predicates. A binary predicate has the following form: "var1 R var2", where R is a relational operator such as "<"; while a unary predicate can be written as F(x), where F is a boolean function such as "Even(x)" (see figure 2).

The heuristic "Handle-Executability" verifies if each non-executable path can be made executable and uses the procedure "Extract-Cycle(P,T)" to find the shortest cycle, if it exists, to be inserted in a non-executable path in order to make it executable. For this purpose, we find the first non-executable transition T in the path P. Two cases may arise: If the transition T cannot be executed because some unary predicate is not satisfied, we find a transition t_k, if it exists, among the transitions preceding transition T, which has the same predicate with a different value. An influencing cycle containing t_k is generated (if it exists) and inserted in the path P before transition T. If the predicate is not a unary predicate, we find out, using symbolic evaluation, what the variable causing the non-executability is, and whether it should be increased or decreased for the transition t_k to be executable. This variable must be an influencing one and transitions which update the variable must exist. If this is not the case, an empty cycle is returned, and the path is discarded. If the variable in the predicate is an influencing variable, we search among the transitions preceding T, for a transition t_k which updates properly the variable, generate a cycle containing this variable and insert it in the path. If a path cannot be made executable, it is discarded.

To illustrate the heuristic, suppose that in the EFSM of figure 1, both variables n and b have the value 2. The shortest preamble for t_{11} is (t_1, t_2, t_3, t_4, t_9, t_{11}), but t_{11} is not executable because its predicate "counter>2" becomes "1>2" after interpretation. Our tool finds that the influencing variable is "counter" and that among the transitions preceding t_{11}, t_9 is an influencing transition which may be adequate, because it increases the variable "counter". The cycle (t_{10},t_9) is generated and inserted twice after transition t_9. The path becomes (t_1, t_2, t_3, t_4, t_9, t_{10}, t_9, t_{10}, t_9, t_{11}).

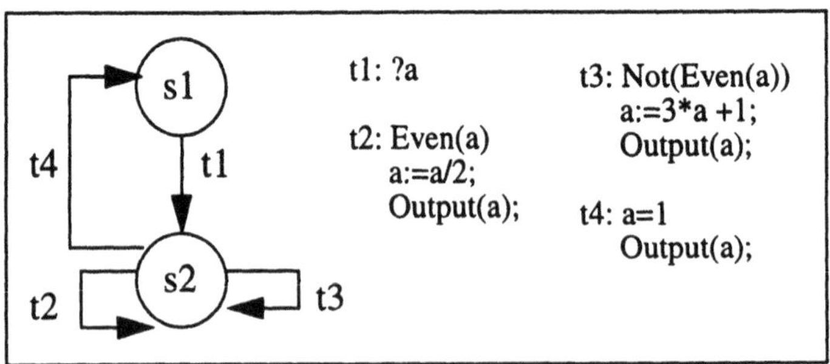

Figure 2. EFSM with unbounded loops and unary predicates.

Figure 2 presents an example of an EFSM with unbounded loops. Each loop is a self-loop with a unary predicate. For this example, since transitions t_2 and t_3 are not bounded, (Chanson, 1993) (Ramalingom, 1995) cannot generate any executable test case for this example.

In table 3, the executable test cases (without state identification) and test sequences for the EFSM in figure 2 are presented. Each test case is relative to one value for the input parameter a.

Table 3. Executable test cases for the EFSM in figure 2

Input parameter	*Executable test case*	*Input/Output sequence*
1	t1, t4	?1!1
5	t1, t3, t2, t2, t2, t2, t4	?5! 16! 8! 4! 2! 1!1
100	t1, t2, t2, t3, t2, t2, t3, t2, t3, t2, t2, t2, t3, t2, t3, t2, t2, t3, t2, t2, t2, t3, t2, t2, t2, t2, t4	?100!50!25!76!38!19!58!29! 88!44!22!11!34!17!52!26!13,!40, 20!10!5!16!8!4!2!1!1
125	-	-

For the EFSM in figure 2, our tool failed to generate any executable test case for a=125. But when we increased the value of the variable Max-Trial (in the procedure Process) , a solution was found. Giving our tool more time to let it find a solution does not mean that a solution will be found. In these cases, out tool cannot decide if a solution exists. After the generation of executable paths, input/output sequences are generated. The inputs will be applied to the IUT, and the observed outputs from the IUT will be compared to the outputs generated by our tool. A conformance relation can then be drawn.

6 RESULTS

Table 4 presents the final executable test cases (without state identification) generated by our tool on the EFSM in figure 1. In many cases, the tool had to look for the influencing cycle to make the test case executable. With state identification, the first executable path will look like: (t_1, t_2, t_3, t_5, t_4, t_8, t_7, $\mathbf{t_{15}}$, t_{16}). The last two paths are added to cover the transitions t_{13}, t_{14}, t_{15} and t_{17} which were not covered by the other paths.

Table 4. Executable test cases for the EFSM of Figure 1

No	*Executable Test Cases*	*Test Purposes*
1	t1, t2, t3, t5, t4, t8, t7, t16,	number, counter, no_of_segment
2	t1, t2, t3, t5, t4, t8, t9, t10, t7, t16	number, counter, no_of_segment, blockbound
3	t1, t2, t3, t5, t4, t9, t10, t8, t7, t16	number, counter, no_of_segment, blockbound
4	t1, t2, t3, t4, t8, t9, t10, t9, t10, t9, t11, t16	number, counter, no_of_segment, blockbound
5	t1, t2, t3, t5, t4, t8, t9, t10, t9, t10, t9, t11, t16	number, counter, no_of_segment, blockbound
6	t1, t2, t3, t5, t4, t9, t10, t9, t10, t9, t11, t16	number, counter, blockbound
7	t1, t2, t3, t5, t4, t9, t12, t4, t7, t16	number, counter, blockbound
8	t1, t2, t3, t4, t6, t4, t7, t16	number, counter, no_of_segment
9	t1, t2, t3, t4, t6, t5, t4, t7, t16	number
10	t1, t2, t3, t4, t8, t7, t16	number, counter, no_of_segment
11	t1, t2, t3, t4, t8, t9, t10, t7, t16	number, counter, no_of_segment, blockbound
12	t1, t2, t3, t4, t9, t10, t6, t4, t7, t16	number, counter, no_of_segment, blockbound
13	t1, t2, t3, t4, t9, t10, t6, t5, t4, t7, t16	number, counter, blockbound
14	t1, t2, t3, t4, t9, t10, t8, t7, t16	number, counter, no_of_segment, blockbound
15	t1, t2, t3, t4, t9, t12, t4, t7, t16	number, counter, blockbound
16	t1, t2, t3, t4, t9, t12, t5, t4, t7, t16	number, counter, blockbound
17	t1, t2, t3, t4, t9, t10, t9, t10, t9, t11, t16	number, counter, blockbound
18	t1, t2, t3, t4, t9, t10, t6, t4, t9, t10, t9, t11, 16	number, counter, blockbound
19	t1, t2, t3, t4, t9, t10, t6, t5, t4, t9, t10, t9, t11, t16	number, counter, blockbound
20	t1, t2, t3, t4, t9, t10, t8, t6, t4, t9, t10, t9, t11, t16	number, counter, no_of_segment, blockbound
21	t1, t2, t3, t4, t9, t10, t8, t6, t5, t4, t9, t10, t9, t11, t16	number, counter, no_of_segment, blockbound
22	t1, t2, t3, t4, t9, t10, t8, t9, t10, t9, t11, t16	number, counter, no_of_segment, blockbound
23	t1, t2, t3, t4, t9, t12, t4, t9, t10, t9, t11, t16	number, counter, blockbound
24	t1, t2, t3, t4, t9, t12, t5, t4, t9, t10, t9, t11, t16	number, counter, blockbound
25	t1, t2, t3, t4, t8, t7, t13, t14, t15, t16	-
26	t1, t17	-

The sequence of input/outputs is extracted from the executable test cases, and applied to test the IUT. For output parameters with variable (such as the output "dt"), symbolic evaluation is used to determine the value of the variable number which has an output use (see Table 3 for an example).

In order to compare our tool to other methods, we implemented an algorithm which generates all the du-paths (like in (Chanson, 1993)), to which we added Cycle Analysis to handle the executability problem instead of loop analysis. We shall call this algorithm "Ch+". Note that "Ch+" verifies the executability after all the du-paths are generated.

Table 5. Results obtained by Ch+ and by our tool

EFSM	*Ch+*		*Our tool*			
	du-paths	Exec	du-paths	discarded du-paths		Exec
fig 1	81	26	60	29	16	26
fig 2	9	1	1	0	0	1
INRES	54	25	24	4	4	22

In table 5, the results obtained by Ch+ and by our tool on three EFSMs are summarized. The third EFSM is a simplified version of the INRES protocol. It has four states, fourteen transitions, four loops two of which are influencing self-loops.

The first column of discarded "du-paths by our tool" specifies the total number of discarded paths during du-path generation. The second column specifies the number of paths that were discarded by the tool without trying to make them executable, because equivalent paths already existed. "Exec" stands for executable.

7 CONCLUSIONS AND FUTURE WORK

As me mentioned earlier, for the EFSM in figure 1, our tool discarded only twenty nine paths (during du-paths generation) while Ch+ discarded fifty five (after generating all the du-paths). Verifying the executability of the du-paths during their generation enables to generate only those paths which are more likely to be executable. Our method generates executable test cases for EFSM-specified systems by using symbolic evaluation techniques to evaluate the constraints along each transition, so only executable test sequences are generated. Also, our method discovers more executable test cases than the other methods and enables to generate test cases for specifications with unbounded loops.

This work is supported by an NSERC Strategic grant STR0167072.

8 REFERENCES

Bochmann, G. v. Das, A. Dssouli, R. Dubuc, M. Ghedamsi, A. and Luo, G. (1991) Fault models and their use in Testing. Proc. IFIP Intern. Workshop on Protocol Test Systems (invited paper), pp. (II-17)-(II-32).

Chanson, S. T. and Zhu, J.(1993) A Unified Approach to Protocol Test Sequence Generation. In Proc. IEEE INFOCOM.

Huang, C.M. Lin, Y.C. and Jang, M.Y. (1995) An Executable Protocol Test Sequence Generation Method for EFSM-Specified Protocols. International Workshop on Protocol Test Systems (IWPTS), Evry, 4-6 September.

Ramalingom, T. Das, A. and Thulasiraman, K.(1995). A Unified Test Case Generation Method for the EFSM Model Using Context Independent Unique Sequences. International Workshop on Protocol Test Systems (IWPTS), Evry, 4-6 September.

Sarikaya, B. and Bochmann, G.v. (1986) Obtaining Normal Form Specifications for Protocols. In Computer Network Usage: Recent Experiences, Elsevier Science Publishers.

Ural, H. and Yang. B. (1991) A Test Sequence Selection Method for Protocol Testing. IEEE Transactions on Communication, Vol 39, No4, April.

Weyuker, E.J. and Rapps, S. (1985) Selecting Software Test Data using Data Flow Information. IEEE Transactions on Software Engineering, April.

9 BIOGRAPHY

Chourouk Bourhfir is a Ph-d student in the Département d'Informatique et de Recherche Opérationnelle (DIRO), Université de Montréal. She received the M.Sc. degree in Computer Science in Université Laval, Canada in June 1994. Her research interests include modeling and automatic test generation.

Rachida Dssouli is professor in the DIRO, Université de Montréal. She received the Doctorat d'université degree in computer science from the Université Paul-Sabatier of Toulouse, France, in 1981, and the Ph.D. degree in computer science in 1987, from the University of Montréal, Canada. She is currently on Sabbatical at NORTEL, Ile des Soeurs. Her research area is in protocol engineering and requirements engineering.

El Mostapha Aboulhamid, received his Ing. Degree from INPG, France, in 1974, The M.Sc. and Ph.D. degrees from Universite de Montréal in 1979 and 1985 respectively. Currently, he is an associate Professor at Université de Montréal. His current research interests include Hardware software codesign, testing, modeling and synthesis.

Nathalie Rico is currently working in Nortel, Ile des soeurs, Montréal. She is the manager of the DMS Network Applications group.

A method to derive a single-EFSM from communicating multi-EFSM for data part testing

I. Hwang, T. Kim, M. Jang, J. Lee, S. Lee, H. Oh[*]*, and M. Kim*[**]
Department of Electronic Engineering, Yonsei University
Seodaemun-Gu Shinchon-Dong 134, Seoul, Korea, 120-749
Phone: +82-2-361-2864, Fax: +82-2-312-4584
E-mail: {his, jyl}@nasla.yonsei.ac.kr
[*]*Electronics and Telecommunications Research Institute(ETRI)*
Yusong-Gu Kajong-Dong 161, Taijon, Korea, 305-350
E-mail: hsohs@pec.etri.re.kr
[**]*Korea Telecom Research Laboratories*
Socho-Gu Umyun-Dong 17, Seoul, Korea, 137-792
E-mail: mckim@sava.kotel.co.kr

Abstract

The necessity that test sequences are automatically generated from a protocol specification for the purpose of testing data flow in the implementation has been emphasized because the cost of the most existing data part testing strategies is prohibitively high. However, existing automatic test generation methods based on single-module structure are not applicable to real protocol having multi-module one. In this paper, we propose a method which transforms multi-module model into an equivalent single-module. Since the proposed method uses the reachability analysis technique, it can minimize semantic loss of the specification during the transformation process.

Keywords

Conformance testing, data part testing, test generation, single-module

Testing of Communicating Systems, M. Kim, S. Kang & K. Hong (Eds)
Published by Chapman & Hall © 1997 IFIP

1 INTRODUCTION

Correct implementation of the protocol and the interoperability of the heterogeneous system have been emphasized as computer communication is widely used. Protocol conformance testing is carried out to check the conformance of a protocol implementation under test(IUT) to the protocol specification that it implements. Conformance to a communication protocol or service is considered to be prerequisite for the correct interoperability of open system.

In conformance testing, tester cannot observe or control the insides of the IUT and testing is done by applying test cases to the IUT and observing the output from it. Thus, test coverage is determined by the test cases and they should test most part of the protocol. Recently, necessity of generating test cases automatically has been emphasized due to the complexity of the communication protocols and a large amount of automatic test case generation methods have been proposed(Lee and Lee 1991)(Chanson and Zhu 1993)(Li *et al.* 1994)(Chin *et al.* 1997).

From a given protocol specification written in formal description techniques(FDTs) such as Estelle(ISO 1989a), LOTOS(ISO 1989b), and SDL(ITU-T 1993), finite state machine(FSM) or extended FSM(EFSM) model is obtained. Then, test cases are generated based on this FSM model to test control flow of the protocol or on EFSM model to test both control and data flow of the protocol.

Early work on test case generation has been based on a single-module FSM or EFSM model. However, most protocols have multi-module structure and existing test case generation methods are not applicable to these protocols. Therefore, transformation of the multi-module structure into an equivalent single-module is required. In this paper, we propose a method to obtain a single-module from multi-module protocol using reachability analysis technique. Since the proposed method simulates the behaviour of the protocol, semantic loss of the protocol during the transformation process can be minimized.

Section 2 presents related work and the outline of this work. In section 3, preliminaries necessary for the proposed method and the transformation algorithm are presented. Section 4 contains empirical results and section 5 discusses test case generation methods from EFSM model. Finally section 6 concludes this paper.

2 RELATED WORK

Conformance testing is done by applying proper input to the IUT and observing the output from it. In other words, disregarding the internal structure of the implementation, testing follows black box test which is carried out just at the interface. It can be divided into two categories; control part testing and data part testing. Control part testing is to test the control flow of the protocol based on FSM model to examine that transitions of the FSM are implemented correctly; to check whether there exist transition errors which are errors in the output function and transfer errors which are the case when next state is not the one expected after transition. Data part testing is to test the data flow of the protocol based on EFSM model. Test sequences are generated referring to data flow graph, and methods to generate test sequences are divided into two categories; using functional program test technique(Sarikaya *et al.* 1987) and using data flow analysis technique(Ural and Yang 1991)(Miller and Paul 1992)(Chanson and Zhu 1993)(Ramalingom *et al.* 1996).

In general, implementing environment, even though for the same protocol, may be different case by case. For this reason, most specifications do not fully describe functions of some procedures, which are implemented to be suitable to different environments. However, in the case that a function of a procedure is not described, this procedure could not be tested and data flow testing is also meaningless. Therefore, the minimum features to be implemented in a procedure should be described in the specification and tester must check the minimum items which are general to all environments. In this paper, protocol specification is assumed to be full specification where every part of the protocol is described.

During the testing process the entire protocol becomes an IUT, which is regarded as a black box. Since most protocols consist of multi-module, test sequences that are capable of testing multi-module should be generated. However, existing test sequence generation methods are for either single-FSM or single-EFSM so that generated test sequences are for single-module. Therefore, we must transform the given protocol model into an equivalent single-module to generate test sequences. A test architecture for a multi-module IUT is given in Figure 1(Linn 1990).

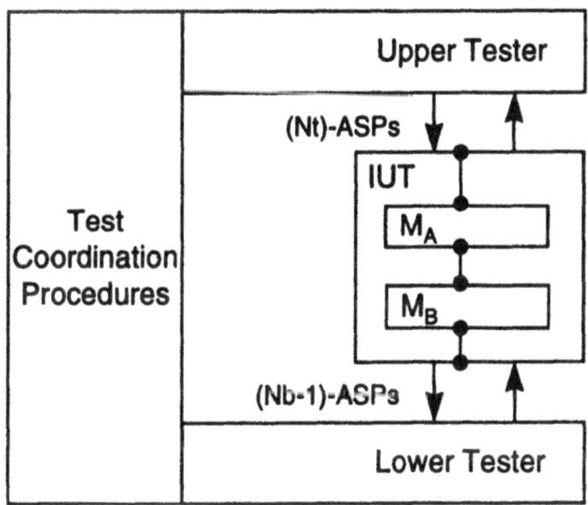

Figure 1 Test architecture for a multi-module IUT.

In Figure 1, messages given to or received from the IUT are called ASPs(Abstract Service Primitives). Tester cannot observe or control the interactions between M_A and M_B, and these interactions are called internal interactions. Interactions between M_A or M_B and the outside of the IUT are called external interactions which can be observed and controlled by the tester.

In (Sarikaya and Bochmann 1986), a single-module is obtained by textual replacement where internal communication is eliminated. For example, assume that an output interaction is transmitted to module M_B in a transition t_i of module M_A through internal channel invisible from the outside, and is received in a transition t_j of M_B. In this case, internal communication can be removed by substituting the output part of t_i with the action part of t_j. At the same time, the conditional statement of t_i should be properly modified according to the conditional statement in t_j.

In Estelle, firable condition of a transition is determined complicately. In the case that the protocol has hierarchical structure, by the parent/children priority principle, whether a transition in a child module can be fired depends on the firable transition in the parent module even though input interaction and conditional statement are satisfied(Dembinski and Budcowski 1989). However, with the method proposed in (Sarikaya and Bochmann 1986), it is difficult to consider this situation and there may occur some semantic loss during the transformation process.

In this paper, a method is proposed that simulates the behaviour of the protocol to obtain a single-module. This method adapts reachability analysis technique(West 1978) and a single-module structure is obtained as a result by simulating the behaviour of the protocol for all given external inputs. We assume that parameter values are considered when test sequences are generated. Therefore, we do not consider the parameter values of the external input during the transformation process and generated model includes all the possible cases according to the parameter values. Since the proposed method uses simulation, it can fully represent the behaviour of the protocol. It is assumed that non-determinism that makes the behaviour of the protocol complicated does not exist or can be eliminated by the tester(Lee and Lee 1991).

3 THE PROPOSED METHOD

3.1 Protocol Modelling

In Formal Methods in Conformance Testing(FMCT), a module of a protocol is modelled by an Input-Output State Machine(IOSM)(ISO 1995). IOSM is a 4-tuple $M =< S, L, T, s_0 >$ where S is a non empty finite set of states, L is a non empty finite set of interactions, $T \subseteq S \times ((\{?, !\} \times L) \cup \{\tau\}) \times S$ is transition relation, and s_0 is the initial state of IOSM. Each element of transition relation T corresponds to a transition and has observable action such as input(?a) and output(!a), and internal action τ.

IOSM represents a protocol as an observable-based machine. However, it is not suitable to the automation of test case generation for data part testing based on formal model testing because the internal action τ is defined abstractly. Furthermore, it is difficult to identify which module is the destination when a module communicates with a lot of modules. In this paper, we introduce a communicating EFSM where the action block is simplified and interaction points are clearly defined. An interaction point is an interface of the communication.

Definition 1 Communicating ESFM model is a 9-tuple $M =< Q, q_0, I, O, V, P, A, \delta, IP >$ where

- Q is a non empty finite set of states;
- $q_0 \in Q$ is the initial state of M;
- I is a non empty finite set of input interactions;
- O is a non empty finite set of output interactions;
- V is a set of variables;

- P is a set of parameters;
- A is a set of actions;
- δ is a transition relation $\delta : Q \times A \longrightarrow Q$;
- IP is a non empty finite set of interaction points.

Each element of an action set A is a 4-tuple (*input*, *predicate*, *output*, *compute_block*). *input*, *output* are 3-tuple $(ip_1,\ i,\ p_i)$, $(ip_2,\ o,\ p_o)$, respectively, where $ip_1,\ ip_2 \in IP$, $i \in I$, $o \in O$, $p_i,\ p_o \in P$. *predicate* is a Pascal-like predicate expressed in terms of the variables and parameters. *compute_block* is a computation block which consists of linear functions $f : V \times P \longrightarrow V \times P$.

Note that input, output interactions and interaction points are defined in the action block of the CEFSM model. When a module communicates with a lot of modules, we can clearly identify which CEFSM is concerned with current communication due to the interaction points. Also note that the computation block is composed of linear equations to expedite the automation of test case generation and constraint solving. *compute_block* can be simplified using existing methods(Sarikaya and Bochmann 1986)(Miller and Paul 1992). In order to model hierarchical structure, CEFSM can be extended to 10-tuple by adding a property *ParentM* that represents the parent module of it.

A communicating system is a set of communicating EFSMs exchanging messages through FIFO(First In First Out) channels. In most cases, communicating system is made up of more than two communicating machines. In this paper, however, we will use only communicating systems composed of two communicating machines. It can be easily extended to general systems.

Definition 2 Communicating system is a 4-tuple $S =< CM_1,\ CM_2,\ C_{12},\ C_{21} >$ where

- $CM_i =< Q_i,\ q_{0i},\ I_i,\ O_i,\ V_i,\ P_i,\ A_i,\ \delta_i,\ IP_i >$, $i = 1, 2$ is CEFSM model;
- C_{ij}, $1 \leq i \neq j \leq 2$ is FIFO channel connecting interaction point ip_i and ip_j, where $ip_i \in IP_i$, $ip_j \in IP_j$.

The set M_{ij} is a set of messages from CM_i to CM_j and messages contained in a channel C_{ij} are represented as $c_{ij} \in (M_{ij})^*$. c_{ij} is the output message of CM_i and the input message of CM_j at the same time. As an example, when CM_i sends a message $c_{ij} = output_i = (ip_i,\ o_i,\ p_{oi})$ to CM_j, CM_j receives this messsage c_{ij} as an $input_j = (ip_j,\ i_j,\ p_{ij})$, where $i_j = o_i$, $p_{ij} = p_{oi}$.

The model obtained from communicating system through the transformation is called a global model. Global model is a directed graph $G = (V,\ E)$ where V is a set of global states and E corresponds to a set of global transitions. Global state and global transition are defined as follows.

Definition 3 A global state of a global model is a 2-tuple $g =< q_1,\ q_2 >$ where $q_i \in Q_i$ is the current state of CM_i.

Definition 4 A global transition of a global model is a pair $t = (i,\ \alpha)$ where $\alpha \in A_i$.

A global transition $t = (i,\ \alpha)$ is said to be firable in $g = < q_1,\ q_2 >$ if and only if the following two conditions are satisfied where $\alpha = (input,\ predicate,\ output,\ compute\_block)$ and c_{12}, c_{21} are messages contained in channel C_{12}, C_{21}, respectively.

- a transition relation $\delta_i(q_i,\ \alpha)$ is defined.
- $input = \epsilon$ and $predicate = True$ or
  $input = a$ and $c_{ji} = aw,\ w \in (M_{ji})^*$ and $predicate = True$.

After the global transition t is fired, the system goes to global state $g' = < q_1',\ q_2' >$ and messages contained in each channel are $c_{12}',\ c_{21}'$ where

- $q_i' = \delta_i(q_i,\ \alpha),\ q_j' = q_j$.
- if $input = \epsilon$ and $output = \epsilon$, then $c_{12}' = c_{12},\ c_{21}' = c_{21}$
  if $input = \epsilon$ and $output = b = (ip_i,\ o_i,\ p_{oi})$, then $c_{ij}' = c_{ij}b,\ c_{ji}' = c_{ji}$
  if $input \neq \epsilon$ and $output = \epsilon$, then $c_{ij}' = c_{ij},\ c_{ji}' = w$
  if $input \neq \epsilon$ and $output = b = (ip_i,\ o_i,\ p_{oi})$, then $c_{ij}' = c_{ij}b,\ c_{ji}' = w$.

A global model corresponds to an EFSM model. Therefore, we can get a single-EFSM from communicating multi-EFSM through the transformation of communicating system into global model.

3.2 Transformation Algorithm

Assume that a protocol is represented as in Figure 2. If module M_A and M_B are entities in (N)-layer, user A and user B are (N+1)-entities. Protocol specification written in formal method describes all the entities user A, user B, M_A, M_B, and transmission media. Each entity is a module or can be composed of multi-module as in Figure 2.

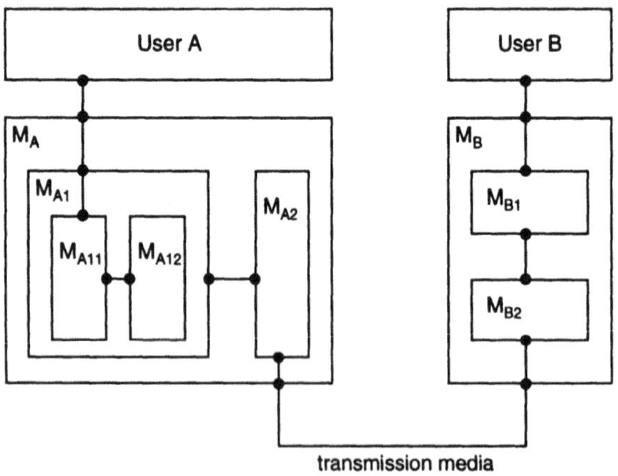

Figure 2 Protocol having hierarchical structure.

In order to test this protocol, IUT should be selected at first. In most cases, a protocol is designed to operate in at least one layer. Since the target of the testing is the implementation, minimum size of the IUT is the (N)-entity, M_A or M_B. Each of user A, user B, and transmission media can also be an IUT. An IUT may extend to more than one layer and M_A combined with transmission media can also be an IUT. In protocol specification, however, only definitions of interactions and simple functions to exchange interactions with other layers are defined for user A, user B, and transmission media. Therefore, we can assume that all the substantial functions of the protocol are in (N)-layer and it is sufficient to test only (N)-entity.

After selecting the module M_A as an IUT, we simulate the behaviour of the module M_A including all the child modules. During the simulation dynamic changing of the protocol structure may cause the size of the global model to get very large, so reduction of the size is needed. In this paper, we restrict the scope to the protocol that does not change its internal structure dynamically after initialization.

(Chun 1991) proposed a method that combines all modules one after another to generate a single-module. For example, the peer module M_{A11} and M_{A12} are combined using reachability analysis and single-module $M_{A11} + M_{A12}$ is obtained at first stage. In the second stage, parent module M_{A1} and the child module $M_{A11} + M_{A12}$ are combined, and then module $M_{A1} + M_{A11} + M_{A12}$ is generated. After repeating this procedure, all modules are combined and finally a single-module is obtained.

This method cannot fully reflect the behaviour of the protocol as in (Sarikaya and Bochmann 1986). As we have mentioned in section 2, the action of the child module can be affected by the parent module. However, if modules are combined one after another, it is difficult to consider this situation and the behaviour of the generated single-module may not be same to that of the protocol. In our method, actions of all the modules are simulated simultaneously so that the semantic loss during the transformation process can be minimized.

The obtained global model through the transformation should include all the behaviour of the communicating system and produce the same output for a given input. All the possible inputs are given to the protocol and reachability analysis technique is used to simulate the behaviour of the protocol. Simulation begins at the initial global state of the protocol*. For a given global state, we give all the possible external inputs to the protocol and check whether each transition is firable. Since we have assumed that the external input parameter values are determined during the test case generation, the value of the *predicate* that depends only on the external parameter is always *True*. For a firable transition, a directed edge from the current global state to the next global state is generated where the action block of the edge is same to that of the transition.

When all the external inputs are considered for the initial global state, the next global states moved from the initial state are taken into account. For each next global state, we check if there exist firable transitions. In this stage, we must consider the messages contained in the internal channel. Assume that there exists a message c_{ji} in the internal channel C_{ji} and a firable transition that receives the message c_{ji} as an input also exists. We can assume that the time required to process the internal interactions in a module is much shorter than the time interval between the external events because external events can be controlled by the tester. Then, there's no need

---

*Communicating system is initialized by the tester to consider the dynamic configuration of the protocol. More than one global states can be generated during the initialization phase.

to consider the external inputs and only transitions that receive no input or internal input are firable.

In the proposed method, messages in the channel are not contained in the global state and they only affect the firable condition of the transitions. This is the main difference with the reachability analysis technique where messages in the channel are containd in the global state. Essentially, the proposed method aims at obtaining a single-EFSM, not single-FSM. We just provide all the possible behaviour of the protocol and do not consider concrete values of the external parameters. Thus, state explosion that is the critical problem in verification technique where new states are generated for all input values does not occur, and the number of the generated global states is always finite. Transformation is completed when there is no global state to check. Algorithm for the transformation is as follows.

Algorithm

- Input: Communicating system S
- Output: Global model $G=(V, E)$

```
begin
initialize S
/* Communicating system S is initialized by the tester. Dynamic configuration
   of S is considered, and then some global states and edges are generated. */
for all generated global states g_k and edges e_k do
    begin
    V = V ∪ {g_k}
    E = E ∪ {e_k}
    end
V_CUR = V      /* global states to visit in current step */
V_NEXT = ∅     /* global states to visit in the next step */
while (V_CUR ≠ ∅) do
    begin
    for all global states g ∈ V_CUR do
        for all transitions t defined in g do
            begin
            if (firable transition t_i by null input exists)
            /* if at least one transition is firable without input, there's
               no need to consider any input. */
               for all transitions t_i
                   Process_This_Transition(t_i)
            else if (firable transition t_i by internal event exists)
            /* if at least one transition is firable by internal event,
               there's no need to consider any external input. */
               for all transitions t_i
                   Process_This_Transition(t_i)
            else
            /* all external inputs are considered. */
               for all transitions t_i firable by external event
                   Process_This_Transition(t_i)
```

```
            end
        V_CUR = V_NEXT
        V_NEXT = ∅
        end
end
```

```
/* Process_This_Transition() decides whether this transition can be a new
   global edge. */
Process_This_Transition(t)
begin
    Create an edge e labelled by t.
    if (e ∉ E)
        begin
        E = E ∪ {e}
        if (next global state g' ∉ V)
            begin
            V = V ∪ {g'}        /* new global state */
            V_NEXT = V_NEXT ∪ {g'}
            end
        else
            V_NEXT = V_NEXT ∪ {g'}        /* existing global state */
        end
    else
        discard e        /* visited edge */
end
```

In (Chun 1991), global states having queue contents are divided into three categories; stable state, unstable state, and transient state.

- stable state: a state having no firable transition or no message in the internal queue
- unstable state: a state having firable transitions by internal events
- transient state: a state having firable transitions without any input

Tester cannot control the behaviour of the protocol in unstable and transient state. So, removing these states makes no difference to the behaviour of the protocol when we observe it from the outside. Since the global state in the proposed model does not include the queue contents, global states should be classified in different way.

- If there exists a firable transition without any input, this state is a transient state.
- If a state is not a transient state and there exists at least one firable transition that makes the system go to this state without giving any output to internal channel, this state is a stable state.
- If a state is not a transient state and there exists at least one firable transition by internal event, this state is an unstable state.

A global state may have properties of both stable and unstable state, and the removal of unstable state does not mean the removal of the global state. Unstable state can be removed by combining two transitions that communicate through internal channel, and this global state can be removed if it doesn't have the property of stable state. We can combine two transitions using symbolic execution technique(Clark and Richardson 1985).

4 EMPIRICAL RESULTS

In this paper, we have applied the proposed method to the Class 0 transport protocol(TP0). Modular structure of the TP0 is shown in Figure 3.

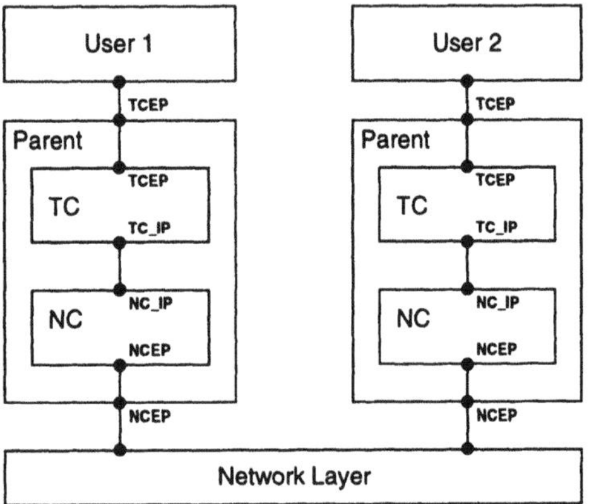

Figure 3 Modular structure of the TP0.

In Figure 3, module Parent initializes and releases the module TC, NC and internal channels, and delivers interactions that were sent from TC or NC to the external modules. TC makes TPDU(Transport Protocol Data Unit) for each input primitive and NC makes NSDU(Network Service Data Unit) for each TPDU to communicate with network layer.

As the target protocol of the testing is transport protocol, the IUT is module Parent, and three modules, Parent, TC, and NC should be combined to generate test sequences. In TP0, module Parent changes the dynamic configuration of the structure, so this should be excluded in the transformation process. However, Parent is in charge of establishment and release of the connection, so we will include this module partially to test the overall flow of the protocol. After the connection is established, we restrict the function of the module Parent to delivering interactions between module TC, NC and external modules. Then, TCEP of TC and NCEP of NC become the external interaction points where tester can observe or control the interactions

Transformation process is divided into two steps. First, we simulate the behaviour of the protocol and the global model is generated. At this time, communication is classified into internal one and external one. In the case of TP0, TCEP and NCEP are external interaction points and TC_IP, NC_IP are internal interaction points.

When TC receives disconnection request from the upper layer, TC makes a TPDU containing this information and sends it to NC. NC receives this TPDU and sends it to the lower layer after encoding it. In this process, communication between TC and NC is an internal communication that tester cannot observe or control. Figure 4 shows this procedure.

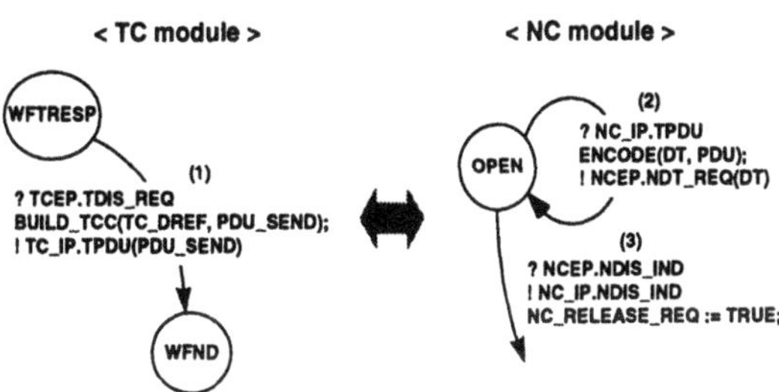

Figure 4 Internal communication in TP0.

Assume that TC, NC are in WFTRESP, OPEN state, respectively, and there's no message in the internal channel connecting TC_IP and NC_IP. The global state of this system is then, (WFTRESP, OPEN) and three transitions (1), (2), and (3) are related to this state. Since there's no message in the internal channel, transition (1), (3) are firable and global transitions labelled (1), (3) are generated. However, we will consider just transition (1) to simplify the explanation. At first, transition (1) is fired and global transition labelled (1) is generated. After the transition (1) is fired, the system goes to the global state (WFND, OPEN) and the message TPDU exists in the internal channel. Although two transitions, (2) and (3) are concerned with this state, only transition (2) is firable because (2) receives internal input while (3) receives external one. After the transition (2) is fired, the internal channel is emptied and the transition (3) becomes firable. Generated global model from Figure 4 is shown in Figure 5(a).

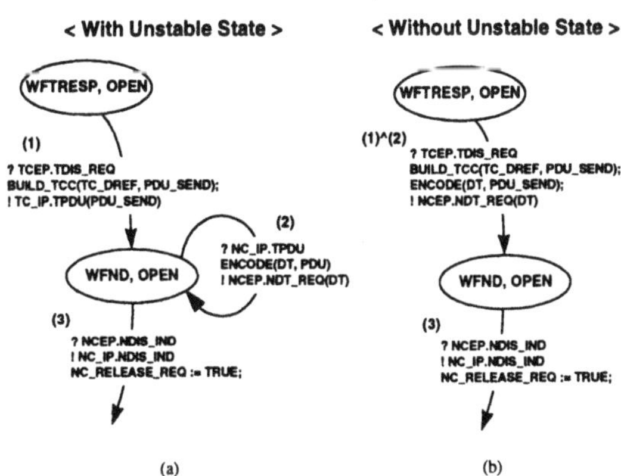

Figure 5 Global model with unstable state and without unstable state.

As a second step, transient states and unstable states are eliminated. In the (WFND, OPEN) state in Figure 5(a), transition (2) is firable by internal event and there's a transition that makes the system go to (WFND, OPEN) state without giving any internal event. Thus, (WFND, OPEN) state has the properties of both stable and unstable state. Unstable property of the state (WFND, OPEN) can be removed by combining two transitions, (1) and (2) which communicate through internal channel. Figure 5(b) shows the result after combining two transitions.

When there's no global state to consider, transformation ends and finally we can get the global model of the communicating system. Figure 6 shows the global model of TP0. In this figure, dotted line represents the dynamic changing of the protocol structure that should be processed manually. It is important to note that 4_1, 5_1, and 6_2 states are unremoved unstable states or transient states due to the specification where some part of the protocol are not described. The action blocks of the transitions are available in (Hwang 1997).

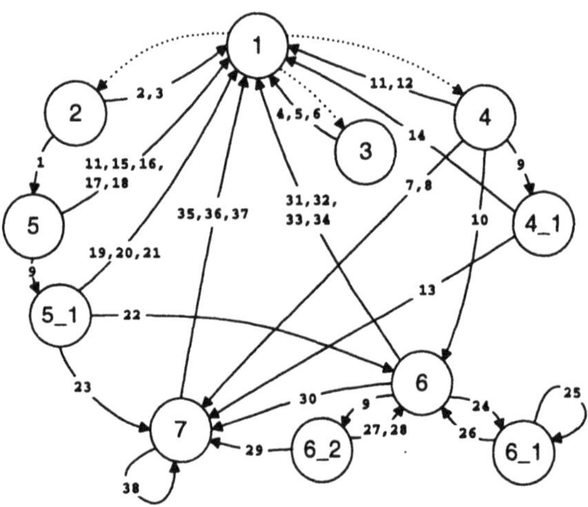

Figure 6 Global model of TP0.

5 TEST CASE GENERATION

Currently, there are two approaches to generate test sequences from EFSM. One is to generate test cases in respect to test purposes(Guerrouat and König 1996). A test case is derived from a test purpose which represents a control flow of the protocol. In this approach, only control flow is checked and data part of the protocol is used for generating executable test cases. Since test purposes are usually informal and may express different kinds of conformance requirements, it is impossible to automate this procedure. Therefore, formalization of the test purposes and automatic test case generation from the formally described test purposes are required. In (Guerrouat and König 1996), test cases are generated automatically for restricted classes of test purposes using a knowledge based techniques.

In the second approach, test sequences are generated automatically using algorithms. In order to test the control flow of the protocol, one can get FSM from the EFSM by exhaustive simulation with input parameter values. In most cases, however, it is not feasible to generate FSM for all input values, so compromise between the size of the generated model and the accuracy of the automation is required. To test the data flow of the protocol, some test sequence generation algorithms based on data part testing criteria are applied to EFSM(Chanson and Zhu 1993)(Li *et al.* 1994)(Ramalingom *et al.* 1996)(Chin *et al.* 1997). In general, the size of the generated test sequences is very large when they are generated automatically even though the target protocol is simple. As the size of the protocol increases, the size of the generated test sequences increases more rapidly. Thus, it is impractical to generate test sequences for complex protocol and the size of the target protocol should be reduced.

Protocol model can be divided into sub-graphs based on some criteria. (Park *et al.* 1994) proposed a method to reduce the size of the problem by applying test purposes to the protocol. Since a test purpose represents a control flow, it can be a sub-graph of the protocol and test sequences are generated based on this sub-graph. In order to automate this procedure, protocol should be divided systematically and manual effort should be minimized.

6 CONCLUSIONS

We have presented a method that transforms communicating multi-module protocol into an equivalent single-module to generate test cases for both control and data part testing. The transformation process is divided into two steps. First, we simulate the behaviour of the protocol for all the possible inputs, and then the global model is generated. As a second step, transient and unstable states are eliminated to reduce the size of the global model. Since the proposed method adapts reachability analysis technique rather than textual replacement, we can minimize the semantic loss of the protocol during the transformation process.

In order to apply the proposed method to the FDTs, a detailed study is needed for each FDT. Definition of the EFSM should be extended according to the FDT's properties and firable conditions should also be modified. Currently, we are extending our study to obtaining an EFSM from the specification written in a FDT and generating feasible test cases from the global model.

ACKNOWLEDGEMENTS

This work has been supported partially by the Electronics and Telecommunications Research Institute(ETRI) of Korea and partially by the Korea Telecom.

7 REFERENCES

Chanson, S.T. and Zhu, J. (1993) A Unified Approach to Protocol Test Sequence Generation. In *Proceedings of the IEEE INFOCOM*, San Francisco, Califor-

nia, 106–114.
Chin, B., Kim, T., Hwang, I., Jang, M., Lee, J., and Lee, S. (1997) Generation of Reliable and Optimized Test Cases for Data Flow Test with a Formal Approach. In *Proceedings of the 11th International Conference on Information Networking*, Taipai, Taiwan.
Chun, W. (1991) Test Case Generation for Protocols Specified in Estelle. *Ph.D. thesis*, Dept. Computer and Information Sciences, University of Delaware.
Clarke, L.A. and Richardson, D.J. (1985) Application of Symbolic Evaluation. *the Journal of Systems and Software*, **5**, 15–35.
Dembinski, P. and Budcowski, S. (1989) Specification Language ESTELLE in *The Formal Description Technique Estelle* (eds. M. Diaz *et al.*), Results of the ESPRIT/SEDOS Project, North Holland.
Guerrouat, A. and König, H. (1996) Automation of test case derivation in respect to test purposes. In *Proceedings of the 9th International Workshop on Testing of Communicating Systems*, Darmstadt, Germany, 207–222.
Hwang, I. (1997) A method to derive a simple single-EFSM from communicating multi-EFSM for data part testing. *M.S. thesis*, Dept. Electronic Engineering, Yonsei University.
URL: http://nasla.yonsei.ac.kr/~ his/Publications/Th_annex.ps
ISO (1989a), *Information Processing Systems - Open Systems Interconnection - Estelle: A Formal Description Technique based on an Extended State Transition Model.* ISO/IEC IS 9074.
ISO (1989b), *Information Processing Systems - Open Systems Interconnection - LOTOS: A Formal Description Technique based on the Temporal Ordering of Observational Behaviour.* ISO/IEC IS 8807.
ISO (1995), ISO/IEC JTC1/SC21 WG7, ITU-T SG10/Q.8 *Formal Methods in Conformance Testing Part, working draft.* ISO Project 1.21.54, ITU-T Z.500. ISO, ITU-T, September, 1995. Output from ISO/ITU-T Meeting in Geneve.
ITU-T (1993) ITU-T Z.100 *CCITT Specification and Description Language.* ITU Recommendation.
Kim, T. (1995) Automatic generation of observation-based and length-optimized test cases for EFSM model in conformance testing. *M.S. thesis*, Dept. Electronic Engineering, Yonsei University.
Lee, D.Y. and Lee, J.Y. (1991) A well defined Estelle specification for the automatic test generation. *IEEE Trans. on Computer*, **COM-40(4)**, 526–542.
Li, X., Higashino, T., Higuchi, M. and Taniguchi, K. (1994) Automatic generation of extended UIO sequences for communication protocols in an EFSM model. In *Proceedings of 7th International Workshop on Protocol Test Systems*, Tokyo, Japan, 225–240.
Linn Jr., R.J. (1990) Conformance Testing for OSI Protocols. *Computer Network and ISDN Systems*, **18**, 203–219.
Miller, R.E. and Paul, S. (1992) Generating Conformance Test Sequences for Combined Control and Data Flow of Communication Protocols. In *Proceedings of 12th International Symposium on Protocol Specification, Testing and Verification*, 1–15.
Park, J.H., Lee, J.Y., Jung, I.Y., and Hong, J.P. (1994) A Conformance Testing Framework for Applying Test Purposes. In *Proceedings of 7th International Workshop on Protocol Test Systems*, Tokyo, Japan, 291–298.

Ramalingom, T., Thulasiraman, K. and Das, A. (1996) Context Independent Unique Sequences Gerneration for Protocol Testing. In *Proceedings of the IEEE INFOCOM*, San Francisco, California, 1141–1148.

Sarikaya, B. and Bochmann, G.v. (1986) Obtaining Normal Form Specification for Protocols. *Computer Network Usage* (eds. L. Csaba *et al.*), 601–612.

Sarikaya, B., Bochmann, G.v. and Cerny, E. (1987) A Test Design Methodology for Protocol Testing. *IEEE Trans. on Software Engineering*, **SE-13**, 518–531.

Ural, H. and Yang, B. (1991) A Test Sequence Selection Method for Protocol Testing. *IEEE Trans. on Communication*, **COM-39**, 514–523.

West, C. H. (1978) An automated technique of communications Protocol Validation. *IEEE Trans. on Communication*, **COM-26**, 1271–1275.

8 BIOGRAPHY

Iksoon Hwang received the B.S. and M.S. degrees in electronic engineering in 1995 and 1997, respectively, from Yonsei University, Seoul, Korea. Currently, he is in the Ph.D. degree course in electronic engineering at Yonsei University. His current interests include protocol engineering, ATM networks, and computer networks.

Taehyong Kim received the B.S. and M.S. degrees in electronic engineering in 1993 and 1995, respectively, from Yonsei University, Seoul, Korea. He is currently a Ph.D. candidate in electronic engineering at Yonsei University. His current research interests include protocol engineering, software engineering, and computer networks.

Minseok Jang received the B.S. and M.S. degrees in 1989 and 1991, respectively, and Ph.D. degree in electronic engineering in 1997 from Yonsei University, Seoul, Korea. His current research interests include protocol engineering, software engineering, especially conformance testing, and computer networks.

Jaiyong Lee received the B.S. degree in electronic engineering in 1977 from Yonsei Universtiy, Seoul, Korea and M.S. and Ph.D. degrees in computer engineering in 1984, and 1987, respectively, from Iowa State University, Ames, Iowa. From 1977 to 1982, he was a research engineer at Agency for Depense Development of Korea. From 1987 to 1994 he was a Associate Professor at Pohang Institute of Science and Technology. He is currently a Professor in the Department of Electronic Engineering, Yonsei University, Seoul, Korea. He is interested in high speed/multimedia communication, multimedia PCS protocol, and conformance testing.

Sangbae Lee received the B.S. from the R.O.K. Air Force Academy of Korea in 1958. the B.E. degree in electrical engineering from the Seoul National University, Seoul, Korea, in 1961, the M.S. degree in from Standford University, California, in 1964, and the Ph.D. degree from the University of Newcastle, England in 1975. He is currently a Professor in the Department of Electronic Engineering, Yonsei University, Seoul, Korea. From 1969 to 1979, he was an Assistant Professor at Seoul National University and from 1982 to 1983, a Visiting Professor at the University of Newcastle, England. He has served as a Chairman of the IEEE Korea Section from 1986 to 1987 and as a Chairman of the Korea Institute of Telematics and Electronics. He is interested in B-ISDN, graph theory, LAN/MAN internetworking, and multimedia communication.

Heangsuk Oh received the B.S. and M.S. in eletronic material engineering from Hanyang University in 1981 and 1983, respectively, and Ph.D. in computer science from Chungbuk University in 1997. His research interests include computer network, formal description techniques, and protocol engineering. From 1993 to the present, he worked as a senior researcher of protocol engineering project where he serves in PEC(Protocol Engineering Center) at ETRI.

Myungchul Kim received B.A. in electronic engineering from Ajou Univ. in 1982, M.S. in computer science from the Korea Advanced Institute of Science and Technology in 1984, and Ph.D. in computer science from the Univ. of British Columbia in 1992. Currently he is with the Korea Telecom Research and Development Group as a managing director, Chairman of Profile Test Specification-Special Interest Group of Asia-Oceania Workshop, and is the Co-Chair of the 10th IWTCS'97. His research interests include protocol engineering on telecommunications and multimedia.

PART FIVE

Test Coverage and Testability

Basing test coverage on a formalization of test hypotheses

O. Charles, R. Groz
FRANCE TÉLÉCOM - CNET,
DTL/MSV, 2 avenue Pierre Marzin, F-22307 LANNION Cedex, FRANCE, Tel: +33 2 96 05 37 90, Fax: +33 2 96 05 39 45,
E-mail: (charleso, groz)@lannion.cnet.fr

Abstract

This paper defines a characterization of protocol conformance test coverage based on test hypotheses. All test selection methods and coverage computation make use of test hypotheses in one way or another. Test hypotheses are assumptions made on the implementation, which justify the verdict of conformity provided by testing; thus they are an important part of the coverage. We propose a model of these hypotheses based on functions on automata, enabling a definition of coverage based on test hypotheses, which we call TH-based coverage.

Keywords

Conformance testing, coverage, test hypothesis

1 INTRODUCTION

Many test coverage measures have been defined in the area of software testing and especially protocol testing. The aim of those definitions is to measure the quality of the test suite. The notion of quality of a test suite for conformance testing can be defined in two complementary ways. On the one hand, it is the ability of the test suite to prove that the implementation conforms to its specification. On the other hand, it is the ability of the test suite to find faults. As a result, we can split coverage definitions into two families. We call the first one specification coverage and the

Testing of Communicating Systems, M. Kim, S. Kang & K. Hong (Eds)
Published by Chapman & Hall © 1997 IFIP

second one fault coverage.

Specification coverage reflects how the test suite probes the specification. These measures are usually expressed as a percentage of elements of the specification that are exercised by a test suite. For example in the case of traditional software testing the usual coverage measure are branch coverage, path coverage and more generally the different criteria typically defined by (Weyuker, 1984), (Rapps, 1985). For protocol testing, (Ural, 1991) defines similar criteria. The metric based theory of coverage (Vuong, 1991) also expresses to what extent the test suite explores the specification. Specification coverage is practical since the test suite is compared to the specification (Groz, 1996) but gives little information on residual faults.

Fault coverage gives the number and the types of faults that can be discovered by a test suite. The types of faults that may have been introduced while implementing are supposed to be known. They are listed in a fault model and the fault coverage is evaluated with respect to this model. The computation is performed either by mutation analysis (Bochmann, 1991) (Dubuc, 1991) (Motteler, 1993) (Sidhu, 1989), by numbering mutants (Yao, 1994), or generating indistinguishable implementations (Zhu, 1994). The test coverage defined in (ISO, 1995) is fault coverage.

In fact, both types of coverage measures rely on test hypotheses (Gaudel, 1992) to reduce the infinite set of possible implementations to a smaller one in which the coverage will actually be computable. For example the number of states of the implementation or the probable faults are supposed to be known, but many other properties may be assumed on the implementation.

In practice, test designers also use (albeit implicitly) test hypotheses when designing test suites. Indeed, if they were to develop test suites discovering all errors, those test suites would be infinite. They also assume that the implementations have good properties to limit the field of their investigation.

Since fault coverage, specification coverage, and test selection all resort to some sort of test hypotheses, we propose a new coverage definition based on hypotheses. The TH-based coverage (coverage based on test hypotheses) of a test suite is defined as the series of hypotheses that must be made on the implementation such that the test suite is perfect. The main advantage of this coverage is that it provides a unifying concept for the other notions of coverage. Furthermore, it provides richer information than the usual percentage computations, and test hypotheses are meaningful for test designers. However, one problem remains: how can we formalise test hypotheses? We have already presented the first ideas of this work in (Charles, 1996). In this paper we present a more elaborate formalism based on functions on automata whereas in the previous paper the hypotheses were given as trace sets. Thus this model gives a better level of abstraction, closer to testing practice.

In section 2 we give a formalism of test hypotheses and test hypotheses coverage. Our goal is not to study systems in order to extract hypotheses (this work has already been done (Bernot, 1991)(Gaudel, 1992)(Phalippou, 1994)) but to catch this concept in a formal and suitable way for TH-based coverage. In section 3 we recall the IOSM model defined by (Phalippou, 1994). In section 4 it is shown that hypotheses can be viewed as trace sets. In section 5 we introduce a data structure — the partial automata — to store the information on the implementation gathered either by testing or by hypotheses. In section 6 we study on a few examples how hypotheses can be viewed as functions on partial automata. This is formalised in section 7. Finally we show in section 8 that our model embodies the natural concept of strength of hypotheses.

2 PRINCIPLES OF TH-BASED COVERAGE

2.1 Basic test suites properties

In this section, we recall the basic formal notions that underlie testing (ISO, 1995). The goal of a test suite is to assess the correctness of an implementation with respect to its specification. The correctness is defined by means of a conformance relation **imp** on $Imp \times Spec$; an implementation $I \in Imp$ conforms to its specification $S \in Spec$ if and only if $\mathbf{imp}(I, S)$.

Let K be the set of all possible tests and $T \subseteq K$ the test suite; we denote $pass(I, T)$ the fact that an implementation I executes successfully a test suite T (an instance of this relation will be defined in section 4.1). (ISO, 1995) defines three properties on test suites:

definition 1 Let $T \subseteq K$ be a test suite and $S \in Spec$ a specification:

- T is exhaustive $=_{def} (\forall I \in Imp)\ (pass(I, T) \Rightarrow \mathbf{imp}(I, S))$
- T is sound $=_{def} (\forall I \in Imp)\ (\neg pass(I, T) \Rightarrow \neg\mathbf{imp}(I, S))$
- T is complete $=_{def} T$ is sound and exhaustive.

2.2 Test hypotheses

Test suites for real systems are seldom exhaustive and therefore are usually incomplete. Those systems are indeed very complex and an exhaustive test suite, assuming one exists, would be infinite. Test designers resort to test hypotheses to build a smaller test suite that keeps the same properties under these hypotheses.

Test hypotheses have been proposed for test selection. (Bernot, 1991) proposes suitable hypotheses to generate finite test sets for implementation that would request infinite ones otherwise. In (Phalippou, 1994) hypotheses for the IOSM model are proposed. In this paper we shall use uniformity hypotheses to illustrate our discussion (see 6.1). Such hypotheses claim that if the implementation behaves correctly for some elements of a given domain, then it behaves correctly on the whole domain. We can also mention regularity, reliable reset, independence or fairness hypotheses. Since our goal is not to study these hypotheses but to give a suitable representation of them in the aim of defining test coverage, we shall not detail any further. However we can mention that test hypotheses are always defined in the same way: «if the implementation is correct for some behaviours, then it is also correct for a larger set». As a result, by a successive composition of the hypotheses, the valid behaviour domain grows.

The idea of TH-based coverage is based on that practical use of hypotheses: the formulation of test hypotheses is an iterative process starting from the test set (containing the only behaviours known to be correct) and ending in the exhaustive test set. According to its iterative nature, the process of formulating test hypotheses cannot take into account all the hypotheses at the same time, but one after the other. In most cases the order in which hypotheses are formulated is significant. This non-commutative property will be reflected in our model.

In other words, looking for the coverage of a test suite comes down to asking: «What is the series of assumptions that must be made such that the test suite is exhaustive for the set of implementations satisfying those assumptions?»

2.3 TH-based coverage

For the moment, let us regard an hypothesis H as a predicate on the set of implementations Imp. We shall see further that in fact H can be the result of the composition of a series of hypotheses $\{H_i\}_{i \in N}$.

definition 2 Let $T \subseteq K$ be a test suite, $S \in Spec$ a specification and H an hypothesis:

- T is exhaustive under hypothesis H
  $=_{def} (\forall I \in Imp)\ (pass\,(I, T) \wedge H\,(I) \Rightarrow \mathbf{imp}(I, S))$.
- T is complete under hypothesis H
  $=_{def} (\forall I \in Imp)\ (pass\,(I, T) \wedge H\,(I) \Leftrightarrow \mathbf{imp}(I, S))$

The hypothesis H expresses to what extend T is exhaustive. This is the basis of what we have called TH-based coverage. Instead of considering the coverage to be a percentage of fired transitions, executed branch conditions, or killed mutants, we propose to define the test coverage as the assumption that must be made on the implementation under test such that the test suite is exhaustive under.

definition 3 Let H be an hypothesis, $T \subseteq K$ a test suite and $S \in Spec$ a specification. H is a coverage of $T =_{def} T$ is exhaustive under hypothesis H.

3 MODEL DEFINITION

In order to see how our general framework for TH-based coverage can be instantiated in the case of protocols and communicating systems, we shall base the rest of this paper on a model suitable for this domain.

3.1 Input-Output State Machines

Input-Output State Machines (IOSM) have been presented in (ISO Annex A, 1995) as a more fundamental model of systems than the standardised syntactic languages such as Estelle or SDL. Despite there exists a lot of models for communicating systems we choose this one because the notion of test hypothesis has been studied in this framework (Phalippou, 1994).

definition 4 An Input-Output State Machine is a 4-tuple $\langle S, L, T, s_0 \rangle$ where:

- S is a finite non-empty set of states;
- L is a finite non-empty set of interactions;
- $T \subseteq S \times ((\{?, !\} \times L) \cup \{\tau\}) \times S$ is the transition relation. Each element from T is a transition, from an origin state to a destination state. This transition is associated either to an observable action (input ?a or output !a), or to the internal action τ.
- s_0 is the initial state of the automaton.

We give also the following definitions and notations.

definition 5 Let $S = \langle S_s, L_s, T_s, s_{0s} \rangle$ and $(\sigma = \mu_1 \ldots \mu_n) \in (\{!, ?\} \times L_s)^*$.

- (s_0, σ, s_n) iff $(\exists (s_i)_{1 \le i < n} \in S_s^n)\ (\forall i, 1 \le i \le n)\ ((s_{i-1}, \mu_i, s_i) \in T_s)$
- (S, σ, s_n) iff (s_0, σ, s_n)
- (s_0, ε, s_1) iff $s_0 = s_1$ or $(\exists n \ge 1)\ (s_0, \tau^n, s_1)$
- $(s_0, \vec{\mu}, s_1)$ iff $(\exists s_2, s_3 \in S_s)\ ((s_0, \varepsilon, s_2) \wedge (s_2, \mu, s_3) \wedge (s_3, \varepsilon, s_1))$
- $(s_0, \vec{\sigma}, s_n)$ iff $(\exists (s_i)_{1 \le i < n} \in S_s^n)\ (\forall i, 1 \le i \le n)\ ((s_{i-1}, \vec{\mu}_i, s_i) \in T_s)$

definition 6 A trace of S is a sequence of observable actions $\sigma \in (\{!, ?\} \times L_s)^*$ such that $(\exists s_n \in S_s)\ (s_{0s}, \vec{\sigma}, s_n)$. The set of all the traces is denoted $Tr(S)$.

The set $Spec$ is chosen to be the set of all IOSM. The test hypothesis (ISO, 1995) allows us to choose also the set of all IOSM as the set Imp.

3.2 Conformance relations

Many conformance relations have been defined on $IOSM \times IOSM$ to suit testing practice. We list here three of them that are suitable for **imp**.

- $I \ge_{tr} S =_{def} Tr(S) \subseteq Tr(I)$
- $I \le_{tr} S =_{def} Tr(I)\ \ Tr(S)$
- $R_5(I, S) =_{def} (\forall \sigma \in T\ \ S))\ (\sigma \in Tr(I) \wedge (O(\sigma, I) = O(\sigma, S)))$ where $O(\sigma, I) = \{a \in L \mid \sigma!a \in Tr(I)\}$ is the set of outputs of I after σ. This relation is used in our automatic test generation tool TVEDA (Clatin, 1995).

4 TEST CONCEPTS

4.1 Test verdict and tests as traces

(ISO, 1995) tells us that during the execution of a test case all that can be observed are the interactions exchanged between the implementation and the tester. A test verdict is assigned thanks to the observed execution trace. But from the coverage point of view the exchanged interactions are more informative than the verdict itself. In other words, the observed trace gives more information about the implementation than the verdict assigned to this trace.

That is the reason why in this paper we shall adopt the view that one successful test case execution reveals exactly one trace of the implementation. Moreover we shall merge test cases and test executions by choosing the set of traces $(\{!, ?\} \times L_S)^*$ as the set of tests K where L_S is the set of interactions declared in the specification. Now we can define successful test case executions and successful test suite executions.

definition 7 Let $t \in (\{!, ?\} \times L_S)^*$ be a test case and let $T \subseteq (\{!, ?\} \times L_S)^*$ be a test suite

- $pass(I, t) =_{def} t \in Tr(I)$

- $pass(I, T) =_{def} T \subseteq Tr(I)$

  Let us remark that because of this definition, the inputs and outputs $(\{!, ?\})$ of test cases are defined from the implementation point of view and no longer from the tester. For example the test case $t = ?a!b$ means that an interaction a is sent by the tester to the implementation and b is received by the tester from the implementation.

4.2 Viewing test hypotheses as trace sets

Remember that an hypothesis H is a coverage of a test suite T iff $(\forall I \in Imp)\ (pass(I, T) \wedge H(I) \Rightarrow \mathbf{imp}(I, S))$. According to definition 7, this is equivalent to: $(\forall I \in Imp)\ (T \subseteq Tr(I) \wedge H(I) \Rightarrow \mathbf{imp}(I, S))$.

Let us instantiate **imp** with the conformance relations listed in 3.2. We can see that they are based in a way or another on a comparison between $Tr(I)$ and $Tr(S)$. Thus, making an hypothesis H covering a test suite T comes down to assuming that some traces are in the implementation and some others are not. For example let us instantiate **imp** with :

- $I \geq_{tr} S$

  H is a coverage of a test suite T iff
  $(\forall I \in Imp)\ (T \subseteq Tr(I) \wedge H(I) \Rightarrow Tr(S) \subseteq Tr(I))$.
  In that case we get : H is a coverage of a test suite T iff
  $(\forall I \in Imp)\ (H(I) \Rightarrow Tr(S) - T \subseteq Tr(I))$.
  Thus, for the conformance relation $I \geq_{tr} S$, making an hypothesis covering a test suite comes down to assuming that the traces of S that are not tested are also in the implementation.
  Proof:
  if: $(\forall I \in Imp)\ (H(I) \Rightarrow Tr(S) - T \subseteq Tr(I))$ implies
  $(\forall I \in Imp)\ (T \subseteq Tr(I) \wedge H(I) \Rightarrow T \subseteq Tr(I) \wedge Tr(S) - T \subseteq Tr(I))$ implies
  $(\forall I \in Imp)\ (T \subseteq Tr(I) \wedge H(I) \Rightarrow Tr(S) \subseteq Tr(I))$. Thus H is a coverage of T.
  only if: Assume there exists I such that $H(I) \wedge\ Tr(S) - T \not\subset Tr(I)$. That implies $Tr(S) \not\subset Tr(I)$. Thus I is not conformant and consequently H is not a coverage of T.

- $I \leq_{tr} S$

  H is a coverage of a test suite T iff
  $(\forall I \in Imp)\ (T \subseteq Tr(I) \wedge H(I) \Rightarrow Tr(I) \subseteq Tr(S))$.
  In that case we get :
  H is a coverage of a test suite T iff $(\forall I \in Imp)\ (H(I) \Rightarrow Tr(I) \subseteq Tr(S))$.
  Thus for this conformance relation, making an hypothesis covering a test suite implies that the traces of I are included in the traces of S, or in other words,

$((\{!,?\} \times L_S)^* - Tr(S)) \cap Tr(I) = \emptyset.$

Proof:

if: trivial

only if: Assume there exists I such that $H(I) \wedge Tr(I) \not\subset Tr(S)$. That implies that I is not conformant and thus H is not a coverage T

We mention that we obtain a similar result with the conformance relations R_5 and R_1 ($R_1(I, S) \Leftrightarrow (\forall \sigma \in Tr(S))(\sigma \in Tr(I) \Rightarrow O(\sigma, I) \subseteq O(\sigma, S))$). R_1 is very close to the relation **ioco** defined on the IOLTS model (Tretmans, 1996). It shows that the results presented in this paper hold for other models than the IOSM.

To sum up this case study, we can say that for the main conformance relations used for testing, formulating a useful test hypothesis comes down to assuming that there exists a set of traces in the trace set of the implementation and another one in the complementary set of the trace set of the implementation. We shall see in further sections that this result will enable us to give a formal representation of hypotheses.

5 PARTIAL AUTOMATA

5.1 Definition

Remember that the goal of test coverage is to make up one's mind about the correctness of an implementation under test knowing that the implementation can execute correctly a test suite T and that it verifies some hypotheses.

Here we define a new data structure to store the information that could be gathered from the implementation either by test or by hypotheses. Moreover this structure must allow us to determine when we have enough information to give a verdict about the correctness of the implementation.

We have seen in the previous section that making hypothesis implies that some traces are included in the implementation and some others are not. We have assumed that both the specification and the implementations can be modelled by automata. Consistently, we shall consider only hypotheses corresponding to regular sets of traces. Therefore, we shall represent these traces on an automaton.

However, this automaton should be able to represent some traces that are known to belong to the implementation and some others that are known not to belong to the implementation. The first point is easy: is it sufficient that some traces of this automaton (what we call from now on partial automaton since it gives a partial view of the implementation under test) are exactly the traces which have been identified as belonging to the IUT. For the second point, the traces «outside» the implementation are also given as traces of the partial automaton but are distinguished from the others by finishing in a particular state of the partial automaton, denoted out. Moreover, we know that if $\alpha \in (\{!,?\} \times L_s)^*$ is not a trace of I then any extension of α is not a trace of I as well. As a result, no transition can go from out to another state and there is a loop-transition on out labelled by each observable interaction. We also know that an implementation modeled as an IOSM can never refuse an input (Phalippou, 1994)(Tretmans, 1996), that is to say it is always possible to send something to the implementation. It means that $\alpha \in Tr(I)$ implies

$(\forall a \in L_s)\ (\alpha?a \in Tr(I))$. So the transitions reaching out from another state are necessary labelled by an output.

definition 8 A partial automaton Ip is an IOSM $\langle S_{Ip}, L_{Ip}, T_{Ip}, s_{0Ip}\rangle$ with a particular state denoted out and verifying

- $(\forall \mu \in \{!,?\} \times L_{Ip})\ (\forall s_j \in S_{Ip})\ ((out, \mu, s_j) \in T_{Ip} \Rightarrow s_j = out)$
- $(\forall \mu \in \{!,?\} \times L_{Ip})\ ((out, \mu, out) \in T_{Ip})$
- $(\forall \mu \in \{!,?\} \times L_{Ip})\ ((\exists s \in S_{Ip})\ ((s, \mu, out) \in T_{Ip}) \Rightarrow \mu \in \{!\} \times L_{Ip})$

definition 9 We denote *IOSMp* the set of partial automata.

Now we have to define which are the implementations that verify the information stored in a partial automaton *Ip*. We call these implementations the candidate implementations because they stand as candidate for being the implementation under test with respect to the constraints stored in the partial automaton. According to the above discussion, each candidate must have the traces of Ip that do not end in out and must not have the others.

definition 10 An implementation I is a candidate w.r.t. the partial automaton Ip, and we note $cand(Ip, I) =_{def} (\forall \sigma \in Tr(Ip))$

$$([(Ip, \sigma, out) \Rightarrow \sigma \notin Tr(I)] \wedge [(Ip, \sigma, s) \wedge s \neq out \Rightarrow \sigma \in Tr(I)]).$$

5.2 Basic ideas on how partial automata work

Before giving some new definitions we give some informal clues on how we shall use partial automata.

The partial automaton embodies test cases and test hypotheses in the shape of traces. Assume the hypotheses to be formulated one after the other, then each hypothesis is a function that adds new traces to the partial automaton. If one of these functions adds one new trace that fall in the out state, it corresponds to an hypothesis that assumes that this trace is not in the IUT. The other added traces correspond to an hypothesis that assumes that these traces are in the implementation.

Thus, at the beginning of the process the partial automaton contains only the traces of the tests, and is enriched as soon as the hypotheses are applied.

5.3 Initial partial automaton

The initial partial automaton is a partial automaton storing nothing but the test set. Since practical test sets are finite, the initial partial automaton has neither loop-transition nor cycles, but there exists many corresponding partial automaton. We choose the minimal tree given by the following algorithm.

definition 11 Let $T = \{\mu_{11}\ldots\mu_{1n_1}, \ldots, \mu_{i1}\ldots\mu_{ij}\ldots\mu_{in_i}, \ldots, \mu_{m1}\ldots\mu_{mn_m}\}$ be a test set. The initial partial automaton of T is the IOSMp $Ip_0 = \langle S_{Ip}, L_{Ip}, T_{Ip}, s_{0Ip}\rangle$ given by the following algorithm:

```
1. S_Ip := {s_0IP, out}, L_Ip:=L,
   T_Ip:= {(out, μ, out)| μ ∈ {!,?} ×L}
```

```
2. for i from 1 to m
   . current-state:= s0Ip
   . for j from 1 to ni
     if (∃s ∈ S_Ip) ((current-state, μ_ij, s) ∈ T_Ip)
     then
       . current-state := s
     else
       . S_Ip := S_Ip ∪ {s_ij}
       . T_Ip := T_Ip ∪ {(current-state, μ_ij, s_ij)}
       . current-state := sij
     end if
```

For example, the initial partial automaton of the test *T={?a!b?c!d, ?a!b?e!f, ?a!d, ?e!b}* is given on figure 1.

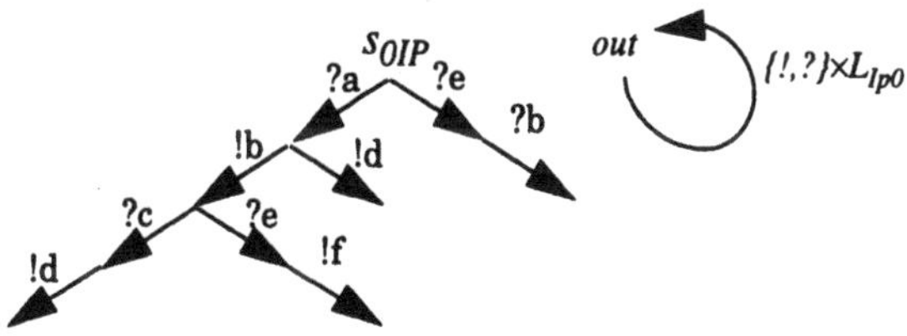

Figure 1 Initial partial automaton of tests suite {?a!b?c!d, ?a!b?e!f, ?a!d, ?e!b}.

6 EXAMPLES OF HYPOTHESES VIEWED AS FUNCTIONS

Along this example and before generalization, we shall use the conformance relation $I \geq_{tr} S$. Thus all significant hypotheses will be formulated as a trace additions (see 4.2 1.). As a result we shall not need the *out* state (see 5.1). So it will not appear on the figure of this example but it must be kept in mind that it exists.

Let us consider the specification *S* given in figure 2. A conforming implementation of *I* is a system that must at least respond *!x* after receiving *?a*, *?b* or *?c* and then going back to its initial state after receiving *?r* (*?r* for reset).

Testing an implementation with *T={?b!x?r}* is far from being sufficient to check that *I* conforms to *S* w.r.t. the conformance relation $I \geq_{tr} S$. All that can be said of *I* is that *I* passes *T* and then the elements of *T* are traces of *I*. Than can be summed up in the initial partial automaton Ip_0 (see figure 2).

Let us consider both following hypotheses:

- The implementation is uniform on the set of interactions $\{?a, ?b, ?c\}$ that is to say its behaviour is the same for these three interactions ((Bernot, 1991)(Phalippou, 1994)).
- The reset is correctly implemented, that is to said it actually brings back the implementation to its initial state.

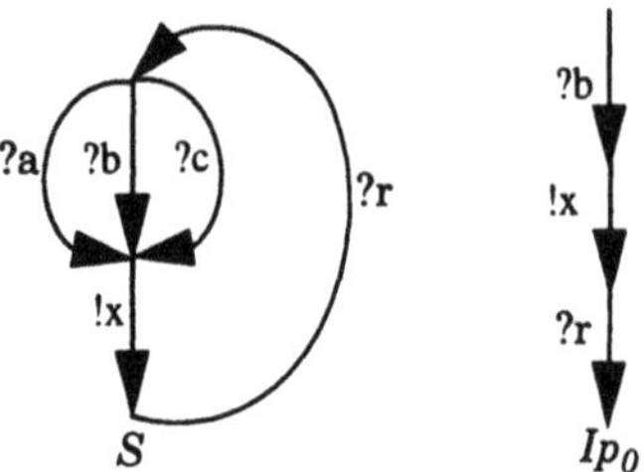

Figure 2 Specification S and initial partial automaton of T={?b!x?r}.

6.1 Uniformity hypothesis

From the trace point of view, the uniformity hypothesis on $\{?a, ?b, ?c\}$ tells us that if there exist $\alpha, \beta \in (\{!, ?\} \times L_s)^*$ and $\mu \in \{?a, ?b, ?c\}$ such that $\alpha\mu\beta \in Tr(I)$, then $(\forall\mu' \in \{?a, ?b, ?c\})\ (\alpha\mu'\beta \in Tr(I))$.

In our example we know that $?b!x?r \in Tr(I)$. Thanks to this hypothesis, we can assume that we also have $?a!x?r \in Tr(I)$ and $?c!x?r \in Tr(I)$. These two new traces must be added to the partial automaton since they are new pieces of information. Ip_1 (see figure 3 a)) is the resulting partial automaton. We can notice that the gap between Ip_1 and Ip_0 is bridged by adding two new transitions labelled respectively *?a* and *?c*.

It is easy to imagine that for other examples the mechanism of transformation of a partial automaton into another partial automaton that takes the uniformity hypothesis into account will always duplicate the missing transitions. Thus we can represent this by a function on partial automata that performs the addition of transitions. For example, the uniformity hypothesis on *{?a,? b, ?c}* can be given as a function $U^{[abc]} : IOSMp \rightarrow IOSMp$ defined by $Ip' = U^{[abc]}(Ip)$ where:

$S_{Ip'} = S_{Ip}$, $L_{Ip'} = L_{Ip}$, $s_{0_{Ip'}} = s_{0_{Ip}}$ and

$$T_{Ip'} = T_{Ip} \cup \{ (s_i, ?a, s_j), (s_i, ?b, s_j), (s_i, ?c, s_j) \mid (s_i, ?a, s_j) \in T_{Ip} \vee (s_i, ?b, s_j) \in T_{Ip} \vee (s_i, ?c, s_j) \in T_{Ip} \}$$

1.1 Reliable reset hypothesis

The reliable reset hypothesis assumes that each time a *?r* interaction is sent to the implementation, it normally goes back to its initial state. Actually on our example that implies that the test set *T={?b!x?r}* is equivalent to *T'={?b!x?r, ?b!x?r?b!x?r, ?b!x?r?b!x?r?b!x?r...}*. The partial automaton Ip_1' of this test set is given on figure 3 b).

As we have already noticed for the uniformity hypothesis, the reliable reset hypothesis can be viewed as a transformation of partial automata. Here we can see on figure 3 b) that we can go from Ip_0 to Ip_1' by looping the transition labelled by

?r. It easy to convince oneself that the well implemented reset hypothesis will always corresponds to this transformation, no matter what the partial automaton is.

As a result we are able to define this hypothesis as a function $Ri: IOSMp \rightarrow IOSMp$ defined by $Ip' = Ri(Ip)$ where

$$S_{Ip'} = S_{Ip} - \{s_j | (\exists s_i \in S_{Ip}) ((s_i, ?r, s_j) \in T_{Ip})\}$$

$$L_{Ip'} = L_{Ip}$$

$$s_{0_{Ip'}} = s_{0_{Ip}}$$

$$T_{Ip'} = T_{Ip} \cup \{(s_i, ?r, s_{0_{Ip'}}) | (\exists s_j \in S_{Ip}) ((s_i, ?r, s_j) \in T_{Ip})\}$$

$$\cup \{(s_{0_{Ip'}}, \mu, s_k) | (\exists s_i, s_j, s_k \in S_{Ip}) ((s_i, ?r, s_j) \in T_{Ip} \wedge (s_j, \mu, s_k) \in T_{Ip})\}$$

$$\cup \{(s_j, \mu, s_{0_{Ip'}}) | (\exists s_i, s_j, s_k \in S_{Ip}) ((s_i, ?r, s_k) \in T_{Ip} \wedge (s_j, \mu, s_k) \in T_{Ip})\}$$

$$-(\{(s_i, ?r, s_j) \in T_{Ip}\} \cup$$

$$\{(s_j, \mu, s_k) \in T_{Ip} | (\exists s_i, s_j, s_k \in S_{Ip}) ((s_i, ?r, s_j) \in T_{Ip} \wedge (s_j, \mu, s_k) \in T_{Ip})\} \cup$$

$$\{(s_j, \mu, s_k) | (\exists s_i, s_j, s_k \in S_{Ip}) ((s_i, ?r, s_k) \in T_{Ip} \wedge (s_j, \mu, s_k) \in T_{Ip})\})$$

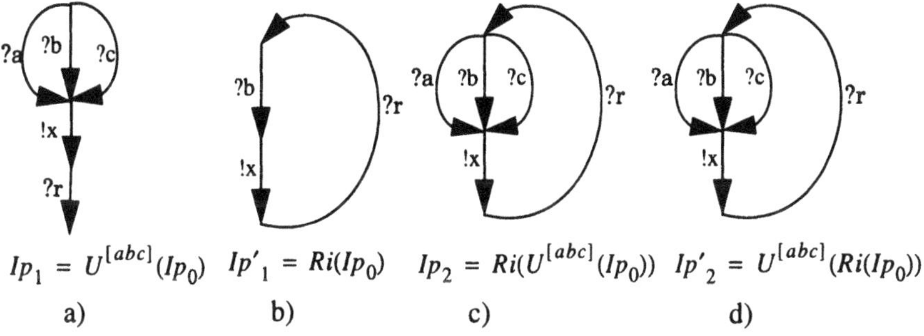

Figure 3 Partial automata.

6.2 Combination of uniformity and reliable reset hypotheses

It's now possible to give the partial automaton which gives (according to the *cand* relation) the set of the implementations that: a) pass the test set *T*, b) satisfy the uniformity hypothesis on *{?a,?b,?c}* c) satisfy the reset hypothesis. Figures 4 c) and d) show $Ip_2 = Ri(U^{[abc]}(Ip_0))$ and $Ip'_2 = U^{[abc]}(Ri(Ip_0))$. We notice that these automata are equal, but it is not a general rule.

The implementations *i* represented by Ip_2 w. r. t. the *cand* relation verifies $(\forall \sigma \in Tr(Ip2))\ (\sigma \in Tr(i))$. On account on their being isomorphic, Ip_2 and *S* have the same trace set. As a result, $(\forall \sigma \in Tr(S))\ (\sigma \in Tr(i))$ and then all the candidate implementations conforms to *S*. We conclude that *T={?b!x?r}* is exhaustive under the uniformity hypothesis on *{?a,?b,?c}* and well implemented reset hypothesis.

7 FORMALIZATION

7.1 Hypotheses and specification

It is important that the specification may be taken into account in the functional definition of test hypotheses. For example, we define uniformity hypothesis on a fixed set *{?a,?b,?c}*. We might wish to define this hypothesis on any set; but at the same time, we would like to apply the hypothesis only in contexts (trace prefixes) where it matches some sort of uniformity within the specification. The point is that the function has to identify the uniform sets and then apply a similar algorithm to $U^{[abc]}$. We can write this function $U: IOSM \times IOSMp \rightarrow IOSMp$ where $Ip' = U(S, Ip)$ if and only if ($S_{Ip'} = S_{Ip}$, $L_{Ip'} = L_{Ip}$, $s_{0_{Ip'}} = s_{0_{Ip}}$ and $T_{Ip'}$ is constructed by the following algorithm

1. $T_{Ip'} = T_{Ip}$
2. For each $s_i \in S_{Ip}$

   If there exist $\alpha \in Tr(Ip)$, $\mu \in L_{Ip}$, $s_j \in S_{Ip}$ such that $(Ip, \alpha, s_i)(s_i, \mu, s_j)$ then,

   (since there exist $s_{s_i}, s_{s_j} \in S_S$ such that $(S, \alpha, s_{s_i})(s_{s_i}, \mu, s_{s_j})$)

   for each $\lambda \neq \mu \in L_S$

   if $(s_{s_i}, \lambda, s_{s_j}) \in T_S$ then add (s_i, λ, s_j) to T_{Ip}

We now have sufficient material to define a functional hypothesis.

definition 12 A functional hypothesis is a function
$F: IOSM \times IOSMp \rightarrow IOSMp$ such that
$(\forall S \in IOSM)\ (\forall Ip \in IOSMp)\ (Tr(Ip) \subseteq Tr(F(S, Ip)))$.

We need to fixe S in $F(S, Ip)$, therefore we note $F_S(Ip)$ for $F(S, Ip)$.

The application of a functional hypothesis to a partial automaton *Ip* reduces the set of candidate implementations. The set of eliminated candidates is
$\{I \in IOSM |\ cand(Ip, I) \wedge \neg cand(F(Ip), I)\}$

7.2 Stop condition

It is high time to sum up all we have seen till now. First we have seen that for the most frequently used conformance relations, making a test hypothesis comes down to assuming that some traces are in the implementation and some others are not. Secondly we introduced an original data structure – the partial automata – to store test sets and hypotheses in the shape of automata. Thirdly, we have defined test hypotheses as functions that enrich partial automata. At the beginning of the process, the partial automaton contains only the test cases (we have called it the initial partial automaton). Then, as the hypotheses are formulated, the corresponding functions are applied to the partial automaton. Now the question is: how can we make sure than we have applied enough hypotheses such that exhaustivity (under these hypotheses) is reached?

Remember that the partial automaton represents the set of candidate implementations, that is to said the set of implementations over which the implementation under test ranges provided it passes the test set and verifies the test hypothesis stored in the partial automaton. As a result, if all the candidate implementations conform, the implementation under test necessarily conforms.

Formally, let T be a test set. At first we compute the corresponding initial partial automaton Ip_0. Let us assume that we have a set of functional hypotheses $\{f_S^{h1}, \ldots, f_S^{hn}\}$. We apply these functional hypotheses one after the other on Ip_0. The resulting partial automaton is $f_S^{hn} \circ \ldots \circ f_S^{h1}(Ip_0)$. The candidate implementations all conform if and only if

$(\forall I \in IOSM)\ (cand(f_S^{hn} \circ \ldots \circ f_S^{h1}(Ip_0), I) \Rightarrow \mathbf{imp}(I, S))$. According to the above discussion we define the exhaustivity condition in the following manner.

definition 13 Let T be a test set and Ip_0 the corresponding initial partial automaton, $f_S^{h1}, \ldots, f_S^{hn}$ a sequence of functional hypotheses.

$(\forall I \in IOSM)\ (cand(f_S^{hn} \circ \ldots \circ f_S^{h1}(Ip_0), I) \Rightarrow \mathbf{imp}(I, S))$ implies T is exhaustive under the functional hypotheses $f_S^{h1}, \ldots, f_S^{hn}$.

Considering a sequence rather than a set of hypotheses reflects the reality since the impact (and even the expression) of new hypotheses may depend on previous ones. As a result, exhaustivity cannot be computed in an «environment» of hypotheses but for an ordered list of hypotheses. This is perfectly reflected by the non-commutative composition of functional hypotheses (as opposed to a purely logical view that would use conjunction of predicates).

This definition is not operational since it is based on the candidate implementations. It can be rewritten by using the definition of *cand* and **imp** in order to obtain a «syntactic» coverage condition (the proofs are omitted but are quite obvious).

- For the relation $I \geq_{tr} S$, $(\forall \sigma \in Tr(S))\ ((Ipn, \sigma, s) \wedge s \neq out)$ implies T is exhaustive under the functional hypotheses $f_S^{h1}, \ldots, f_S^{hn}$.
- For the relation $I \leq_{tr} S$ the condition is $(\forall \sigma \in L^* - Tr(S))\ (Ipn, \sigma, out)$.
- Finally for $R_5(I, S)$ we get

  $(\forall \sigma \in Tr(S))\ (\forall \mu \in O(\sigma, S))\ (\forall \mu' \in L - O(\sigma, S))$

  $(Ipn, \sigma, s) \wedge s \neq out \wedge (Ipn, \sigma\mu, s') \wedge s' \neq out \wedge (Ipn, \sigma\mu', out)$

8 STRENGTH OF FUNCTIONAL HYPOTHESES

8.1 partial order on hypotheses

It is well known that we can be more or less confident in an hypothesis. It depends on many subjective factors and particularly the knowledge we have on the implementation. We dealt we this problem in another paper (Charles, 1996). Here we propose to build a partial order on functional hypotheses based on their distinguishing power. We have seen that applying a functional hypothesis to a partial automaton comes down to reducing the set of candidate implementations. It is obvious that the more candidate implementations are eliminated, the stronger the hypothesis is. For example, imagine the hypothesis assuming the implementation conform whatever the test set. In other words, this hypothesis eliminates all non-conforming implementation from the set of candidates. It is hard to believe without testing that an implementation conform, so it is a very strong hypothesis. Conversely imagine an hypothesis that never reduce the set of candidates : it is the weakest hypothesis because it assumes no further good properties on the implementation.

definition 14 Let F and F' be two functional hypotheses. F is stronger than F' $=_{def} (\forall Ip \in IOSMp)\ (\forall I \in IOSM)\ (cand(F(Ip), I) \Rightarrow cand(F'(Ip), I))$

If we look at the algorithms that define the functional hypotheses, we can see that this definition has a practical meaning. Consider function $U^{[abc]}$ defined in section 6.1. The underlying significance of this hypothesis is : if there exists $\alpha, \beta \in (\{!, ?\} \times L_s)^*$ and $\mu \in \{?a, ?b, ?c\}$ such that $\alpha\mu\beta \in Tr(I)$ then it can be assumed that $(\forall \mu' \in \{?a, ?b, ?c\})\ (\alpha\mu'\beta \in Tr(I))$. That is to say, testing I with one the interactions of the set $\{?a, ?b, ?c\}$ is equivalent to testing with all three. But one may say that this hypothesis is too hazardous and should be better verified before being applied. For example it could be demand before assuming the whole set $\{?a, ?b, ?c\}$ that:

- two interactions of $\{?a, ?b, ?c\}$ are tested or,
- one interaction of $\{?a, ?b, ?c\}$ are tested with at least a two interaction long preamble (β in the discussion above) or,
- two interactions of $\{?a, ?b, ?c\}$ are tested with a two interaction long preamble each, etc...

That defines three new functions on $IOSM \times IOSMp \to IOSMp$ (resp. $U_1^{[abc]}$, $U_2^{[abc]}$, $U_3^{[abc]}$). Their algorithms are not given here since they are light variations of $U^{[abc]}$.

Now it can be proved that $U^{[abc]}$ is stronger than $U_1^{[abc]}$ and $U_2^{[abc]}$, which are stronger than $U_3^{[abc]}$. Thus this relation reflects perfectly the intuition felt in the examples.

Of course, this is only a partial order. We cannot order hypotheses of different types, for example we cannot establish which of the reliable reset hypothesis or the uniformity hypothesis is the stronger. However it is the first step towards a total order and weight assignment that would reflect the strength of hypotheses as used in (Charles, 1996).

8.2 Possible enhancement of TH-based coverage with weight assignment

As we show in (Charles, 1996), ordering and assigning weights to hypotheses according to their strength can improve significantly the TH-based coverage definition. Indeed, for a given test suite *T* there might exist many series of hypothesis that cover *T*. But if we know the weight (ie the strength) of each hypothesis, we can compute a global weight of a series of hypotheses that reflects how we can be confident in that series. As a result we can redefine TH-based coverage of a test suite by restricting to minimal weight series of hypotheses.

We also show in (Charles, 1996) that with a suitable weight assignment to hypotheses and with a suitable way to compute the global weight of series of hypotheses our model embodies the metric based theory of coverage of (Vuong, 1991) and (Curgus, 1993).

9 CONCLUSION

In this paper we have introduced the concept of TH-based coverage. We have focused on the major role of test hypotheses in usual coverage definition and test practice. This has led us to define an original data structure – the partial automata – that represents the set of the implementations – the candidate implementations – passing a test suite and verifying some test hypotheses. Thereafter we have seen that the hypotheses could be viewed as functions on partial automata. This has given us a practical way to handle test hypotheses for TH-based coverage. Finally we have introduced the notion of strength of hypotheses and explained how this could enhance the TH-based coverage definition.

It may seem paradoxical that we propose in this paper a formal and abstract test coverage measure while we claim in (Groz, 1996) that the very poor transition coverage is sufficient. In fact both approaches aims at giving pieces of information understandable and practical to test designers. In (Groz, 1996) we explained how relating the test suites to specifications through a visual tool turns to be more informative than a coverage given as a figure. In this paper we keep on thinking that test coverage must be richer than a percentage and must incorporate the know-how of test designers. Since the know-how of test designers is expressed as test hypotheses, why not mix both approaches in a visual tool that would link a test suite to a specification by means of test hypotheses?

10 REFERENCES

Bernot G., Gaudel MC. , Marre B. , *Software testing based on formal specifications: a theory and a tool.* Software Engineering Journal, November 1991.

Bochmann G. v. , Das A. , Dssouli R. , Dubuc M. , Ghedamsi A. , Luo G. , *Fault models in testing*, Proceedings of IWPTS, Leidschendam, October 1991.

Charles O. , Groz R. , *Formalisation d'hypothèses pour l'évaluation de la couverture de test.* Proceedings of CFIP 96, Rabat, Morocco.

Clatin M. , Groz R. , Phalippou M. , Thummel R. , *Two approaches linking a test generation tool with verification techniques*, IWPTS 95, Evry, France.

Curgus J. , Vuong S. T. *A Metric Based Theory of Test Selection and Coverage.* Proceedings of PSTV XIII, Liege, Belgium, 1993.

Dubuc M. , Dssouli R. , Bochmann G. v. , *TESTL: A Tool for Incremental Test Suite Design Based on Finite State Model,* IWPTS, Leidschendam, 1991.

Gaudel MC. , *Test Selection Based on ADT Specifications.* Proceedings of IWPTS 92, Montréal, Canada.

Groz R. , Charles O. , Renévot J. , *Relating Conformance Test Coverage to Formal Specifications.* Proceedings of FORTE 96, Kaiserslautern, Germany.

ISO/IEC/JTC1/SC21/P.54 - ITU-T SG10 Q.8, *Formal Methods in Conformance Testing,* Working Draft, March 1995. To become rec. Z.500 from ITU-T.

Motteler H. , Chung A. , Sidhu D., *Fault Coverage of UIO-based Methods for Protocol testing.* Proceedings of IWTCS 93, Pau, France.

Phalippou M. , *Relations d'implantation et hypothèses de test sur des automates à entrées et sorties,* université de Bordeaux I, 1994.

Rapps S. , Weyuker E. J. . *Selecting Software Test Data Using Data Flow Information.* IEEE Trans. on Software Engineering, vol. SE-11, NO. 4, April 1985.

Sidhu D. , Leung T. , *Formal Methods for Protocol Testing : A Detailed Study,* IEEE transactions on software engineering, vol. 15, n. 4, april 1989.

Tretmans J. , *Test Generation with Inputs, Outputs, and Repetitive Quiescence.* CTIT Technical Report series No. 96-26, ISSN 1381-3625. To be published in Software — Concept and Tools. 1996.

Ural H. , Yang B. , *A Test Sequence Selection Method for Protocol Testing,* IEEE Transactions on Communications, vol. 39, no 4, April 1991, pp. 514-523.

Vuong S. T. , Curgus J. . *On Test coverage Metrics for Communication Protocols.* Proceedings of IWPTS IV, Leinschendam. The Netherlands. 1991.

Weyuker E. J. , *The Complexity of Data Flow Criteria for Test Data Selection,* Information Processing Letters 19 (31 August 1984), pp 103-109.

Yao M. , Petrenko A. , Bochmann G. V. , *A Structural Analysis Approach to the Evaluation of Fault Coverage for Protocol Conformance Testing.* Proceedings of FORTE 94, Berne, Switzerland, 1994.

Zhu J. , Chanson S. T. , *Toward Evaluating Fault Coverage of Protocol Test Sequences.* Proceedings of PSTV XIV, Vancouver, Canada, 1994.

11 BIOGRAPHY

Olivier Charles is a Ph.D. student under the joint supervision of Roland Groz, André Schaff and Rachida Dssouli. He graduated from the Ecole Supérieure d'Informatique et Applications de Lorraine, University of Nancy.

Roland Groz received a Ph.D. in computer science from University of Rennes I in 1989. He currently manages a department in France Télécom - CNET devoted to research in the area of validation, specification and prototyping .

9

Design for testability: a step-wise approach to protocol testing

Hartmut König[a], Andreas Ulrich[b], Monika Heiner[a]
[a] Department of Computer Science, BTU Cottbus, PF 101344, 03013 Cottbus, Germany, e-mail: {koenig, mh}@informatik.tu-cottbus.de
[b] Department of Computer Science, University of Magdeburg, PF 4120, 39016 Magdeburg, Germany, e-mail: ulrich@cs.uni-magdeburg.de

Abstract

We present an approach to support the design for testability aspect of communication protocols. It combines the ad-hoc techniques partitioning and instrumentation known from integrated circuit testing. A protocol specification is divided into modules of reasonable size. This module structure is preserved in the implementation. Extra test points are added to observe inter-module communication. The test procedure consists of several steps. In the first step, modules are tested separately by applying a powerful test method, whereas following integration tests of modules exploit additional information provided by observers. The application of less sophisticated test methods is propagated for these steps. We show that this testing approach extends testability while fault detection capability is maintained.

Keywords

Network protocols; requirements / specifications; testing.

Testing of Communicating Systems, M. Kim, S. Kang & K. Hong (Eds)
Published by Chapman & Hall

1 MOTIVATION

Due to the limited power of verification, testing has always been an important method in practice to validate the correctness of communication protocols. Nevertheless, the test of communication protocols has been proven to be difficult and expensive. Reasons are the complexity of communication protocols that makes exhaustive tests impossible as well as the need for complementary tests, e.g. development tests during the implementation phase of a protocol, conformance test to prove the compliance of the implementation with the specification or a protocol standard, interoperability test to demonstrate the ability of implementations to work together, performance test to measure, whether the implementation provides the specified efficiency, and robustness test to prove, whether the implementation behaves stable in erroneous situations.

Up to now, testing aspects are usually not considered during protocol design and protocol implementation. To make sophisticated test methods more efficient and applicable in practical testing, the test process itself has to be reconsidered. This demand is especially enforced by new requirements from high performance communication that require new protocols and communication architectures as well as new implementation techniques [Clar 90]. To make protocol implementations more testable, dedicated techniques and methods have to be applied already during the design phase in order to reduce efforts and costs of testing. In addition, testing aspects should be taken into consideration during the whole protocol development process. Therefore, *design for testability* (DFT) has become an important research topic in protocol engineering.

Testability, in general, is a property of an object that facilitates the testing process [Vuon 94]. It can be obtained in two ways: (1) by introducing special observation features that give additional information about the (internal) behavior of the object, and (2) by a systematic design for testability. The choice of the DFT strategy depends on two factors: the goals of the testing process, and the kind of application.

DFT has been applied in integrated circuit (IC) technology already for a long time. The techniques used there can be divided into two categories [Will 82]: ad-hoc techniques and structured approaches. Ad-hoc techniques solve the testing problem for a given design. They are not generally applicable to all designs. Examples of ad-hoc techniques are partitioning and extra test points. Structured approaches, on the other hand, are generally applicable techniques that are based on a certain design methodology with fixed design rules.

DFT is still a new topic in protocol engineering. It is obvious that some of the approaches worked out in the IC area are also tried to be applied in protocol engineering. First proposals, such as the introduction of points of observation [Dsso 91, 95], can be categorized as ad-hoc techniques according to the classification introduced above. Structured approaches have been not known, yet.

According to [Will 82], DFT comprises a collection of techniques that are, in some cases, general guidelines and, in other cases, precise design rules. Consequently, there will be not only a single approach, but several ones. For the protocol area, this means that the objective of DFT should be to develop a set of approaches that can be applied depending on the test context, the associated cost of implementing them, and the return on investment. Therefore, DFT research should not be limited to a certain test category. It should have a general view and consider all methods that improve the ability of detecting faults during testing and decreasing cost. A selection of specific DFT techniques is needed bearing in mind the benefits they will bring in a given test context.

Starting from this position, we present an testing approach to support DFT of communication protocols that combines the ad-hoc techniques of partitioning a protocol specification into module structures and adding extra test points to observe inter-module communication. The idea of the approach presented in this paper is to use instrumentation not only for getting additional information about the behavior of the implementation under test but also to use this information to decrease the testing efforts by reducing the length of the test suite. The proposed testing procedure is a step-wise one. In the first step, the modules are tested separately by applying a powerful test method, whereas for the following integration tests of the modules (in one or more steps) the application of a less sophisticated test method is propagated to decrease test efforts while fault detection capability is maintained.

The rest of the paper is organized as follows. Section 2 gives a short overview of the proposed testing procedure that is evaluated in more detail in Section 3. Section 4 is dedicated to aspects of multi-module testing and concurrency. Section 5 relates our work to existing ones, and finally, Section 6 concludes the paper.

2 A STEP-WISE TESTING APPROACH – OVERVIEW

The step-wise testing approach proposed in this paper follows the ad-hoc approach in integrated circuit testing [Will 82]. In particular, we use partitioning and adding of extra test points. According to these techniques, we propose to partition a protocol specification into a set of modules of reasonable size which can be executed sequentially and/or in parallel. Such a structuring is natural for protocol design. Most formal description techniques (FDTs) support a certain module structure in the specification, but structuring is usually not used to support testing.

We suppose that the module structure is preserved in the implementation. But we do not make any assumption that the specified inter-module communication is correctly implemented. The inter-module communication, however, is traced by extra test points used as points of control and observation (PCOs) or only as points of observation (POs).

Supposing such a module structure, testing can be executed step-wise in the following manner (cf. Figure 1):

1. *Module testing*: Each module is tested separately. This test is a black-box test. The extra test points associated to the module serve as PCOs. The modules can be considered as software ICs [Hoff 89].

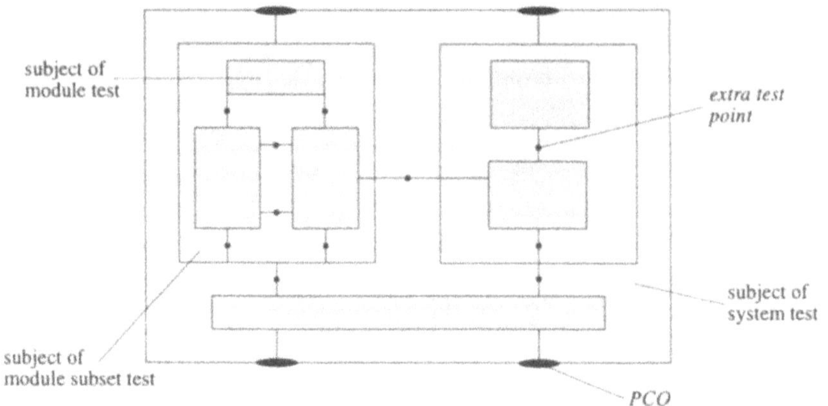

Figure 1 Subjects of test steps for a protocol entity.

2. *Module subset testing*: Reasonable subsets of modules are tested together. The used test method is grey-box testing. Extra test points between modules are POs to observe internal communication between modules.
3. *System testing*: The complete system is tested by integration of all modules and subsets of modules using again grey-box testing as described in the second step.

Steps 2 and 3 are *integration tests* [Myer 79] that test the correct cooperation of the modules, i.e. the correct implementation of the inter-module communication. Step 2 is optional and can be omitted, or may be repeated several times with changing subsets of modules.

The step-wise testing procedure takes advantage of the modularization within the protocol entity. First, each module is tested separately (e.g. by applying the W-method). After that, subsets of modules are tested, and eventually the whole system. Due to the testing efforts already done at module testing level, application of less sophisticated test generation methods is suggested at module subset or system level (e.g. the T-method). The simplification is motivated by the types of faults that can still appear at the second or third testing level (see Section 3.2). The necessary information to find faults that are usually not detectable by a transition tour will be derived from the observation of inter-module communication.

Applying this test strategy we have to show two things: (1) whether the proposed testing approach increases testability, and (2) whether a less sophisticated test generation method in combination with grey-box testing guarantees still high fault coverage. The feasibility of these requirements is discussed in Section 3.

To measure the degree of testability T, we apply the measure introduced in [Petr 94] for finite state machines (FSMs) under the complete coverage assumption:

$$T = \frac{o}{mn^2 p^{m-n+1}} \tag{1}$$

where m is the number of states in the implemented FSM, n is the number of states in the reference FSM, p denotes the number of inputs, and o the number of outputs. In the case that the number of states of the reference FSM equals to the number in the implemented FSM, i.e. $m = n$, the formula is simplified to:

$$T = o/(n^3 p) \qquad (2)$$

The measure is proposed to evaluate FSM based module structures, in order to compare different designs with respect to testability. It assumes that testability is inversely proportional to the amount of testing efforts. The latter is proportional to the length of the test suite needed to achieve full fault coverage in a predefined fault domain. Further, it is obvious that an implementation becomes more testable, if more outputs can be observed during testing.

The reduction of the length of a test suite has a larger impact on the increase in testability, since it more effectively cuts test efforts. Consequently, to estimate the increase in testability, we have to show that the average total test suite length of the step-wise testing procedure is shorter than the length of a test suite from the unstructured testing approach.

3 ADVOCATING THE STEP-WISE TESTING APPROACH

In this section, we want to discuss the feasibility of the step-wise testing approach. We suppose that the protocol specification is given in the form of interacting modules as depicted in Figure 1. In order to perform systematic tests, test suites must be derived that are complete to a chosen fault model. A test suite is complete if it can distinguish all faulty implementations among all implementations in the chosen fault model. For example, a complete test suite is produced by the W-method [Chow 78] under the assumption that the number of states in the implementation equals to the one in its specification [Petr 94]. Therefore, we apply the W-method as test generation method for module testing and show that under certain prerequisites the test suite of the less powerful transition tour method (T-method) [Sidh 89] is complete in case of integration test.

For the sake of simplicity, we consider only module testing and system testing. The necessity to introduce further module subset test steps depends on the complexity of the specification. It does not principally change the discussion here, because the procedure is the same as in the system test. It has only to be taken into account for evaluating a concrete test situation.

3.1 Assumptions and basic notations

To follow the sequel of the paper, we introduce some necessary assumptions on the protocol specification as well as some basic notations.

First, we suppose a formal protocol specification as a parallel composition $\mathfrak{J} = M_1 \parallel \ldots \parallel M_k$ of interacting modules. Each module realizes a certain part of the protocol. It is described by a sequential automaton (finite state machine, FSM). Modules communicate with each other solely via interaction points. The communication pattern used is synchronous communication and non-blocking send based on interleaving semantics. Transmitting messages and their receipt through interaction points are referred to *actions*.

To distinguish the different kinds of communication, we denote all inputs and outputs of the protocol implementation from/to the environment as *external*, analogously all inputs and outputs belonging to the inter-module communication as *internal*. Events appearing only inside a module are not considered.

In our discussion, we need to distinguish three types of automata: the module automaton M, the composite automaton CA, and the entity automaton EA.

Module automaton (M)

The module automaton specifies the expected behavior of the module within the protocol entity. It is modeled as a finite state machine.

A *finite state machine (FSM)* M is defined by a quadruple $(S, A, \rightarrow, s_0)$, where S is a finite set of states; A is a finite set of actions (the alphabet) consisting of a subset of inputs A_I and a subset of outputs A_O; $A_I \cup A_O = A$; $\rightarrow \subseteq S \times A_I \times A_O \times S$ is a transition relation; and $s_0 \in S$ is the initial state.

A transition $(s_1, a, b, s_2) \in \rightarrow$ with input a and output b is also written as $s_1 - a_i / b_o \rightarrow s_2$. A *trace* denotes a sequence of actions a_i transferring M from state s to state s' and traversing a set of intermediate states: $s - a_1/b_1 \rightarrow s_1 - a_2/b_2 \rightarrow s_2 - \ldots \rightarrow s'$. With no loss of generality, we assume that each component FSM is initially connected.

Composite automaton (CA)

The composite automaton specifies the behavior of a subset of modules and of the complete protocol entity. The joint behavior of the multi-module system $\mathfrak{J} = M_1 \parallel \ldots \parallel M_n$ can be described by means of a so-called composite machine defined over $A_{\mathfrak{J}} \subseteq A_1 \cup \ldots \cup A_n$, the (global) alphabet of system $\mathfrak{J}$ that is defined by the parallel composition operator $\parallel$. According to the semantics of this operator, components execute shared actions that require rendezvous of a matching input/output pair of two component FSMs along with local actions that are executed by a component and its environment only.

A *composite automaton* of a given concurrent system $\mathfrak{J}$ of k FSMs $M_i = (S_i, A_i, \rightarrow_i, s_{0i})$ is the quadruple $(S_{\mathfrak{J}}, A_{\mathfrak{J}}, \rightarrow_{\mathfrak{J}}, s_{\mathfrak{J}})$, where $S_{\mathfrak{J}}$ is a global state space, $S_{\mathfrak{J}} \in S_1 \times \ldots \times S_k$; $A_{\mathfrak{J}} \subseteq A_1 \cup \ldots \cup A_k$ is the set of actions (the global alphabet), $s_{\mathfrak{J}} = (s_{01}, \ldots, s_{0k})$ is the initial global state. The transition relation $\rightarrow_{\mathfrak{J}}$ is given by the following three transition rules assuming P and Q are two given FSMs, s_P, s_P' and s_Q, s_Q' are states in P and Q, and a, x, b are actions in the corresponding subsets of inputs or outputs of action sets A_P or A_Q.

- If $s_P\text{–}a/x\rightarrow s_P'$ and $x \notin A_{Q_I}$ then $(s_P, s_Q)\ \text{–}a/x\rightarrow_{\mathfrak{Z}} (s_P', s_Q)$.
- If $s_P\text{–}a/x\rightarrow s_P'$ and $s_Q\text{–}x/b\rightarrow s_Q'$ then $(s_P, s_Q)\ \text{–}a/x/b\rightarrow_{\mathfrak{Z}} (s_P', s_Q')$.
- If $s_Q\text{–}x/b\rightarrow s_Q'$ and $x \notin A_{P_O}$ then $(s_P, s_Q)\ \text{–}x/b\rightarrow_{\mathfrak{Z}} (s_P, s_Q')$.

The notation of a global transition $s_{\mathfrak{Z}}\ \text{–}a/x/b\rightarrow_{\mathfrak{Z}} s_{\mathfrak{Z}}'$ illustrates that after input a has occurred, internal action x between two modules is exchanged and output b is produced finally.

Entity automaton (EA)

The entity automaton specifies the global, observable behavior of the protocol entity. It can be derived formally from *CA* by restricting the global alphabet $A_{\mathfrak{Z}}$ to the set of actions observable by the environment of the protocol, i.e. internal communication between modules is suppressed in the description of *EA*. The notion of an entity automaton is introduced here merely for the purpose of comparison.

3.2 Fault model

Now we discuss the types of faults that may appear in a faulty multi-module implementation. We assume that the specification has been verified to be correct. That means, there are no deadlocks or unreachable states in the specification.

Fault model of the module automaton

In our test approach, the module test is a black-box test, in which a test suite is applied that is complete to the fault model of a single module. We suppose in the following discussion that the single modules have been successfully tested and that they behave as specified.

Fault model of the composite automaton

At the level of integration tests the following faults are still possible*:

- *Data flow faults:* Data exchanged between modules may be faulty. This is a common implementation fault. Testing related to data flow is still a partly unsolved issue for which only specialized solutions have been found [Guer 96]. The observation of the inter-module communication can in part detect data faults. This may be in some cases useful because inter-module communication often consists of simple data structures as, for instance, signals that inform about

* Faults in module interactions that do not appear in an appropriate sequence or even incorrect sequences of actions can be considered as design faults of the protocol. They can be found by static analysis of the composite automata concerning communication inconsistencies (e.g. on the basis of Petri net analysis [Hein 92] [Ochs 95]). Synchronization faults due to a change in the communication pattern from synchronous to asynchronous communication or vice versa are not considered here since once the communication principle has been selected in the design phase of the protocol, it should remain the same throughout the design trajectory.

a state achieved in the sending module or transfer data as credit information. According to our approach false outputs, i.e. data of a wrong type, are detected during module test. Faults in the data flow caused by false values of components of the data are not considered in our discussion.

- *Coupling faults among modules*: Inter-module communication can be implemented by different means, e.g. procedure calls, shared variables, communication channels or others. It is also often a source for faulty implementations. Coupling faults appear if interaction points of the modules are erroneously connected with each other, i.e. the output of a module is sent to a wrong module that is, however, able to consume this event performing a corresponding input event. This type of fault must be detected during integration test. A coupling fault can be reduced to a state fault in the composite automaton.

3.3 Feasibility of the approach

To justify the step-wise testing approach, we have to show that

- the average total length of the test suite for the step-wise approach is shorter than the length of the test suite derived from the entity automaton;
- the fault coverage of the step-wise approach is the same as for the conventional approach based on the single entity automaton, i.e. all possible faults that can be detected in the conventional approach shall be detected by the step-wise approach, too.

Let A be a finite state automaton, $length(W(A))$ is the test suite length of the W-method applied to A, $length(T(A))$ is the test suite length of the T-method applied to A. The conjecture is that the following equation holds for a suitable number k of modules in the specification:

$$length(W(EA)) > \sum_{i=1}^{k} length(W(M_i)) + length(T(CA)) \quad (3)$$

The formula means that the total length of the test suite applied in the step-wise approach is shorter than the length of the test suite that would be derived from the entity automaton EA.

According to the formula, we have to show that the total length of test suites in a step-wise approach is generally shorter than the length of the test suite derived from the monolithic entity automaton. We demonstrate that this statement holds for the case of an equal number of states in the implementation and the specification.

To test the entity automaton, the W-method is applied since it produces a complete test suite in the fault model of implementations with an equal number of states. The number of states in the entity automaton EA can be estimated in the worst case by $n_{EA} \leq \prod n_i$, where n_i is the number of states of module M_i. This estimation also assumes that automaton reduction applied when constructing the entity automaton does not contribute to a reduction in the number of states, i.e. all global states in the reduced composite automaton are distinguishable. Thus, the length of the W-method is bounded to $O(p_{EA} \cdot n_{EA}^3) = O(n_{EA}^4) = O(n_1^4 \cdot \ldots \cdot n_k^4)$ if we assume that the number of transitions is nearly the same as the number of states.

On the other side of the formula, we have a finite sum of the length of the test suites for all single modules (W-method) plus the length of the test suite for the composite automaton (T-method): $O(p_1 \cdot n_1^3) + \ldots + O(p_k \cdot n_k^3) + O(p_{CA} \cdot n_{CA}) = O(n_1^4) + \ldots + O(n_k^4) + O(n_{CA}^2)$. The number of states n_{CA} in the composite automaton is also bounded to the product of the number of states of individual modules: $n_{CA} \leq n_1 \cdot \ldots \cdot n_k$.

Since the length of the T-method is reduced by the power of 2 compared to the W-method, and a sum of numbers greater than 1 is always less than their product, it follows that the total length of test suites in the step-wise approach is shorter. It implies that testability increases according to the testability metric from [Petr 94] quoted in Section 2. In addition, testability will be further improved by the number of events additionally observed at points of observation.

Now we turn to the second requirement of our approach. We have to show that the transition tour in combination with the use of extra test points is a complete test suite for integration test. As known, a transition tour is only capable to indicate output faults (caused by erroneous inter-module communication), but not to detect wrong states. However after the module test has been carried out successfully, i.e. the correctness of the module implementations was verified, we can assume that wrong states in the composite automaton can only occur as result of coupling errors. Therefore, we must show that the transition tour together with the observation of inter-module communication will be capable to detect wrong coupling of modules.

The detection of these faults depends on the way how the observation of the inter-module communication can be performed. A pragmatic approach for realizing this observation would be to implement the extra test points in a such a way that the gates of the modules send the information to the observer, which data have passed. Thus, wrong data and coupling errors can be very easily detected, because the way the transition tour has taken in the integration test can be traced. But it would require that the implementations of the modules support extra test points. This approach influences the implementation and is therefore not feasible. We suppose in the following that the extra test points do not influence the module implementation. They can only “see“ the data sent by the modules.

A coupling fault may appear, if there exist two equivalent traces Tr_{i1} and Tr_{i2} between a state s_{im} in module *m* and a state s_{jn} in module *n* such that the transition tour can follow another way. Since we know from the module test that the local actions are correctly implemented, the selection of another way can only be forced by a wrong coupling between modules. If we can prove that all traces between two CA states that include inter-module communication are distinguishable, then coupling faults in the composite automaton will lead to sequences of internal and external outputs that do not correspond to traces of the specified automaton.

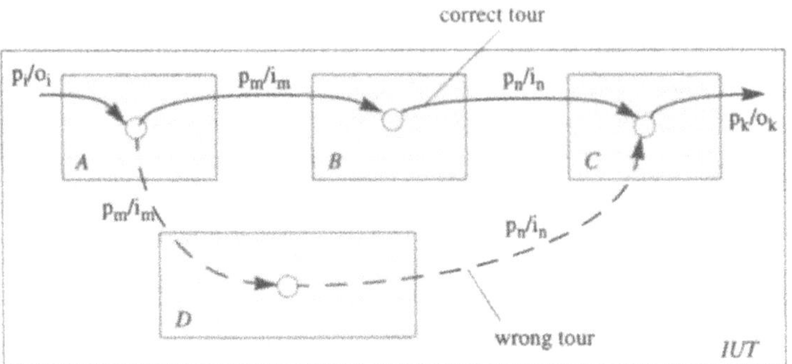

Figure 2 Wrong tour due to coupling error.

Let us now suppose that there exist a coupling fault between two modules and that the observed trace Tr_{i1} between s_{im} and s_{jn} coincides by chance with another trace Tr_{i2} between the two CA states. This is only possible if the states of the module automata passed by Tr_{i1} possess a same transition as Tr_{i2}. Figure 2 depicts this situation. In this case, a transition tour cannot detect without any further information the wrong coupling. To exclude this situation, we have two choices:

1. To make the states of the modules that are involved in inter-module communication distinguishable at the receiving side. This can be done by analyzing the specification for such states in advance and to introduce an additional loop transition back to the same state in the specification and implementation of these states. The transition tour executes the additional transition to validate that it has reached the correct state.
2. To use distinguishable messages for inter-module communication, i.e. the shared actions in A_3 are unique. In this case a data error will be observed.

If such a measures are accepted for DFT purposes, a transition tour is a complete test suite for integration test.

Example

To illustrate the above discussion, we consider the XDT protocol [Koen 96]. XDT (*eXample Data Transfer*) is an example protocol used for teaching protocol engineering. It provides a connection-oriented data transfer service based on the *go-back-N* principle. In our discussion we only consider the sender part. The sender starts with an implicit connection set up *(XDATrequ)*, which is indicated to the service user by a *XCONconf* when finished successfully, otherwise the attempt is stopped by an time-out (*to_t1*). After that the service user can continuously send (*XDATrequ*). The sending may be interrupted *(XBRKind*, *XBRKend*), when the buffer for storing the *DT*-PDU copies is full. The sender repeats the transmission of a *DT*-PDU and the following (already sent) ones (*go-back-N*), when the transmission of a *DT* is not confirmed by an *ACK*-PDU within a certain time (*to_t2*). The connection is released (*XDISind*) after confirming the successful transmission of the last *DT*-PDU. The transmission can be aborted with an *ABO*-PDU by the receiver (indicated to the service user by a *XABOind*), when the PDU sequence is not reestablished in a reasonable time. The FSM of the sender entity is depicted in Figure 5 in the appendix.

To estimate the testability of the sender FSM we use the measure from [Petr 94] (see Section 2). The sender FSM has 5 states, 8 inputs and 5 outputs. The upper bound of the length of the test suite when applying the W-method will be $8*5^3 = 1000$, and the testability degree is $5/1000 = 0.005$.

We now divide the specification according to their logical function in 3 modules *M1*, *M2* and *M3* (see Figure 3 and Figure 4 in the Appendix). Module *M1* performs the connection set-up, *M2* the data transfer and *M3* supervises the acknowledgments. It also initiates the *go-back-N* mechanism and accepts the *ABO*-PDU. For inter-module communication the internal events *i1*, *i2*, *i3*, *i4*, *i5*, *i6* are introduced. The upper test suite lengths for each of the 3 modules, when applying the W-method, are $3*2^3 = 24$ (*M1*), $8*4^3 = 512$ (*M2*) and $5*2^3 = 40$ (*M3*). The upper length of the transition tour is $8*2*4*2 = 128$ (with 8 external inputs). The maximum length of the test suite is therefore 704 test events. The testability degree is $11/704 = 0.0156$ (with 11 internal and external outputs), i.e. the testability increases remarkably. The length of the transition tour for the system test can be even further reduced, because module *M1* terminates before the other two modules start. This knowledge from the specification could be also exploited in the step-wise test approach.

4 CONCURRENT MODULE STRUCTURES

In this section, we discuss the application of the step-wise testing approach for a protocol specification and its corresponding implementation, in which modules are executed concurrently. The assumption of true concurrency is realistic for protocol implementations. However, testing implementations based on multi-module specifications is complicated by a number of problems that are unique to the nature of concurrent systems. Under these problems the most important ones are the occurrence of concurrent events during testing; the reproduction of test runs with the same test data; and state explosion that occurs when the system is being analyzed.

A conventional approach to test suite generation starts from a monolithic, single automaton, i.e. from the entity automaton in our case. Since the entity automaton is usually not given in advance, it must be constructed, e.g., by computing the product of the module automata using interleaving semantics rules to obtain the composite automaton and reducing the composite automaton eventually to obtain its reduced automaton that equals to the entity automaton. The generation of a transition tour from the interleaving model of an entity automaton has its limitations since concurrent events are serialized. Due to a lack of controllability during testing, this approach is not feasible. The resulting order of concurrent events in a test run could not be predicted. The order of events is, however, essential to assess whether an implementation is correct.

If we apply the proposed step-wise testing approach, we are able to use the structure information given as a set of communicating modules during test suite generation. In [Ulri 95], we extended the notion of a *transition tour* [Sidh 89] and applied it as a test suite for distributed systems. A transition tour is defined for a single automaton as the shortest path that covers all transitions in the automaton at least once. In the context of distributed systems a transition tour is extended to a *concurrent transition tour* (CTT) such that all transitions in all modules of the system are visited at least once on the shortest possible path through the system. A CTT takes concurrency among actions of different modules into account.

A CTT is depicted graphically as a time event sequence diagram where nodes are events and the directed arcs define the causality relation between events. It can be considered as a set of local transition tours TT_i through the single modules of the system by taking into account synchronization constraints, i.e. $CTT = (TT_1, \ldots, TT_k)$. Its construction, however, does not necessarily follow from this definition. A feasible construction algorithm of a CTT is presented in [Ulri 97].

The actual length of a concurrent transition tour depends on the degree of concurrency among the modules. The lowest bound of the length is determined by the least common multiple of completed cycles of single transition tours through the modules if no branching occurs at all. In the worst case, the length of the concurrent transition tour equals to the length of a transition tour derived from the interleaving model, i.e. $length(CTT) \leq length(TT)$. Thus, using concurrent test sequences instead of interleaved based ones reduces test efforts further.

5 RELATED WORK

Design for testability is a relatively new approach in protocol engineering. It aims at decreasing the efforts in protocol testing and supporting a better detection of faults in implementations. The testability of protocols may be influenced by many factors in the context of design, implementation and testing. Dssouli and Fournier have, therefore, first proposed to introduce DFT as a development step in the protocol development process [Dsso 91]. A general framework for DFT for protocols was given by Vuong, Loureiro, and Chanson in [Vuon 94].

Grey-box testing is considered as the preferred approach to increase testability. Theoretical aspects of grey-box testing have been pioneered by Yao, Petrenko, Bochmann, and Yevtushenko [Yao 94] [Yevt 95]. A metric for testability based on finite state machines under the complete fault coverage assumption was proposed by Petrenko, Dssouli, and König [Petr 94]. Most approaches that follow this way use means to instrument the implementation with extra test points in order to observe the behavior of the implementation under test. A framework for this approach is proposed by Dssouli, Karoui, Petrenko, and Rafiq in [Dsso 95]. A generic scheme to automatically instrument a formal specification is described by Kim, Chanson, and Yoo in [Kim 95].

A similar incremental approach to structural testing was first proposed by Koppol and Tai in [Kopp 96]. Here, the incremental approach is used to alleviate state explosion during the derivation of test cases for a concurrent system using interleaving semantics. They establish test derivation on structural test coverage criteria, e.g. the coverage of every transition in the modules of the system at least once, instead of providing a fault model, and they do not discuss the degree of testability of their approach.

The work on a concurrent transition tour as a test suite for distributed systems [Ulri 95, 97] can be regarded as an alternative approach to test derivation to alleviate state explosion. It has been advocated by approaches on trace analysis [Yang 92] [Kim 96]. These approaches assume that valid sets of traces through the modules, i.e. valid execution sequences of the system, are already given, but do not provide methods to derive them according to a certain fault coverage. Since a concurrent transition tour requires a grey-box approach in testing to avoid nondeterminism in distributed systems, the test method proposed in this paper follows immediately.

6 CONCLUSIONS

We have presented an approach to support design for testability for communication protocols. The approach combines a step-wise test procedure with grey-box test principles. Applying the approach, we have to consider two further aspects.

First, an appropriate module structure of the protocol specification has to be found. Its design depends often on subjective decisions made by a designer. However, protocols themselves support modularization in most cases. They usually consist of several protocol phases represented by separated (partial) services. These phases can be designed as different subsets of modules and implemented and tested separately. Such a modularization is also supported by the standardized FDTs.

In addition, a test architecture has to be provided that supports the step-wise testing approach. Extra test points must be designed in such a manner that they can be used as PCOs for module tests and POs for integration tests. Their inclusion should be automated as proposed in [Kim 95].

Nondeterminism is a real issue in testing concurrent systems as it was shortly pointed out in Section 4. This problem is aggravated further since additional forms of nondeterminism may exist in a concurrent system, due to nonobservability of internal interactions or data races, even if all its modules behave deterministically. In this case, only a grey-box testing approach and further measures must be taken into account to guarantee a deterministic test run [Tai 95].

Up to now, the step-wise testing approach has been elaborated and justified for concurrent modules communicating synchronously. However, work on an extension of the current method to asynchronous communication is needed. Furthermore, any impact of data flow on the internal behavior of modules has been neglected. A more sophisticated grey-box test procedure is needed to trace the influence of data exchanged over communicating modules. Suggestions for related techniques that are probably applicable in the area of protocol engineering are already known from software engineering of parallel processes (see e.g. [Lebl 87]).

Our approach also facilitates interoperability test because separate accessible modules can be tested against each other. The additional information obtained from POs supplements the test data recorded by a test monitor. Thus, these tests are useful in particular for locating faults when the interoperability test was not successful.

7 REFERENCES

[Chow 78] Chow, T. S.: *Testing Software Design Modeled by Finite-State Machines;* IEEE Trans. on Software Engineering (1978) 3,178-187.

[Clar 90] Clark, D. D.; Tennenhouse, D. L.: *Architectural Considerations for a New Generation of Protocols;* Proc. ACM SIGCOMM, 1990, 200-208.

[Dsso 91] Dssouli, R.; Fournier, R.: *Communication Software Testability;* In Davidson, I.; Litwack, W. (eds:): Protocol Test Systems III. North Holland, 1991, pp. 45-55.

[Dsso 95] Dssouli, R.; Karoui, K., Petrenko, A.; Rafiq, O.: *Towards testable communication software;* Proc. IWPTS'95, Paris, 1995, pp. 239-255.

[Guer 96] Guerrouat, A.; König, H.; Ulrich, A.: *SELEXPERT - A knowledge-based tool for test case selection;* In Formal Description Techniques VIII, Chapman&Hall, 1996, pp. 313-328

[Hein 92] Heiner, M.: *Petri Net Based Software Validation, Prospects and Limitations;* ICSI-TR-92-022, Berkeley/CA, 3/1992.

[Hoff 89] Hoffman, D.: *Hardware testing and Software ICs;* Proc. Pacific NW Software Quality Conference, Portland, 1989, pp. 234-244.

[Kim 95] Kim, M.; Chanson, S. T.; Yoo, S.: *Design for testability of protocols based on formal specifications;* Proc. IWPTS'95, Paris, 1995, 257-269.

[Kim 96] Kim, M. C.; Chanson, S. T.; Kang, S. W.; Shin, J. W.: *An approach for testing asynchronous communicating systems*; 9th IWTCS'96, Darmstadt, Germany; 1996.

[Koen 96] König,H.: The XDT Protocol. Technical Report 04-96. BTU Cottbus, Fakultät Mathematik,Naturwissenschaften und Informatik,1996.

[Kopp 96] P. V. Koppol, K. C. Tai: *An incremental approach to structural testing of concurrent software*; International Symposium on Software Testing and Analysis (ISSTA'96); San Diego, California; 1996; pp. 14–23.

[Lebl 87] Leblanc, T.; Mellor-Crummey, J. M.: *Debugging Parallel Programs with Instant Replay;* IEEE Trans. on Computers 36 (1987) 4, 471-482.

[Myer 79] Myers, G. J.: *The art of software testing;* John Wiley & Son, 1979.

[Ochs 95] Ochsenschläger, P.; Prinoth, R.: *Modelling of distributed systems*; Vieweg, 1995, (in German).

[Petr 94] Petrenko, A.; Dssouli, R.; König, H.: *On Evaluation of Testability of Protcol Structures;* In Rafiq, O. (ed.): Protocol Test Systems VI, North Holland,1994, pp. 111-123.

[Sidh 89] Sidhu, D. P.; Leung, T. K.: *Formal methods for protocol testing: a detailed study;* IEEE Trans. on Software Eng. 15 (1989) 4, 413–426.

[Tai 95] Tai, K. C.; Carver, R. H.: *Testing of distributed programs*; in A. Zomaya (ed.): Handbook of Parallel and Distributed Computing; McGraw Hill; 1995.

[Ulri 95] Ulrich, A.; Chanson, S. T.: *An approach to testing distributed software systems;* Proc. 15th PSTV'95; Warsaw, Poland; 1995.

[Ulri 97] Ulrich, A.: *A description model to support test suite derivation for concurrent systems*; in M. Zitterbart (ed.): Kommunikation in Verteilten Systemen, GI/ITG-Fachtagung; Springer Verlag, 1997.

[Vuon 94] Vuong, S. T.; Loureiro, A. A. F.; Chanson, S. T.: *A Framework for the Design for Testability of Communication Protocols;* In Rafiq, O. (ed.): Protocol Test Systems VI, North Holland, 1994, pp. 89-108.

[Will 82] Williams, T.W.; Parker, K. P.: *Design for Testability - A Survey;* IEEE Trans. on Computers C-31 (1982) 1, 2-15.

[Yang 92] R. D. Yang, C. G. Chung: *Path analysis testing of concurrent programs*; Information and Software Technology; 34(1992)1, 43–56.

[Yao 94] Yao, M., Petrenko, A., Bochmann, G.: *A structural analysis approach to evaluating fault coverage of software testing in respect to the FSM model*; Proc. 7th FORTE'94; Bern, Switzerland; 1994.

[Yevt 95] Yevtushenko, N.; Petrenko, A.; Dssouli, R.; Karoui, K.; Propenko, S.: *On the Design for testability of Communication Protocols;* Proc. IWPTS'95, Paris, 1995.

8 APPENDIX

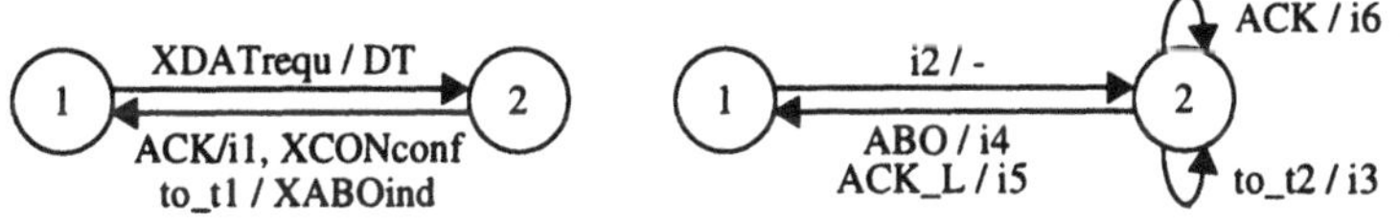

p = {XDATrequ, ACK, to_t1}
o = {DT, XABOind, XCONconf, i1}

p = {i2, ABO, ACK, ACK_L, to_t2}
o = {i3, i5, i6}

Figure 3 FSMs *M1* and *M3*.

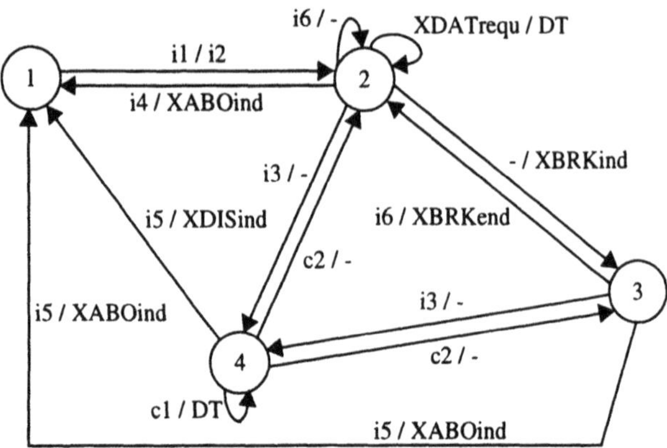

p = {i1, i3, i4, i5, i6, c1, c2, XDATrequ}
o = {DT, XBRKind, XBRKend, XDISind, XABOind, i2}

Figure 4 FSM *M2*.

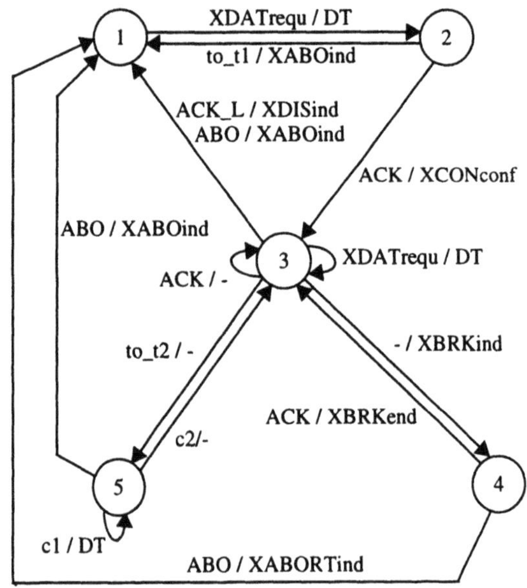

p = {XDATrequ, ACK, ACK_L, ABO, to_t1, to_t2, c1, c2}
o = {DT, XDISind, XABOind, XBRKind, XBRKend}

Figure 5 FSM of the XDT sender.

PART SIX

Theory and Practice of Protocol Testing

Developments in testing transition systems

Ed Brinksma, Lex Heerink and Jan Tretmans
Tele-Informatics and Open Systems group, Dept. of Computer Science
University of Twente, 7500 AE Enschede, The Netherlands
{brinksma,heerink,tretmans}@cs.utwente.nl

Abstract

This paper discusses some of the developments in the theory of test generation from labelled transition systems over the last decade, and puts these developments in a historical perspective. These developments are driven by the need to make testing theory applicable to realistic systems. We illustrate the developments that have taken place in a chronological order, and we discuss the main motivations that led to these developments. In this paper the claim is made that testing theory (slowly) narrows the gap with testing practice, and that progress is made in designing test generation algorithms that can be used in realistic situations while maintaining a sound theoretical basis.

1 INTRODUCTION

Testing and verification Testing and verification are complementary techniques that are used to increase the level of confidence in the correct functioning of systems as prescribed by their specifications. While verification aims at proving properties about systems by formal manipulation on a mathematical model of the system, testing is performed by exercising the real, executing implementation (or an executable simulation model). Verification can give certainty about satisfaction of a required property, but this certainty only applies to the model of the system: any verification is only as good as the validity of the system model. Testing, in practice being based on observing only a small subset of all possible instances of system behaviour, is usually incomplete: testing shows the presence of errors, not their absence. Since test-

Testing of Communicating Systems, M. Kim, S. Kang & K. Hong (Eds)
Published by Chapman & Hall © 1997 IFIP

ing can be applied to the real implementation, it is useful in those cases when a valid and reliable model is not present.

There is an apparent paradox between the attention that verification and testing get in usage and research. Whereas most of the research in the area of distributed systems is concentrated on verification, testing is the predominant technique in practice. People from the realm of verification very often consider testing as inferior, because it can only detect some errors, but it cannot prove correctness; on the other hand, people from the realm of testing consider verification as impracticable and not applicable to realistically-sized systems.

Protocol conformance testing Protocol conformance testing is concerned with checking protocol implementations against their specifications by means of experimentation. Tests are derived from the protocol specification, then applied to the implementation under test, and, based on observations made during the execution of the tests, a verdict about the correct functioning of the implementation is given. Since conformance testing is a mainly manual, laborious and time-consuming process, automating the testing process has always received much attention. To automate the generation of test cases the protocol specification must be in a form amenable to manipulation by tools. Natural language specifications do not serve this purpose; formal languages do. The availability and increasing use of formal methods has resulted in theories, methods and pragmatics for the (semi-)automatic derivation of tests from formal specifications. In the area of test execution there are currently commercial protocol-tester tools available that can execute tests for many different protocols. For such tools to work properly it is important that test cases can be specified precisely and unambiguously. The standardised test specification language TTCN [22, part 3] is widely used for this purpose.

Conformance testing and formal methods Starting point for protocol conformance testing based on formal methods is a formal specification, e.g., a specification written in one of the currently standardised formal description techniques Estelle [20], LOTOS [21], or SDL [10]. Correctness and validity of this specification is assumed, and is not considered as part of conformance testing. Furthermore, there is an implementation, referred to as the implementation under test (IUT), which is treated as a black box, exhibiting external behaviour. The IUT is a physical, real object that is in principle not amenable to formal reasoning. We can only deal with implementations in a formal way, if we make the assumption that any real implementation has a formal model with which we could reason formally. This formal model is only assumed to exist, but it need not be known a priori. This assumption is referred to as the *test hypothesis* [3, 39, 23]. The test hypothesis allows to reason about implementations as if they were formal objects, and, consequently, to express conformance of implementations with respect to specifications by means of a formal relation between such models of implementations and specifications.

Such a relation is called an *implementation relation* [8, 23]. Conformance testing now consists of performing experiments to decide whether the unknown model of the implementation relates to the specification according to the implementation relation. The experiments are specified in test cases. Given a specification, a test generation algorithm must produce a set of such test cases, called a test suite. The test suite must be sound, i.e., it must give a negative verdict only if the implementation is incorrect. Additionally, the test suite must be as complete as possible, i.e., if the implementation is incorrect, it must have a high probability to give a negative verdict.

Many different approaches to algorithmic test generation, based on different protocol specification formalisms, have been undertaken. Two main approaches can be distinguished: those based on Finite State Machines (FSM) and those based on Labelled Transition Systems (LTS). FSM-based protocol testing has been inspired by functional hardware testing and is based on modelling the behaviour of a protocol as a Mealy machine (Finite State Machine FSM) [5, 16, 27, 26, 30, 37, 46].

Goal and overview LTS-based testing has its basis in the formal theory of testing equivalences for labelled transition systems and process algebras, which is based on the formalisation of the notion of test and observation in [13, 12], and which continues with [1, 33, 24, 17].

The goal of this paper is to describe the developments in the theory for test generation for labelled transition systems, as they have led to the current status. We will show that the approach that started from practice and the one that started from theory are now at the point of meeting each other, leading to practical test generation algorithms that have a sound theoretical basis. One indication for this claim is that the algorithm implemented in TVEDA can be given a theoretical basis in the theory of refusal testing [33, 24] by adding to this theory a distinction between input and output actions. This was shown using the theory of Input/Output Transition Systems (IOTS) in [41]. The model of IOTS can be used very well to describe SDL and TTCN processes. Recent results [19] also link the notion of channel (as in SDL) or Point of Control and Observation (PCO) into the LTS-based testing theory.

Section 2 introduces LTS and fixes notation, and section 3 introduces testing concepts for LTS as described by, e.g., [13, 12]. Next, section 4 presents a testing theory for LTS that uses these concepts, and shows how tests can be constructed that are able to check correctness of implementations. Since this theory assumes that implementations communicate in a symmetric manner with their environment, which is unrealistic in practice, a more refined testing theory, based on IOTS, is presented in section 5. Section 6 discusses a refinement of the IOTS model that takes the distribution of PCOs of implementations into account. This theory can serve as an unified model in which both the traditional testing theory of section 3, and the refined theory of section 5, can be expressed. Section 7 ends with conclusions and further work.

2 LABELLED TRANSITION SYSTEMS

In this paper we will concentrate on a testing theory for labelled transition systems. We will use this formalism to model the behaviour of specifications, implementations and tests. A labelled transition consists of nodes and transitions between nodes that are labelled with actions. Formally, a (labelled) transition system (LTS) over L is a quadruple $\langle S, L, \rightarrow, s_0 \rangle$ where

- S is a (countable) set of states;
- L is a (countable) set of observable actions;
- $\rightarrow \subseteq S \times (L \cup \{\tau\}) \times S$ is a set of transitions; and
- $s_0 \in S$ is the initial state.

The special action $\tau \notin L$ represents an unobservable, internal action. We restrict to (strongly) convergent transition systems, i.e., transition systems that are not able to perform an infinite sequence of internal transitions. The class of all convergent transition systems over L is denoted by $\mathcal{LTS}(L)$, and the set of all finite words over L is denoted by L^*. In order to describe the sequences of actions in L and $\mathcal{P}(L)$ that can be performed from a given state (where $\mathcal{P}(\cdot)$ denotes the powerset operator on sets) we use the following abbreviations, with $p = \langle S, L, \rightarrow, s_0 \rangle$ a labelled transition system such that $s, s' \in S, \lambda, \lambda_i \in \mathcal{P}(L) \cup L \cup \{\tau\}, \alpha, \alpha_i \in \mathcal{P}(L) \cup L$ and $\sigma \in (\mathcal{P}(L) \cup L)^*$.

$$
\begin{array}{lll}
s \xrightarrow{\lambda} s' & =_{def} & \begin{cases} (s, \lambda, s') \in \rightarrow, & \text{if } \lambda \in L \cup \{\tau\} \\ s = s' \text{ and } \forall \mu \in \lambda \cup \{\tau\}, \forall s'' : \neg(s \xrightarrow{\mu} s''), & \\ & \text{if } \lambda \in \mathcal{P}(L) \end{cases} \\
s \xrightarrow{\lambda_1 \cdot \lambda_2 \cdot \ldots \cdot \lambda_n} s' & =_{def} & \exists s_0, s_1, \ldots, s_n : s = s_0 \xrightarrow{\lambda_1} s_1 \xrightarrow{\lambda_2} \ldots \xrightarrow{\lambda_n} s_n = s' \\
s \xrightarrow{\lambda_1 \cdot \lambda_2 \cdot \ldots \cdot \lambda_n} & =_{def} & \exists s' : s \xrightarrow{\lambda_1 \cdot \lambda_2 \cdot \ldots \cdot \lambda_n} s' \\
s \stackrel{\epsilon}{\Longrightarrow} s' & =_{def} & s = s' \text{ or } s \xrightarrow{\tau \cdot \ldots \cdot \tau} s' \\
s \stackrel{\alpha}{\Longrightarrow} s' & =_{def} & \exists s_1, s_2 : s \stackrel{\epsilon}{\Longrightarrow} s_1 \xrightarrow{\alpha} s_2 \stackrel{\epsilon}{\Longrightarrow} s' \\
s \stackrel{\alpha_1 \cdot \alpha_2 \cdot \ldots \cdot \alpha_n}{\Longrightarrow} s' & =_{def} & \exists s_0, s_1, \ldots, s_n : s = s_0 \stackrel{\alpha_1}{\Longrightarrow} s_1 \stackrel{\alpha_2}{\Longrightarrow} \ldots \stackrel{\alpha_n}{\Longrightarrow} s_n = s' \\
s \stackrel{\sigma}{\Longrightarrow} & =_{def} & \exists s' : s \stackrel{\sigma}{\Longrightarrow} s'
\end{array}
$$

Self-loop transitions of the form $s \xrightarrow{A} s$ where $A \subseteq L$ are called refusal transitions. In this case A is called a refusal of s. Such a refusal transition explicitly encodes the inability to perform any action in $A \cup \{\tau\}$ in state s. A failure trace consists of a sequence over refusal transitions $\xrightarrow{A}$ with $A \subseteq L$ and 'normal' transitions $\xrightarrow{\mu}$ with $\mu \in L \cup \{\tau\}$ where an abstraction from internal actions τ is made. For readability we do not distinguish between a labelled transition system and its initial state, e.g., $p \stackrel{\sigma}{\Longrightarrow} =_{def} s_0 \stackrel{\sigma}{\Longrightarrow}$ where s_0 is the initial state of labelled transition system p. If $p \stackrel{\sigma}{\Longrightarrow}$ where $\sigma \in L^*$ then σ is called a trace of p. For $p \in \mathcal{LTS}(L)$ we will use the following definitions.

1. $f\text{-}traces(p) =_{def} \{\sigma \in (\mathcal{P}(L) \cup L)^* \mid p \stackrel{\sigma}{\Longrightarrow}\}$
2. $traces(p) =_{def} \{\sigma \in L^* \mid p \stackrel{\sigma}{\Longrightarrow}\}$
3. p **after** σ **refuses** $A =_{def} \exists p' : p \stackrel{\sigma}{\Longrightarrow} p'$ and $\forall \mu \in A \cup \{\tau\} : \neg(p' \stackrel{\mu}{\longrightarrow})$
4. p **after** σ **deadlocks** $=_{def}$ p **after** σ **refuses** L
5. $der(p) =_{def} \{p' \mid \exists \sigma \in L^* : p \stackrel{\sigma}{\Longrightarrow} p'\}$
6. $init(p) =_{def} \{\mu \in L \cup \{\tau\} \mid \exists p' : p \stackrel{\mu}{\longrightarrow} p'\}$
7. P **after** $\sigma =_{def} \{p' \mid \exists p \in P : p \stackrel{\sigma}{\Longrightarrow} p'\}$ where P is a set of states
8. p is *deterministic* iff $\forall \sigma \in L^* : \mid \{p\}$ **after** $\sigma \mid \leq 1$
9. p has *finite behaviour* iff $\exists N \in \mathbf{N} : \forall \sigma \in traces(p) : |\sigma| \leq N$

In testing, an external observer experiments on an implementation in order to unravel its (unknown) behaviour. A test specifies the behaviour of an observer, and we assume that tests are modelled as LTS. Tests can be run, or executed, against implementations. From the execution of a test against an implementation observations can be made. These observations are then compared with the expected observations that can be obtained by running the same test against the specified behaviour, and a verdict (success or failure) is assigned. Failure should indicate that there is evidence that the implementation did not behave correct, otherwise success should be assigned. Section 5 treats test execution in more detail.

3 TESTING RELATIONS FOR TRANSITION SYSTEMS

In order to decide the correctness of implementations a clear correctness criterion is needed: when is an implementation considered correct with respect to its specification? In the context of labelled transition systems many proposals for such correctness criteria in the form of implementation relations have been made [17]. One of the first significant implementation relations was *observation equivalence* [29]. Observation equivalence is defined as a relation over states of transition systems by means of (weak) bisimulation relations. Informally, two systems $p, q \in \mathcal{LTS}(L)$ are called observation equivalent, denoted by $p \approx q$, if for every trace $\sigma \in L^*$ every state that is reachable from p after having performed trace σ is itself observation equivalent to some state of q that is also reachable after having performed trace σ, and similarly with p and q interchanged. Observation equivalence intuitively captures the notion of equivalent external behaviour of systems; two systems are observation equivalent if they exhibit "exactly the same" external behaviour. See [29] for a formal definition of observation equivalence.

Instead of relating behaviours intensionally in terms of relations over states and transitions between states, it is also possible to relate system behaviour in an extensional way; what kind of systems can be distinguished from each from each other by means of experimentation? [13, 12] were first in comparing system behaviour in this way by explicitly modelling the behaviour of experiments, and relating the observations that can be made when these experiments are applied to systems. In general, for a set of experiments $\mathcal{U}$, and a set of observations $obs(u,p)$ that experiment $u \in \mathcal{U}$ may cause when system p is tested, they define a so-called *testing relation* over systems by relating the observations $obs(u,i)$ and $obs(u,s)$ that are made when experiments $u \in \mathcal{U}$ are carried out against the systems i and s. Formally, such testing relations are defined as follows

$$i \textbf{ conforms-to } s \quad =_{def} \quad \forall u \in \mathcal{U} : obs(u,i) \sqsubseteq obs(u,s) \tag{1}$$

where **conforms-to** denotes the testing relation that is defined. By varying the set of experiments $\mathcal{U}$, the set of observations obs and the relation $\sqsubseteq$ between these sets of observations, different testing equivalences can be defined. [13, 12] discuss, and compare, several different testing relations by varying the set of observations obs and the relation $\sqsubseteq$ between these sets of observations. The theory described in [13, 12] forms the basis for testing theories for transitions systems. We will discuss three instances of such testing relations that are relevant for the remainder of this paper, viz., observation equivalence, testing preorder and refusal preorder, and use a formalisation following [39] that slightly differs from the original formalisation given in the seminal work of [13, 12].

Observation equivalence [1] shows that observation equivalence can be characterised in an extensional way (i.e., following the characterisation of equation (1)), under the assumption that at each stage of a test run infinitely many local copies of the internal state of the system under test can be made, and infinitely many experiments can be conducted on these local copies. Intuitively, this means that at each stage of a test run the implementation must be tested against all possible operating environments. These assumptions are quite strong and too difficult to meet in practice. Therefore, observation equivalence is, in general, too fine to serve as a realistic implementation relation, and weaker notions of correctness between implementations and specifications have to be defined.

Testing preorder In testing preorder it is assumed that the behaviour of external observers can, just as the behaviour of implementations and specifications, be modelled as transition systems (that is, $\mathcal{U} \equiv \mathcal{LTS}(L)$) and these observers communicate in a synchronous and symmetric way with the system under test [13, 12]. From an observer u and system under test p, the binary infix operator $||$ creates a transition system $u \mid\mid p$ that models the behaviour of u experimenting on p in a synchronous way. The transitions that $u \mid\mid p$ can perform are defined by the smallest set of transitions induced by the following inference rules

$$\frac{u \xrightarrow{\tau} u'}{u \parallel p \xrightarrow{\tau} u' \parallel p} \qquad \frac{p \xrightarrow{\tau} p'}{u \parallel p \xrightarrow{\tau} u \parallel p'} \qquad \frac{u \xrightarrow{a} u', p \xrightarrow{a} p'}{u \parallel p \xrightarrow{a} u' \parallel p'} \ (a \in L)$$

Using $\parallel$ a testing preorder on transition systems [13] can be defined in an extensional way following equation (1). Intuitively, an implementation i is testing preorder related to specification s, denoted as $i \leq_{te} s$, if for every external observer u that is modelled as a transition system, each trace that $u \parallel i$ can perform is preserved by $u \parallel s$, and each deadlock of $u \parallel i$ is preserved by $u \parallel s$. Formally, testing preorder $\leq_{te}$ is defined by

$$i \leq_{te} s \quad =_{def} \quad \forall u \in \mathcal{LTS}(L): \quad \begin{array}{r} obs_t(u,i) \subseteq obs_t(u,s) \\ \text{and } obs_c(u,i) \subseteq obs_c(u,s) \end{array}$$

where $obs_t(u,p) =_{def} \{\sigma \in L^* \mid (u \parallel p) \stackrel{\sigma}{\Longrightarrow}\}$ and $obs_c(u,p) =_{def} \{\sigma \in L^* \mid (u \parallel p)$ **after** σ **deadlocks**$\}$. The relation $\leq_{te}$ can be intensionally characterised by $i \leq_{te} s$ iff $\forall \sigma \in L^*, \forall A \subseteq L : i$ **after** σ **refuses** A implies s **after** σ **refuses** A. Testing preorder allows implementations to be "more deterministic" than their specification, but it does not allow that implementations "can do more" than is specified; in this sense the specification not only prescribes what behaviour is allowed, but also what behaviour is not allowed! The relation $\leq_{te}$ serves as the basic implementation relation in many testing theories for transition systems.

Refusal preorder Refusal preorder can be seen as a refinement of testing preorder, and is defined extensionally in the theory of refusal testing [33]. Instead of administrating the successful actions that are conducted on an implementation by an observer, refusal testing also takes the unsuccessful actions into account. The difference between refusal preorder and testing preorder is that observers can detect deadlock, and act on it, i.e., in refusal preorder observers are able to continue after observation of deadlock. Formally, we model this as in [24] by using a special deadlock detection label $\theta \notin L$ (i.e., $\mathcal{U} \equiv \mathcal{LTS}(L \cup \{\theta\})$, cf. equation (1)) that is used to detect the inability to synchronise between the observer u and system under test p. The θ-action is observed if there is no other way to continue, i.e., when p is not able to interact with the actions offered by u. The transition system $u \rceil\!| \, p \in \mathcal{LTS}(L \cup \{\theta\})$ that occurs as the result of communication between a deadlock observer $u \in \mathcal{LTS}(L \cup \{\theta\})$ and a transition system $p \in \mathcal{LTS}(L)$ is defined by the following inference rules.

$$\frac{u \xrightarrow{\tau} u'}{u \rceil\!| \, p \xrightarrow{\tau} u' \rceil\!| \, p} \qquad \frac{u \xrightarrow{a} u', p \xrightarrow{a} p'}{u \rceil\!| \, p \xrightarrow{a} u' \rceil\!| \, p'} \ (a \in L)$$

$$\frac{p \xrightarrow{\tau} p'}{u \rceil\!| \, p \xrightarrow{\tau} u \rceil\!| \, p'} \qquad \frac{u \xrightarrow{\theta} u', u \stackrel{\tau}{\not\longrightarrow}, p \stackrel{\tau}{\not\longrightarrow}, init(u) \cap init(p) = \emptyset}{u \rceil\!| \, p \xrightarrow{\theta} u' \rceil\!| \, p}$$

Observations made by an observer u by means of the operator $]|$ now may include the action θ. The testing preorder induced for observers in $\mathcal{LTS}(L \cup \{\theta\})$ is called refusal preorder, and is defined in the style of equation (1):

$$i \leq_{rf} s \quad =_{def} \quad \forall u \in \mathcal{LTS}(L \cup \{\theta\}) : \qquad \begin{array}{r} obs_c^\theta(u,i) \subseteq obs_c^\theta(u,s) \\ \text{and } obs_t^\theta(u,i) \subseteq obs_t^\theta(u,s) \end{array}$$

where $obs_c^\theta(u,p) \; =_{def} \; \{\sigma \in (L \cup \{\theta\})^* \mid (u]|\, p) \textbf{ after } \sigma \textbf{ deadlocks}\}$ and $obs_t^\theta(u,p) \; =_{def} \; \{\sigma \in (L \cup \{\theta\})^* \mid (u]|\, p) \stackrel{\sigma}{\Longrightarrow}\}$. Informally, $i \leq_{rf} s$ if, for every observer $u \in \mathcal{LTS}(L \cup \{\theta\})$, every sequence of actions that may occur when u is run against i (using $]|$) is specified in $u]|\, s$; i is not allowed to accept, or reject, an action when communicating with u, if this is not specified by s. Refusal preorder is strictly stronger than testing preorder, i.e., $\leq_{rf} \subset \leq_{te}$. Refusal preorder is characterised by inclusion of failure traces: $i \leq_{rf} s$ iff $\textit{f-traces}(i) \subseteq \textit{f-traces}(s)$.

We emphasize that implementation relations that abstract from the nondeterministic characteristics of protocols (e.g., trace preorder or trace equivalence) are, in general, *not* sufficient to capture the intuition behind correctness of systems. Even if protocols are defined as deterministic automata, their joint operation with underlying layers, such as operating systems, generally will behave in a nondeterministic manner.

4 CONF TESTING

As shown in section 3 [13, 12] define a correctness criterion (in terms of a testing relation) by providing a set of experiments ($\mathcal{U}$), a notion of observation (obs), and a way to relate observations of different systems ($\sqsubseteq$) (equation (1)). In test generation the opposite happens: for some implementation relation a set of tests $\mathcal{U}$ has to be designed that is able to distinguish between correct and incorrect implementations by comparing the observations that the implementation produces with the expected observations when the same test is applied to the specification. The first testing theory that treats the problem of test generation in this way is [6, 7].

In [6, 7] a method is presented to derive test cases from a specification that is able to discriminate between correct and incorrect implementation with respect to the implementation relation **conf**. The relation **conf** can be seen as a liberal variant of $\leq_{te}$. The difference with $\leq_{te}$ is that the implementation may do things that are not specified; in **conf** there is no need to perform any robustness tests! Since in **conf** there is no need to check how the implementation behaves for unspecified traces, test generation algorithms for **conf** are better suited for automation than test generation algorithms for $\leq_{te}$. In particular, for a finite behaviour specification this means that only a finite number of traces have to be checked. Formally, the relation **conf** is defined as $\leq_{te}$ restricted to the traces of the specification.

$$i \textbf{ conf } s \;=_{def}\; \forall u \in \mathcal{LTS}(L): \quad obs_t(u,i) \cap traces(s) \subseteq obs_t(u,s) \text{ and } obs_c(u,i) \cap traces(s) \subseteq obs_c(u,s)$$

In literature, this relation is usually known in its intentional characterisation: i **conf** s iff $\forall \sigma \in traces(s), \forall A \subseteq L^*$: i **after** σ **refuses** A implies s **after** σ **refuses** A. Informally, the **conf** relation indicates that an implementation is correct with respect to its specification if, after executing a specified trace, the implementation is not able to reach an unspecified deadlock when synchronised with an arbitrary test process. [6, 7] develops a theory for the construction of a so-called *canonical tester* from a specification. The canonical tester $T(s)$ of s is a process that preserves the traces of s (i.e., $traces(T(s)) = traces(s)$) and that is able to decide unambiguously whether an implementation i is **conf**-correct with respect to specification s, i.e.,

$$\forall i \in \mathcal{LTS}(L) : i \textbf{ conf } s \textit{ iff } i \textit{ conf-passes } T(s)$$

where i *conf-passes* $T(s) =_{def} \forall \sigma \in L^*$: $(i \parallel T(s))$ **after** σ **deadlocks** implies $T(s)$ **after** σ **deadlocks**. This is done by running $T(s)$ against implementation i until it deadlocks, and checking that every deadlock of $i \parallel T(s)$ can be explained by a deadlock of $T(s)$; if $T(s)$ did not end in a deadlock state, evidence of non-conformance with respect to **conf** has been found. The elegance of **conf**-testing is nicely illustrated by the fact that the canonical tester of a canonical tester is testing equivalent with the original specification; $T(T(s)) \approx_{te} s$ (where $\approx_{te}$ is the symmetric reduction of $\leq_{te}$) [6].

In [2, 45] a procedure to construct canonical testers has been implemented for finite Basic LOTOS processes, that is, from a finite behaviour LOTOS specification s without data a tester $T(s)$ is constructed that is again represented as a finite behaviour Basic LOTOS process. [35] has extended this to Basic LOTOS processes with infinite behaviour. A procedure for the construction of tests from a specification related to the theory of canonical testers in such a way that these tests preserve the structure of the specification is sketched in [34]. In [25] a variant of the theory of canonical testers is discussed for a transitive version of the **conf** relation. [15] derives, and simplifies, canonical testers using refusal graphs. Figure 1 presents an example of a process and its canonical tester.

The theory of canonical testers is applicable to situations where the system under test communicates in a symmetric and synchronous manner with an external observer; both the observer and the system under test have to agree on an action in order to interact, and there is no notion of initiative of actions. Since asynchronously communicating systems can be modelled in terms of synchronously communicating systems by explicitly modelling the intermedi-

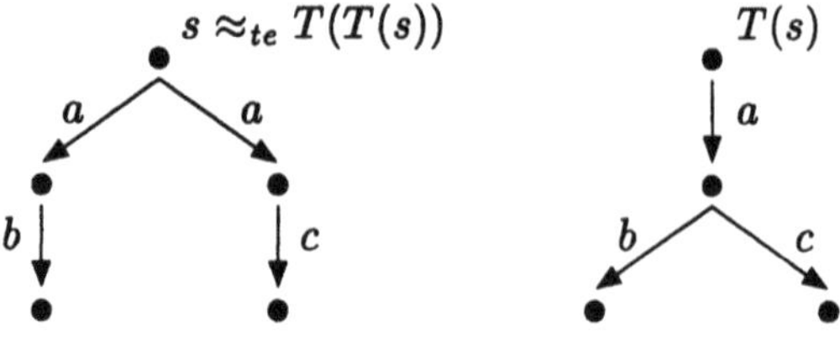

Figure 1 Canonical testers.

ate communication medium between these two systems **conf**-testing can also be applied to asynchronously communicating systems (e.g., the queue systems discussed in section 5). Consequently, **conf**-testing is widely applicable to a large variety of systems.

However, the theory of canonical testers also has some difficulties that restricts its applicability in practice. We will mention the two important ones in our view. The first difficulty has to do with the large application scope of the theory of canonical testers. In general, the more widely applicable a theory becomes, the less powerful this theory becomes for specific situations. In particular, communication between realistic systems is, in practice, often asymmetric. By exploiting the characteristics of such asymmetric communication, a more refined testing theory can be developed. The next section discusses in detail how this can be done.

Another drawback of the theory of canonical testers is its difficulty to handle data in a symbolic way. Since in most realistic applications data is involved, it is necessary to deal with data in a symbolic way in order to generate canonical testers in an efficient way. In [14, 39] some problems with the derivation of canonical testers for transition systems that are specified in full LOTOS (i.e., LOTOS with data) have been identified, such as an explosion in the data part of the specification. In particular, the derivation of canonical testers in a symbolic way is complicated by the fact that not only the data domains and the constraints imposed on the data values that are communicated need to be composed in a correct way, but also the branching structure of the specification (and thus of the canonical tester itself) needs to be taken into account. The problem is that the test generation algorithm for **conf** uses powerset constructions that are, in principle, able to transform countable branching structures into uncountable branching structures.

5 CHANGING THE INTERFACES

Several approaches have been proposed to model the interaction between implementations and their environment more faithfully, e.g., by explicitly considering the asymmetric nature of communication with the aim to come to a testing theory that is better suited for test generation in realistic situations. Moreover, since the standardised test notation TTCN [22, part 3] uses inputs and outputs to specify tests, theories that incorporate such asymmetric communication allow the generation of tests in TTCN. In this section we present a short overview of some of the approaches that have been proposed in this area, and we will elaborate on one of them.

Apply asynchronous theory to transition systems Much research has been done in systems that communicate in an asynchronous manner (e.g., [4]), and some languages used in protocol conformance testing are based on asynchronous paradigms (e.g., SDL [10] , Estelle [20], TTCN [22, part 3]). [9] gives a short overview of translation between labelled transition systems and Mealy machines, which can be used as an underlying semantic model for, e.g., SDL [10]. In particular, research has been done in transforming transition systems without inputs and outputs into FSMs with inputs and outputs, and deriving tests for these FSMs (e.g., [18]). However, many of these developments lack a solid, formal basis, and their use in practice is restricted.

Queue systems In [42] asynchronous communication between an implementation and its environment is modelled explicitly by the introduction of an underlying communication layer. This layer essentially consists of two unbounded FIFO queues, one of which is used for message transfer from the implementation to the environment, and the other for message transfer in the opposite direction (figure 2). Such systems are called queue systems.

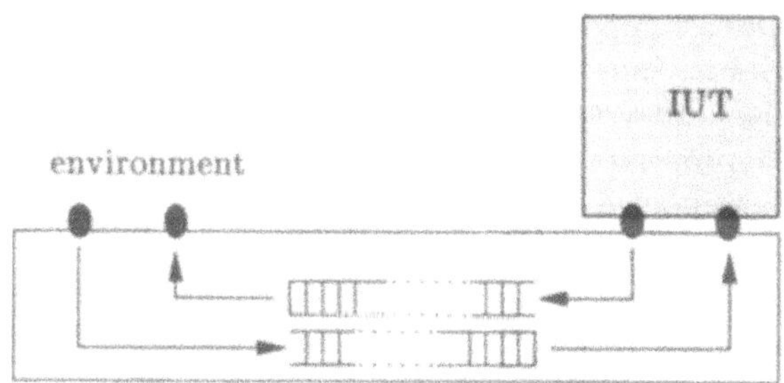

Figure 2 Architecture of a queue system.

In order to formalise the notion of queue systems the set of labels L is partitioned in a set of input labels L_I and a set of output labels L_U (i.e., $L = L_I \cup L_U, L_I \cap L_U = \emptyset$). Input labels are supplied from the environment via the input queue to the IUT, and, similarly, output labels run via the output queue. In particular, [42] is interested in what kind of systems can be distinguished from each other in the asynchronous setting sketched above, and how this compares to the synchronous setting. They therefore define a new implementation relation $\leq_{te}^{Q}$ that captures whether two systems are $\leq_{te}$-related when tested through the queues. Formally,

$$i \leq_{te}^{Q} s \quad =_{def} \quad \mathcal{Q}(i) \leq_{te} \mathcal{Q}(s)$$

where $Q(p)$ denotes the transition system that is induced when p is placed in an environment where communication runs via two queues as sketched above.

They also define classes of asynchronous implementation relations called *queue preorders* $\leq_{Q}^{\mathcal{F}}$ as preorders that disallow the implementation to produce unspecified outputs (where the inability to produce outputs is considered observable) after having performed arbitrary trace in some specified $\mathcal{F} \subseteq L^*$, i.e.,

$$i \leq_{Q}^{\mathcal{F}} s \quad =_{def} \quad \forall \sigma \in \mathcal{F} : \mathcal{O}_i(\sigma) \subseteq \mathcal{O}_s(\sigma) \tag{2}$$

where $\mathcal{O}_p(\sigma) =_{def} \{x \in L_U \mid \mathcal{Q}(p) \stackrel{\sigma \cdot x}{\Longrightarrow}\} \cup \{\delta \mid \mathcal{Q}(p)\, \mathbf{after}\, \sigma\, \mathbf{refuses}\, L_U\}$ and $\delta \notin L$. By restricting the set $\mathcal{F}$ to sets of traces that depend on the specification s asynchronous **conf**-like relations can be defined, and their properties can be investigated. [44] presents an algorithm that is able to derive a complete test suite for such classes of queue implementation relations.

The asynchronous testing theory for queue systems can be seen as an attempt to narrow the gap between testing based on synchronous theories (such as the theory for canonical testers, section 4) and testing based on asynchronous theories via inputs and outputs (e.g., testing based on systems specified in SDL [10]). However, queue systems are restricted in their use; the theory is only appropriate for systems that explicitly communicate with each other via two unbounded FIFO queues, and other communication architectures (such as having more than two queues, allowing media to be non-FIFO, etc.) cannot be described in this model. Fortunately, the requirement that systems communicate with each other via unbounded FIFO queues turns out not to be necessary in order to apply the ideas discussed before: the only essential requirements are that the set of actions can be partitioned in a set of input actions L_I and a set of output actions L_U, and that implementations can never refuse input actions, whereas the environment is always prepared to accept output actions (where input actions and output actions are viewed from the perspective of the system under test). By considering in figure 2 the input queue as part of the implementation, and the output queue as part of the environment, queue systems are just a special case of systems satisfying this requirement. This observation has triggered research on systems that are never able to refuse input actions. We discuss three of such (marginally) different system models: input/output automata (IOA), input/output state machines (IOSM), and input/output transition systems (IOTS).

Input/Output Automata (IOA) Formally, a transition system p where the set of labels L is partitioned in a set of input labels L_I and a set of output labels L_U (i.e., $L = L_I \cup L_U$ and $L_I \cap L_U = \emptyset$), and that satisfies

$$\forall p' \in der(p), \forall a \in L_I : p' \xrightarrow{a}$$

is called an input/output automaton (IOA) [28]. By explicitly distinguishing between inputs and outputs, implementations and their observers are allowed to communicate in a complementary manner; observers control and supply the input actions, while implementations control and produce output actions. [36] applies the ideas from [13] to implementations that are assumed to be modelled as IOA.

Input/Output State Machines (IOSM) [32] introduces a model called (complete) input/output state machines (IOSM) that differs from IOA by requiring that IOSM must have a finite number of states. This model is used as a semantic underpinning for test derivation in the tool TVEDA [11].

Input/Output Transition Systems (IOTS) According to [40, 41] an input/output transition system (IOTS) is a transition system that marginally differs from IOA and IOSM. Like in IOA the set of labels is partitioned in a set of input labels L_I and a set of output labels L_U, but the difference is that instead of requiring that inputs are always strongly enabled, we require for IOTS that inputs are weakly enabled, i.e., $p \in \mathcal{LTS}(L_I \cup L_U)$ is IOTS iff

$$\forall p' \in der(p), \forall a \in L_I : p' \stackrel{a}{\Longrightarrow} \qquad (3)$$

The above condition is strictly weaker than the one imposed on IOA. Consequently, test theory for IOTS is more general than for IOA. Note that queue systems can be seen as subclass of IOTS: every implementation in a queue context satisfies the condition imposed on IOTS, but not vice versa.

Although IOA, IOSM and IOTS differ marginally, we concentrate here on the most liberal one, namely IOTS, and discuss testing theory for implementations that can be modelled as IOTS in the same way as [40, 41]. We denote the universe of IOTS with input set L_I and output set L_U by $\mathcal{IOTS}(L_I, L_U)$.

Inputs and outputs are complementary: inputs for IUT are outputs from the perspective of the environment, and outputs produced by the IUT are inputs for the environment (figure 3). By convention, we will use the terms inputs and outputs always from the perspective of the IUT. Many existing implementations satisfy the test assumption that inputs are always enabled (that is, they can be modelled as an IOTS), and that inputs are initiated and controlled by the environment, whereas outputs are initiated and controlled by the implementation. From now on we will assume that implementations can

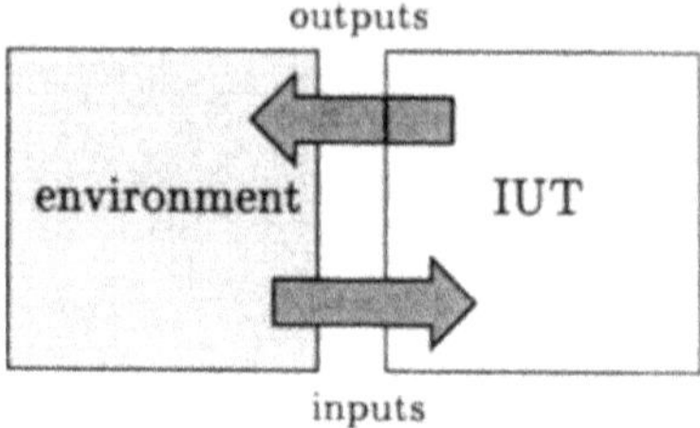

Figure 3 Asymmetric communication between IUT and its environment.

be modelled as members of $\mathcal{IOTS}(L_I, L_U)$. However, if the implementation is not able to refuse inputs initiated by the environment, then it is reasonable to assume that the environment is not able to refuse outputs produced by the implementation. If we allow the environment to also observe the inability of implementations to produce any output by means of θ (see section 3), then this means that the behaviour of the environment can be modelled as a member of $\mathcal{IOTS}(L_U, L_I \cup \{\theta\})$. By instantiating the set of observers with $\mathcal{IOTS}(L_U, L_I \cup \{\theta\})$ and the set of implementations with $\mathcal{IOTS}(L_I, L_U)$, *input/output refusal preorder*, $\leq_{ior}$ [41], is defined following the extensional characterisation given in equation (1))

$$i \leq_{ior} s \quad =_{def} \quad \forall u \in \mathcal{IOTS}(L_U, L_U \cup \{\theta\}) : \quad obs_c^\theta(u,i) \subseteq obs_c^\theta(u,s) \text{ and } obs_t^\theta(u,i) \subseteq obs_t^\theta(u,s) \qquad (4)$$

Since implementations are, by assumption, always prepared to accept input actions and the environment is always prepared to accept output actions, the only way to deadlock for these kind of systems is if the environment does not provide an input action, and the IUT does not produce an output action. The inability to produce outputs is an important characteristic of implementations that is observable by observers that are equipped with a θ-label. Following terminology introduced in [43] we call a state *quiescent* if no output action or internal transition can be produced from this state; $\delta(s) =_{def} s \xrightarrow{L_U}\!\!\!\!\!/\ \, s$. Observing quiescence can be made explicit by means of a special event with label $\delta \notin L$; δ can be observed if the implementation is in a quiescent state. [41] proves that $\leq_{ior}$ can also be characterised intensionally in terms of inclusion between the sets of output actions, including δ, that the implementation and the specification can perform. Formally, $i \leq_{ior} s$ iff after all failure traces in $\sigma \in (L \cup \{L_U\})^*$ the outputs produced by the implementation are specified, and the implementation may only refuse to produce outputs if the specification does so, viz.,

$$i \leq_{ior} s \quad \textit{iff} \quad \forall \sigma \in (L \cup \{L_U\})^* : out(\, i \textbf{ after } \sigma\,) \subseteq out(\, s \textbf{ after } \sigma\,)$$

where $out(S) =_{def} \{x \in L_U \mid \exists s \in S : s \stackrel{x}{\Longrightarrow}\} \cup \{\delta \mid \exists s \in S : \delta(s)\}$ for S a set of states. A failure trace in $(L \cup \{L_U\})^*$ is called a *suspension trace*; $s\text{-}traces(p) =_{def} f\text{-}traces(p) \cap (L \cup \{L_U\})^*$.

Since checking $out(\, i \textbf{ after } \sigma\,) \subseteq out(\, s \textbf{ after } \sigma\,)$ for all suspension traces is hard to achieve by means of testing, the above characterisation can be relaxed by checking this condition for fewer suspension traces. In general, for each $\mathcal{F} \subseteq (L \cup \{L_U\})^*$ an implementation relation $\textbf{ioco}_{\mathcal{F}}$ can be defined that only checks the condition $out(\, i \textbf{ after } \sigma\,) \subseteq out(\, s \textbf{ after } \sigma\,)$ for $\sigma \in \mathcal{F}$, viz.,

$$i \ \textbf{ioco}_{\mathcal{F}}\ s \quad =_{def} \quad \forall \sigma \in \mathcal{F} : out(\, i \textbf{ after } \sigma\,) \subseteq out(\, s \textbf{ after } \sigma\,) \tag{5}$$

Note the correspondence in structure between equation (5) and equation (2).

Validating a system by means of testing involves, in practice, checking how the system reacts to stimuli from the environment. The relation $\textbf{ioco}_{\mathcal{F}}$ captures this intuitive notion of correctness [41]: correct implementations may only give reactions that are specified. From now on we focus on the generation of tests for implementation relation $\textbf{ioco}_{\mathcal{F}}$ with $\mathcal{F} \subseteq s\text{-}traces(s)$.

For testing implementations in $\mathcal{IOTS}(L_I, L_U)$ it suffices to restrict the class of tests to a specific subclass of $\mathcal{IOTS}(L_U, L_I \cup \{\theta\}) \subseteq \mathcal{LTS}(L \cup \{\theta\})$ in order to check whether systems are $\leq_{ior}$-related or not. In particular, [41] shows that it suffices to restrict to deterministic members with finite behaviour of $\mathcal{LTS}(L_U, L_I \cup \{\theta\})$, such that either a single input action is supplied, or all output actions, including θ, can be observed. There is no need to introduce additional nondeterminism in the test, and, since all errors occur within a finite depth of a transition system, they can be found using a finite series of experiments. Formally, a test case t for $\textbf{ioco}_{\mathcal{F}}$ is a labelled transition system over $L_I \cup L_U \cup \{\theta\}$ such that (i) t is deterministic and has finite behaviour, (ii) there exists two states **pass**, **fail** such that $init(\textbf{ pass }) = init(\textbf{ fail }) = \emptyset$, and (iii) for all states $t' \in der(t)$ with $t' \neq$ **pass** , **fail** we have $init(t') = \{a\}$ for some $a \in L_I$, or $init(t') = L_U \cup \{\theta\}$. The universe of tests over L_U and L_I is denoted as $TESTS(L_U, L_I)$, and a test suite T is a set of tests: $T \subseteq TESTS(L_U, L_I)$. We denote test cases with a LOTOS-like syntax: $t := a; t \mid \sum T \mid \textbf{pass} \mid \textbf{fail}$ where $\emptyset \neq T \subseteq TESTS(L_U, L_I)$. The semantics of these expressions is the obvious one: $a; T$ is able to do action a, after which it behaves as t, $\sum T$ behaves make a choice between the behaviours of T, and **pass**, **fail** cannot perform any action at all. Instead of $\sum\{B_1, B_2\}$ we also write $B_1 + B_2$.

In order to give an indication about the (in)correctness of implementations based on observations made after execution of a test case, a verdict (success or failure) is assigned to implementations. For brevity we will identify the verdicts success and failure with the states **pass** and **fail**, respectively. The execution of a test is modelled in terms of test runs. A test run $\sigma \in L^*$ of test t and implementation i is a trace that $t\,]|\, i$ can perform such that test

t ends in **pass** or **fail**: $\exists i' : t \rceil\!| \, i \overset{\sigma}{\Longrightarrow}$ **pass** $\rceil\!| \, i'$ or $t \rceil\!| \, i \overset{\sigma}{\Longrightarrow}$ **fail** $\rceil\!| \, i'$. An implementation i is said to fail test t if there exists a test run of $t \rceil\!| \, i$ that ends in **fail**, i.e., i *fails* $t =_{def} \exists \sigma \in L^*, \exists i' : t \rceil\!| \, i \overset{\sigma}{\Longrightarrow}$ **fail** $\rceil\!| \, i'$. Dually, an implementation passes test t if it does not fail test t: i *passes* $t =_{def} \neg(i$ *fails* $t)$. We shall say that an implementation passes a set of test cases T, denoted as i *passes* T, if it passes all tests in test suite T. The failing of a test suite is defined conversively. To link the passing and failing of an implementation to the correctness and incorrectness of this implementation, respectively, the verdicts **pass** and **fail** in the test case have to be assigned carefully. Ideally, test cases are designed in such a way that correct implementations always pass this set of tests (soundness), and incorrect implementations always fail this set of tests (exhaustiveness). Since exhaustiveness is difficult (if not impossible) to reach in practice we require soundness when designing test suites, and strive for exhaustiveness; erroneous behaviour is likely to be detected by the test suite. A test suite that is both sound and exhaustive is called complete.

Now we can give a test generation algorithm that is able to produce test cases in $TESTS(L_U, L_I)$ from a specification $s \in \mathcal{LTS}(L_I \cup L_U)$ with respect to implementation relation $\mathbf{ioco}_{\mathcal{F}}$, and where it is assumed that implementations can be modelled as members of $\mathcal{IOTS}(L_I, L_U)$. The algorithm is inspired by the one presented in [41] and given in figure 4. In the algorithm we use the notation $\overline{\sigma}$ for a trace in which all occurrences of δ are replaced by the deadlock detection symbol θ that is used to observe this output deadlock, and vice versa: $\overline{\sigma}$ leaves other actions unchanged.

The algorithm is parameterised over a set of suspension traces $\mathcal{F}$ and a specification $s \in \mathcal{LTS}(L_I \cup L_U)$. For each suspension trace in $\mathcal{F}$ the algorithm produces a test case that is able to check that the implementation produces an valid output(cf. equation (5)). The algorithm keeps track of the current states of the specification that are exercised by means of the variable S, which is initialised with $\{s_0\}$ **after** ϵ (where s_0 is the initial state of specification s). Tests are constructed by recursive application of three different steps. Step 1 is used to terminate a test case by assigning **pass**. Step 2 supplies an input $a \in L_I$, that is specified by some trace in $\mathcal{F}$, to the implementation, updates the set of possible current states S of the specification and the set of suspension traces $\mathcal{F}$ that need to be verified, and recursively proceeds. In step 3 the output actions that the implementation produces are checked for validity: a **fail** is assigned if the implementation produces an output that cannot be produced by the specification, and we have already executed a trace in $\mathcal{F}$, i.e., $\epsilon \in \mathcal{F}$. In this case there is evidence that the implementation violates equation (5). If the implementation produces an unspecified output for which no checking is required ($\epsilon \notin \mathcal{F}$) a **pass** is assigned: there is no evidence of incorrectness with respect to $\mathbf{ioco}_{\mathcal{F}}$. In case the implementation produces a specified output then checking needs to be continued, i.e., the algorithm recursively proceeds, where S and $\mathcal{F}$ are updated accordingly.

Input:	specification $s \in \mathcal{LTS}(L_I \cup L_U)$
Input:	set of failure traces $\mathcal{F} \subseteq (L \cup \{L_U\})^*$
Output:	test case $\Pi_{\mathcal{F},S} \in TESTS(L_U, L_I)$.

Initial value: $S = \{s_0\}$ **after** ϵ, where s_0 is the initial state of s.

Apply one of the following non-deterministic choices recursively.

1. (* terminate the test case *)

$$\Pi_{\mathcal{F},S} := \textbf{pass}$$

2. (* supply an input to the implementation *)
Take $a \in L_I$ such that $\mathcal{F}' \neq \emptyset$, then

$$\Pi_{\mathcal{F},S} := a; \Pi_{\mathcal{F}',S'}$$

where $\mathcal{F}' = \{\sigma \mid a{\cdot}\sigma \in \mathcal{F}\}$ and $S' = S$ **after** a

3. (* check the next output of the implementation *)

$$\begin{aligned}\Pi_{\mathcal{F},S} := \quad & \textstyle\sum\{x;\ \textbf{fail} \mid x \in L_U \cup \{\theta\}, \overline{x} \notin out(S), \epsilon \in \mathcal{F}\} \\ & + \textstyle\sum\{x;\ \textbf{pass} \mid x \in L_U \cup \{\theta\}, \overline{x} \notin out(S), \epsilon \notin \mathcal{F}\} \\ & + \textstyle\sum\{x; \Pi_{\mathcal{F}'_x,S'_x} \mid x \in L_U \cup \{\theta\}, \overline{x} \in out(S)\}\end{aligned}$$

where $\mathcal{F}'_x = \{\sigma \mid \overline{x}{\cdot}\sigma \in \mathcal{F}\}$ and $S'_x = S$ **after** $\overline{x}$

Figure 4 Test generation algorithm.

When executing tests obtained using the algorithm in figure 4, implementations that are **ioco**$_{\mathcal{F}}$-correct will never be considered erroneous, i.e., there is no test run that will lead to a **fail**-state when these tests are executed against **ioco**$_{\mathcal{F}}$-correct implementations (soundness). Moreover, executing *all* (usually infinitely many) test cases that are generated by the algorithm can detect all erroneous implementations (exhaustiveness) [41].

Theorem 1 Let $s \in \mathcal{LTS}(L_I \cup L_U)$ and $\mathcal{F} \subseteq s\text{-}traces(s)$.

1. A test case obtained with the algorithm depicted in figure 4 is sound for s with respect to **ioco**$_{\mathcal{F}}$.
2. The set of all test cases that can be obtained by the algorithm depicted in figure 4 is exhaustive for s with respect to **ioco**$_{\mathcal{F}}$.

The testing theory for IOTS is expected to be more useful, due to the distinction between inputs and outputs, than theories that do not make such explicit distinction. This is also motivated by the existence of the tool TVEDA, that originated from protocol testing experience. TVEDA can derive tests that are similar to tests that can be derived by algorithm 4. In [32] an attempt to provide an theoretical foundation behind TVEDA was given, which resulted in an implementation relation R_1 that is very similar to $\mathbf{ioco}_{traces(s)}$. Moreover, since algorithm 4 abstracts from the branching structure of implementations and only deals with trace structures, it is expected that data aspects are more easy to incorporate than in the algorithm for the construction of canonical testers (section 4): in testing for $\mathbf{ioco}_{\mathcal{F}}$ the explosion from countably branching structures to uncountably branching structures (that is present in the construction of canonical testers) is avoided.

6 CLOSING THE CIRCLE

Although the shift from symmetric to asymmetric communication allows for a more realistic modelling of the testing process, still some criticism can be ventilated towards the asymmetric model. As indicated in [36] the requirement that implementations must be modelled as members of $\mathcal{IOTS}(L_I, L_U)$ is still restrictive; not all implementations satisfy the requirement that inputs are always enabled (e.g., systems that communicate with each other via *bounded* queues; if the queue is full, no input can be accepted any more). Furthermore, observers for IOTS are forced to accept *all* outputs, even if these outputs occur at geographically dispersed places, and thereby a possible distribution of the environment itself is ignored. As, in practice, many distributed implementations communicate with their environment via distributed locations, or PCOs (Point of Control and Observation [22]), the distributed nature of the interfaces should be taken into account when testing these systems. For example, the standardised language SDL [10] explicitly incorporates the different locations through which an implementation communicates with its environment by means of channels, and the standardised test notation TTCN [22, part 3] is also able to express the sending and reception of messages to specific locations. In the IOTS model it is not possible to exploit the distributed nature of interfaces. An example of a system that cannot be described as an IOTS is depicted in figure 5.

In order to overcome these deficiencies recent research has lead to a model that refines the IOTS model, and, at the same time, unifies both the symmetric and asymmetric communication paradigm in a single framework. Basically, this is done by making two refinements to the IOTS model that are sufficient to model systems like the one in figure 5, i.e., (i) distinguishing between different locations, or channels, through which an implementation communicates with its environment, and (ii) weaken the requirement for IOTS that all input

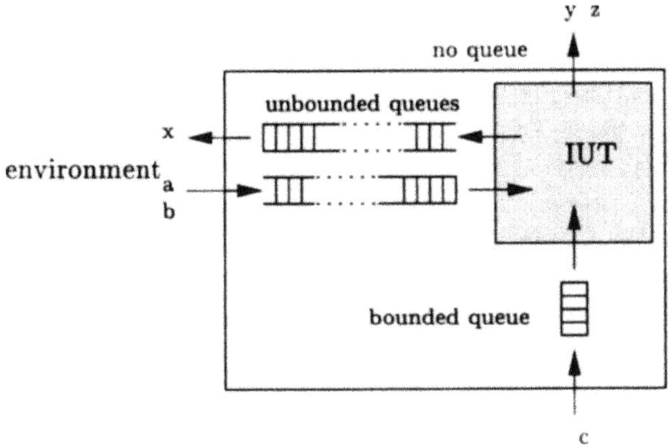

Figure 5 Example of a multi input/output queue system.

actions have to be continuously enabled. We will briefly elaborate on both refinements.

Ad (i) Instead of partitioning the label set L in an input set L_I and an output set L_U, these sets themselves are partitioned in one or more groups (i.e., sets) of actions; $L_I = \bigcup_{1\leq i\leq n} L_I^i$ and $L_U = \bigcup_{1\leq j\leq m} L_U^j$. Each group of actions defines a *channel* where these actions may occur. By distinguishing between the different channels of an implementation an external observer is (potentially) able to observe the inability of an implementation to produce an output at some output channel, while at another output channel the implementation can produce an output. Note that this is not possible in the IOTS model: observers of an IOTS are not able to check that subsets of outputs cannot occur.

Ad (ii) Instead of requiring that input actions must always be enabled, it is required for each input channel that "if an input in a channel is enabled, then all inputs at this channel should be simultaneously enabled", i.e., For each input channel L_I^i we require

$$\forall p' \in der(p), \text{ if } \exists a \in L_I^i : p' \stackrel{a}{\Longrightarrow} \text{ then } \forall b \in L_I^i : p' \stackrel{b}{\Longrightarrow} \tag{6}$$

This requirement is strictly weaker than the one imposed on IOTS where all inputs are always enabled. In particular, this requirement allows us to model communication by means of bounded queues; all inputs in a channel are only enabled if the queue is not full.

Systems that are modelled with these two refinements are called *multi input/output transition systems* (MIOTS) [40, 41]. The class of MIOTS under consideration depends on the specific partitioning of the channels, that is, the set of implementations in MIOTS is parameterised by the location of interfaces through which these implementations communicate with their environment. Such systems can be tested by means of observers that are also modelled as MIOTS (where input channels of the system are output channels for the observer, and vice versa). This yields an implementation relation $\leq_{mior}$ defined similarly as $\leq_{ior}$ (equation (4)), and a characterisation in terms of $\mathbf{mioco}_{\mathcal{F}}$ similarly as $\mathbf{ioco}_{\mathcal{F}}$ (equation (5)), that are parameterised over the distribution of the interfaces of the implementation with the environment. [19] investigates testing theory for MIOTS, and they relate the different instances of $\leq_{mior}$ for the specific distributions of interfaces.

MIOTS allow to relate synchronous and asynchronous testing theories by varying the granularity of the interfaces, and thus close the circle with refusal testing (section 1, [33]). Moreover, the different instances of $\leq_{mior}$ for the specific distributions of the interfaces are related. If all inputs run via a single channel and all outputs run via a single channel, and requirement (6) is strengthened to requirement (3) for inputs on this single input channel, then $\leq_{mior}$ corresponds to $\leq_{ior}$. On the other hand, if each action runs through a separate channel, i.e., the sets L_I and L_U are partitioned in singletons, then $\leq_{mior}$ equals $\leq_{rf}$ [19]. This means that the symmetric testing theory discussed in section 3 and the asymmetric testing theory discussed in section 5 are unified in a single testing framework, and the test algorithm presented in [19] is able to generate tests for $\leq_{rf}$, $\leq_{ior}$, $\leq_{mior}$, $\mathbf{ioco}_{\mathcal{F}}$ and $\mathbf{mioco}_{\mathcal{F}}$.

7 CONCLUSIONS

History In this paper we sketched the developments that have taken place (and still take place) in testing based on labelled transition systems. The seminal work in [13, 12] introduces a testing theory for labelled transition systems based on the assumption that communication between systems and their environments is symmetric. They define, and compare, many testing relations by varying the class of tests, and the class of observations. [33] discusses a refinement of [13, 12] by allowing observers to continue after observation of deadlock. The first mature theory based on [13, 12] that presents an algorithm to derive tests from a specification is presented in [6, 7]. They discuss how to generate a test suite that can distinguish between correct and incorrect implementations (with respect to implementation relation **conf**).

As, in practice, communication between implementations and testers is often asymmetric, many approaches that incorporate such asymmetric communication have been done with the aim to apply testing theory to realistic systems (SDL [10], TTCN [22, part 3]). One of these approaches is [42]. They assume that communication between implementations and testers runs via two unbounded queues, and they define, and analyse, testing relations (so-called queue preorders) for systems that communicate with their environment through these queues. A more general approach is taken by assuming that implementations can be modelled as input/output transition systems (IOTS). An IOTS is a LTS that makes an explicit distinction between input actions and output actions, and assumes that input actions are weakly enabled. In this way it isolates the relevant aspects of queue systems without requiring that communication with the environment is done via queues.

[41] applies the ideas of [33] to IOTS, and defines a testing theory for implementations that can be modelled as IOTS. They assume that the inability to produce output actions, i.e., quiescence, is observable, and define an implementation relation $\mathbf{ioco}_{\mathcal{F}}$ that captures the intuition of correctness in practice. They also present an algorithm that is able to derive a sound and complete set of tests from a specification. These tests resemble tests generated by the tool TVEDA [11] that originated from practical testing experience.

[19] refines the theory of IOTS by taking the distribution of the interfaces of implementations into account. They explicitly model the locations (also: PCOs or channels) where actions can take place, and they require that input actions per input channel are either simultaneously enabled or simultaneously disabled. Such systems are called multi input/output transition systems (MIOTS). For implementations that can be modelled as MIOTS refusal testing [33] is applied, and quiescence is assumed to be observable (cf. [36, 40, 41]). Similar to $\mathbf{ioco}_{\mathcal{F}}$ they define an implementation relation $\mathbf{mioco}_{\mathcal{F}}$ relative to the distribution of interfaces of implementations, and present an algorithm that is able to derive sound and complete test cases for $\mathbf{mioco}_{\mathcal{F}}$. This results in a testing theory that is parameterised over the granularity of interfaces of implementations. [19] shows that specific instances yield the traditional refusal testing theory of [33], and the refusal testing for IOTS [41], and hence incorporates both theories in a single framework.

Future The theory and the algorithm for IOTS/MIOTS can form the basis for the development of test generation tools. In order to use such tools in realistic testing experiments several aspects need elaboration. One of these aspects involves data handling. In many realistic applications data is involved. To deal with data in an efficient way the test generation algorithm has to incorporate such data aspects in a symbolic way; otherwise automation of tests is not feasible due to explosion in the data part. Another aspect concerns the well-known problem of test selection. As test suites grow in size, execution of all of the tests in the test suite becomes too expensive, and selections have to made; which tests are executed, and which are not? (Partial) solutions can be found in defining coverage measures, fault models, strengtening test assumptions, etc. [3, 23, 31]. Experiments in applying the algorithm to realistic problems have to be conducted in order to show the strengths and weaknesses in the testing theory for IOTS. A first trial in which a preliminary version of the theory for IOTS was applied to a simple protocol looks promising [38], but more experiments are needed to draw meaningful conclusions. Finally, the relation between formalisms that incorporate channels (e.g., SDL, TTCN), and MIOTS needs further investigation.

8 REFERENCES

[1] S. Abramsky. Observational equivalence as a testing equivalence. *Th. Comp. Sc.*, 53(3):225–241, 1987.

[2] R. Alderden. COOPER, the compositional construction of a canonical tester. In S.T. Vuong, editor, *FORTE'89*, pages 13–17. North-Holland, 1990.

[3] G. Bernot. Testing against formal specifications: A theoretical view. In S. Abramsky et al., editor, *TAPSOFT'91, Volume 2*, pages 99–119. LNCS 494, Springer-Verlag, 1991.

[4] F. de Boer, J.W. Klop, and J. Rutten. Asynchronous communication in process algebra. In J.W. Bakker et al., editor, *REX Workshop on Semantics: Foundations and Applications.* LNCS 666, Springer-Verlag, 1993.

[5] B. S. Bosik and M. Ü. Uyar. Finite state machine based formal methods in protocol conformance testing: From theory to implementation. *Computer Networks and ISDN Systems*, 22(1):7–33, 1991.

[6] E. Brinksma. On the existence of canonical testers. Memorandum INF-87-5, University of Twente, Enschede, The Netherlands, 1987.

[7] E. Brinksma. A theory for the derivation of tests. In S. Aggarwal et al., editor, *PSTV VIII*, pages 63–74. North-Holland, 1988.

[8] E. Brinksma, R. Alderden, R. Langerak, J. van de Lagemaat, and J. Tretmans. A formal approach to conformance testing. In J. de Meer et al., editor, *Second International Workshop on Protocol Test Systems*, pages 349–363. North-Holland, 1990.

[9] A. Cavalli and M. Philippou. Some issues on testing theory and its applications. In T. Mizuno et al., editor, *Seventh International Workshop on Protocol Test Systems.* Chapman & Hall, 1995.

[10] CCITT. *Specification and Description Language (SDL).* Recommendation Z.100. ITU-T General Secretariat, Geneve, Switzerland, 1992.

[11] M. Clatin, R. Groz, M. Phalippou, and R. Thummel. Two approaches linking test generation with verification techniques. In A. Cavalli et al., editor, *Eight International Workshop on Protocol Test Systems.* Chapman & Hall, 1996.

[12] R. De Nicola. Extensional equivalences for transition systems. *Acta Informatica*, 24:211–237, 1987.

[13] R. De Nicola and M.C.B. Hennessy. Testing equivalences for processes. *Th. Comp. Sc.*, 34:83–133, 1984.

[14] P. Doornbosch. Test derivation for Full LOTOS. Memorandum INF-91-51, University of Twente, Enschede, The Netherlands, 1991.

[15] K. Drira, P. Azéma, and F. Vernadat. Refusal graphs for conformance

tester generation and simplification: A computational framework. In A. Danthine et al., editor, *PSTV XIII.* North-Holland, 1993.

[16] S. Fujiwara et al. Test selection based on finite state models. *IEEE Trans. on Soft. Eng.*, 17(6):591–603, 1991.

[17] R.J. van Glabbeek. The linear time – branching time spectrum II (The semantics of sequential systems with silent moves). In E. Best, editor, *CONCUR'93*, LNCS 715, pages 66–81. Springer-Verlag, 1993.

[18] D. Gueraichi and L. Logrippo. Derivation of test cases for lap-b from a LOTOS specification. In S.T. Vuong, editor, *FORTE'89.* North-Holland, 1990. Also: technical report TR-89-18.

[19] L. Heerink and J. Tretmans. Refusal testing for classes of transition systems with inputs and outputs. CTIT technical report, University of Twente, Enschede, The Netherlands, 1997.

[20] ISO. *Information Processing Systems, Open Systems Interconnection, Estelle - A Formal Description Technique Based on an Extended State Transition Model.* International Standard IS-9074. ISO, Geneve, 1989.

[21] ISO. *Information Processing Systems, Open Systems Interconnection, LOTOS - A Formal Description Technique Based on the Temporal Ordering of Observational Behaviour.* International Standard IS-8807. ISO, Geneve, 1989.

[22] ISO. *Information Technology, Open Systems Interconnection, Conformance Testing Methodology and Framework.* International Standard IS-9646. ISO, Geneve, 1991. Also: CCITT X.290–X.294.

[23] ISO/IEC JTC1/SC21 WG7, ITU-T SG 10/Q.8. *Proposed ITU-T Z.500 and Committee Draft on "Formal Methods in Conformance Testing".* CD 13245-1. ISO – ITU-T, Geneve, 1996.

[24] R. Langerak. A testing theory for LOTOS using deadlock detection. In E. Brinksma et al., editor, *PSTV IX*, pages 87–98. North-Holland, 1990.

[25] G. Leduc. Conformance relation, associated equivalence, and minimum canonical tester in LOTOS. In B. Jonsson et al., editor, *PSTV XI.* North-Holland, 1991.

[26] D. Lee and M. Yannakakis. Testing finite-state machines: State identification and verification. *IEEE Trans. on Comp.*, 43(3):306–320, 1994.

[27] G. Luo, A. Petrenko, and G. von Bochmann. Selecting test sequences for partially-specified nondeterministic finite state machines. Publication N.864, Université de Montréal, Montréal, Canada, 1993.

[28] N.A. Lynch and M.R. Tuttle. An introduction to input/output automata. *CWI Quarterly*, 2(3):219–246, 1989. Also: Technical Report MIT/LCS/TM-373 (TM-351 revised), MIT, U.S.A., 1988.

[29] R. Milner. *Communication and Concurrency.* Prentice-Hall, 1989.

[30] A. Petrenko, N. Yevtushenko, A. Lebedev, and A. Das. Nondeterministic state machines in protocol conformance testing. In O. Rafiq, editor, *Sixth International Workshop on Protocol Test Systems*, pages 363–

378. North-Holland, 1994.

[31] M. Phalippou. Executable testers. In O. Rafiq, editor, *Sixth International Workshop on Protocol Test Systems*, pages 35–50. North-Holland, 1994.

[32] M. Phalippou. *Relations d'Implantation et Hypothèses de Test sur des Automates à Entrées et Sorties.* PhD thesis, L'Université de Bordeaux I, France, 1994.

[33] I. Phillips. Refusal testing. *Th. Comp. Sc.*, 50(2):241–284, 1987.

[34] D. H. Pitt and D. Freestone. The derivation of conformance tests from LOTOS specifications. *IEEE Trans. on Soft. Eng.*, 16(12):1337–1343, 1990.

[35] G. Schoemakers. Generation of canonical testers from recursive LOTOS specifications. Master's thesis, University of Twente, Enschede, The Netherlands, 1994.

[36] R. Segala. Quiescence, fairness, testing, and the notion of implementation. In E. Best, editor, *CONCUR'93*, pages 324–338. LNCS 715, Springer-Verlag, 1993.

[37] D.P. Sidhu and T.K. Leung. Formal methods for protocol testing: A detailed study. *IEEE Trans. on Soft. Eng.*, 15(4):413–426, 1989.

[38] R. Terpstra, L. Ferreira Pires, L. Heerink, and J. Tretmans. Testing theory in practice: A simple experiment. In Z. Brezočnik et al., editor, *COST 247 International Workshop on Applied Formal Methods in System Design*, pages 168–183, Maribor, Slovenia, 1996.

[39] J. Tretmans. *A Formal Approach to Conformance Testing.* PhD thesis, University of Twente, Enschede, The Netherlands, 1992.

[40] J. Tretmans. Conformance testing with labelled transition systems: Implementation relations and test generation. *Computer Networks and ISDN Systems*, 29:49–79, 1996.

[41] J. Tretmans. Test generation with inputs, outputs, and repetitive quiescence. *Software - Concepts and Tools*, 17:103–120, 1996.

[42] J. Tretmans and L. Verhaard. A queue model relating synchronous and asynchronous communication. In R.J. Linn and M.Ü. Uyar, editors, *PSTV XII*, IFIP Transactions, pages 131–145. North-Holland, 1992.

[43] F. Vaandrager. On the relationship between process algebra and input/output automata. In *Logic in Computer Science*, pages 387–398. Sixth Annual IEEE Symposium, IEEE Computer Society Press, 1991.

[44] L. Verhaard, J. Tretmans, P. Kars, and E. Brinksma. On asynchronous testing. In G. Bochmann et al., editor, *Fifth International Workshop on Protocol Test Systems*. North-Holland, 1993.

[45] C. D. Wezeman. The CO-OP method for compositional derivation of conformance testers. In E. Brinksma et al., editor, *PSTV IX*, pages 145–158. North-Holland, 1990.

[46] M. Yannakakis and D. Lee. Testing finite state machines: Fault detection. *Journal of Computer and System Sciences*, 50(2):209–227, 1995.

Checking Experiments with Labeled Transition Systems for Trace Equivalence *

1*Q. M. Tan*[†], *A. Petrenko*[‡] *and G. v. Bochmann*[†]
†*Département d'IRO, Université de Montréal*
C.P. 6128, Succ. Centre-Ville, Montréal, (Québec) H3C 3J7, Canada
E-mail:(tanq,Bochmann)@iro.umontreal.ca Fax:(514)343-5834
‡*CRIM, Centre de Recherche Informatique de Montréal*
1801 Avenue McGill College, Montréal, (Québec) H3A 2N4, Canada
E-mail:petrenko@crim.ca Phone:(514)840-1234 Fax:(514)840-1244

Abstract

We apply the state identification techniques for testing communication systems which are modeled labeled by transition systems (LTSs). The conformance requirements of specifications are represented as the trace equivalence relation and derived tests have finite behavior and provide well-defined fault coverage. We redefine in the realm of LTSs the notions of state identification that were originally defined in the realm of input/output finite state machines (FSMs). Then we present the corresponding test generation methods and discuss their fault coverage.

Keywords

Conformance testing, formal description techniques, test generation, labeled transition systems, communication protocols

*This work was supported by the HP-NSERC-CITI Industrial Research Chair on Communication Protocols, Université de Montréal

Testing of Communicating Systems, M. Kim, S. Kang & K. Hong (Eds)
Published by Chapman & Hall © 1997 IFIP

1 INTRODUCTION

One of the important issues of conformance testing is to derive useful tests for labeled transition systems (LTSs), which serve as a semantic model for various specification languages, e.g., LOTOS, CCS, and CSP. Testing theories and methods for test derivation in the LTS formalism have been developed in [2, 16, 11, 3, 13, 15]. In particular, a so-called **conf** relation and *canonical tester* [2] became the basis for a large body of work in this area.

Unfortunately, the canonical tester approach cannot be taken into account when test generation for real protocols is attempted. The canonical tester has infinite behavior whenever the specification describes an infinite behavior. Moreover, we believe that the **conf** relation alone is too weak as a criterion to accept an implementation. Since this relation does not deal with invalid traces, it allows for a trivial implementation which has a single state with looping transitions labeled with all possible actions, and such an implementation conforms to any LTS specification with the same alphabet with respect to the **conf** relation [14]. Thus even though an implementation is concluded being valid based on **conf**, another relation, such as *trace-equivalence*, has to be tested as well.

Observing and comparing traces of executed interactions is usual means for conformance testing of protocols, and in many cases it is required that an implementation should have the same traces as its specification. In particular, most existing protocols are deterministic, and in the case of determinism several other finer testing semantics, such as failure or failure trace, are reduced to trace semantics. Based on the notion of such experiments and the trace equivalence relation, a number of competing test derivation methods with fault coverage have been elaborated [8, 4, 12, 18, 7, 10, 9] for protocols in the formalism of input/output finite state machines (FSMs), many of which use the state identification techniques to obtain better fault coverage. Compared to FSMs, LTSs are in some sense a more general descriptive model which use rendezvous communication without distinction between input and output; there are various semantics determining whether an implementation conforms to a specification; most existing test derivation methods use the exhaustive testing approach in order to prove the correctness of the implementation in respect to a given conformance relation. Apparently, such an approach is often impractical since it may involve a test suite of infinite length. The approximation approach [11, 16], such as n-testers, which is proposed to solve this problem, provides no fault coverage measure for conformity of the implementation with its specification.

Several attempts have been made to apply the ideas underlying the FSM-based methods to the LTS model [6, 3, 13, 14] for several conformance relations. In particular, this research is directed towards redefining the notions of state identification in the LTS realm for a given relation. However, these attempts are limited to individual or informal applications of the notions of

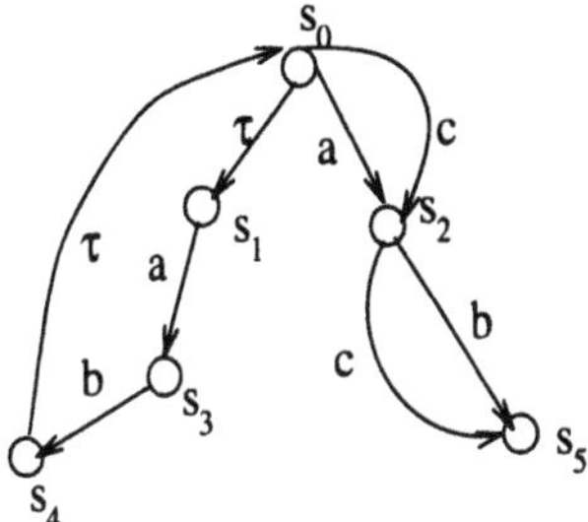

Figure 1 An LTS graph.

state identification underlying the FSM-based methods. In fact, the FSM-based notions can also be applied directly to the LTS model if an appropriate distinguishabilty of states is defined in the LTS model. Therefore, a systematic approach based on the notions of state identification can also be developed in the LTS model such that we could devise alternative and competing techniques that guarantee fault coverage, for constructing useful tests for protocols based on the LTS semantics.

In this paper, we redefine in the LTS model the notions of state identification which were originally used in the FSM realm for trace equivalence. Based on the adapted notions, the corresponding test derivation methods are presented, and it is shown that for an FSM-based method with a notion of state identification we can have a corresponding LTS-based method with a similar notion of state identification, and if the FSM-based method guarantees complete fault coverage then the LTS-analogue also guarantees complete fault coverage.

2 LABELED TRANSITION SYSTEMS

Definition 1 (*Labeled transition system (LTS)*): A labeled transition system is a 4-tuple $< S, \Sigma, \Delta, s_0 >$, where

- S is a finite set of states, $s_0 \in S$, is the initial state.
- Σ is a finite set of labels, called observable actions; $\tau \notin \Sigma$ is called an internal action.
- $\Delta \subseteq S \times (\Sigma \cup \{\tau\}) \times S$ is a transitions set. $(p, \mu, q) \in \Delta$ is denoted by $p{-}\mu{\rightarrow} q$.

An LTS is said to be *nondeterministic* if it has some transition labeled with τ or there exist $p{-}a{\rightarrow} p_1, p{-}a{\rightarrow} p_2 \in \Delta$ but $p_1 \neq p_2$; otherwise it is *deterministic* LTS.

An LTS can also be represented by a directed graph where nodes are states and labeled edges are transitions. An LTS graph is shown in Figure 1.

Given an LTS $\mathsf{S} =< S, \Sigma, \Delta, s_0 >$, the conventional notations are shown in Table 1, as introduced in [2]. In this paper we use $\mathsf{M}, \mathsf{P}, \mathsf{S}, \ldots$ to represent

Table 1 Basic notations for labeled transition systems.

notation	meaning
Σ^*	set of sequences over Σ; σ or $a_1 \ldots a_n$ for such a sequence
$p-\mu_1 \ldots \mu_n \rightarrow q$	$\exists\, p_k,\ 1 \leq k < n$, such that $p-\mu_1\rightarrow p_1 \ldots p_{n-1}-\mu_n\rightarrow q$
$p{=}\varepsilon{\Rightarrow} q$	$p-\tau^n\rightarrow q\ (1 \leq n)$ or $p = q$ (note: τ^n means n times τ)
$p{=}a{\Rightarrow} q$	$\exists\, p_1, p_2$ such that $p{=}\varepsilon{\Rightarrow} p_1 - a \rightarrow p_2 {=}\varepsilon{\Rightarrow} q$
$p{=}a_1 \ldots a_n{\Rightarrow} q$	$\exists\, p_k,\ 1 \leq k < n$, such that $p{=}a_1{\Rightarrow} p_1 \ldots p_{n-1}{=}a_n{\Rightarrow} q$
$p{=}\sigma{\Rightarrow}$	$\exists\, q$ such that $p{=}\sigma{\Rightarrow} q$
$p{\neq}\sigma{\Rightarrow}$	no q exists such that $p{=}\sigma{\Rightarrow} q$
$init(p)$	$init(p) = \{a \in \Sigma \mid p{=}a{\Rightarrow}\}$
p-**after**-σ	p-**after**-$\sigma = \{q \in S \mid p{=}\sigma{\Rightarrow} q\}$; S-**after**-$\sigma = s_0$-**after**-σ
$Tr(p)$	$Tr(p) = \{\sigma \in \Sigma^* \mid p{=}\sigma{\Rightarrow}\}$; $Tr(\mathrm{S}) = Tr(s_0)$

LTSs; $M, P, Q, \ldots$, for sets of states; $a, b, c, \ldots$, for actions; and $i, p, q, s \ldots$, for states. The sequences in $Tr(p)$ are called the *traces* of S for p.

Given $V \subseteq \Sigma^*$, we denote $Pref(V) = \{\sigma_1 \in \Sigma^* \mid \exists \sigma_2 \in \Sigma^*\ (\sigma_1.\sigma_2 \in V)\}$. Given $V_1, V_2 \subseteq \Sigma^*$, we denote $V_1 @ V_2 = \{\sigma_1.\sigma_2 \mid \sigma_1 \in V_1 \wedge \sigma \in V_2\}$. We also write $V^n = V @ V^{n-1}$ for $n > 0$ and $V^0 = \{\varepsilon\}$.

In the case of nondeterminism, after an observable action sequence, an LTS may enter a number of different states. In order to consider all these possibilities, a state subset (multi-state [6]), which contains all the states reachable by the LTS after this action sequence, is used.

Definition 2 (*Multi-state set*): The multi-state set of LTS S is the set $\Pi_S = \{S_i \subseteq S \mid \exists \sigma \in \Sigma^*\ (s_0\text{-}\mathbf{after}\text{-}\sigma = S_i)\}$.

Note that $S_0 = s_0$-**after**-ε is in Π_S and is called the *initial multi-state*. The multi-state set can be obtained by a known algorithm which performs the deterministic transformation of a nondeterministic automaton with trace equivalence [6]. For Figure 1, $\{\{s_0, s_1\}, \{s_2, s_3\}, \{s_2\}, \{s_0, s_1, s_4, s_5\}, \{s_5\}\}$ is the multi-state set. Obviously, each LTS has one and only one multi-state set.

After any observable sequence, a nondeterministic system reaches a unique multi-state. Thus from the test perspective, it makes sense to identify multi-states, rather than single states. This viewpoint is reflected in the FSM realm by the presentation of a nondeterministic FSM as an observable FSM [9], in which each state is a subset of states of the non-observable FSM.

3 CONFORMANCE TESTING

3.1 Conformance Relation

The starting point for conformance testing is a specification in some notation, an implementation given in the form of a black box, and a conformance criterion that the implementation should satisfy. In this paper, the notation of the specification is the LTS formalism; the implementation is assumed to be described in the same model as its specification; a conformance relation, called *trace equivalence*, is used as the conformance criterion. We say that an implementation M conforms to a specification S if M is trace-equivalent to S.

Definition 3 (*Trace equivalence*): The trace equivalence relation between two states p and q, written $p \approx q$, holds iff $Tr(p) = Tr(q)$.
Given two LTSs S and M with initial states s_0 and m_0 respectively, we say that M is trace-equivalent to S, written $\mathsf{M} \approx \mathsf{S}$, iff $m_0 \approx s_0$.

We say that two states are ***distinguishable*** in trace semantics if they are not trace-equivalent. For any two states that are not trace-equivalent we can surely find a sequence of observable actions, which is a trace one of the two states, not both, to distinguish them. We also say that an LTS is *reduced* in trace semantics if all of its states are distinguishable in trace semantics.

3.2 Testing Framework

Conformance testing is a finite set of experiments, in which a set of test cases, usually derived from a specification according to a given conformance relation, is applied by a tester or experimenter to the implementation under test (IUT), such that from the results of the execution of the test cases, it can be concluded whether or not the implementation conforms to the specification.

The behavior of the tester during testing is defined by the applied test case. Thus a test case is a specification of behavior, which, like other specifications, can be represented as an LTS. An experiment should last for a finite time, so a test case should have no infinite behavior. Moreover, the tester should have certain control over the testing process, so nondeterminism in a test case is undesirable [14, 17].

Definition 4 (*Test cases and test suite*): Given an LTS specification $\mathsf{S} =< S, \Sigma, \Delta, s_0 >$, a test case T for S is a 5-tuple $< T, \Sigma_T, \Delta_T, t_0, \ell >$ where:

- $\Sigma_T \subseteq \Sigma$;
- $< T, \Sigma_T, \Delta_T, t_0 >$ is a deterministic, tree-structured LTS such that for each $p \in T$ there exists exactly one $\sigma \in \Sigma_T^*$ with $t_0 = \sigma \Rightarrow p$;
- $\ell : T \rightarrow \{\textbf{pass}, \textbf{fail}, \textbf{inconclusive}\}$ is a state labeling function.

A test suite for S is a finite set of test cases for S.

From this definition, the behavior of test case T is finite, since it has no cycles. Moreover, a trace of T uniquely determines a single state in T, so we define $\ell(\sigma) = \ell(t)$ for $\{t\} = t_0$-**after**-σ.

The interactions between a test case T and the IUT M can be formalized by the composition operator "||" of LOTOS, that is, T || M. When $t_0 \parallel m_0$ after an observable action sequence σ reaches a *deadlock*, that is, there exists a state $p \in T \times M$ such that for all actions $a \in \Sigma$, $t_0 \parallel m_0 = \sigma \Rightarrow p$ and $p \not\xrightarrow{a}$, we say that this experiment completes a *test run*. In order to start a new test run, a global reset is always assumed in our testing framework.

In order to test nondeterministic implementations, one usually makes the so-called *complete-testing assumption:* it is possible, by applying a given test case to the implementation a finite number of times, to exercise all possible execution paths of the implementation which are traversed by the test case [6, 9]. Therefore any experiment, in which M is tested by T, should include several test runs and lead to a complete set of observations $Obs_{(\mathrm{T,M})} = \{\sigma \in Tr(t_0) \mid \exists p \in T \times M, \forall a \in \Sigma \; ((t_0 \parallel m_0) = \sigma \Rightarrow p \neq a \Rightarrow)\}$. Note that for deterministic systems, such as most of real-life protocols, there is no need for this assumption.

Based on $Obs_{(\mathrm{T,M})}$, the success or failure of testing needs to be concluded. The way a verdict is drawn from $Obs_{(\mathrm{T,M})}$ is the *verdict assignment* for T. A pass verdict means success, which, intuitively, should mean that no unexpected behavior is found and the test purpose has been achieved; otherwise, the verdict should be a fail verdict. If we define $Pur(\mathrm{T}) = \{\sigma \in Tr(t_0) \mid \ell(\sigma) = \textbf{pass}\}$ for *the test purpose* of T, then the conclusion can be drawn as follows.

Definition 5 (*Verdict assignment*): Given an IUT M, a test case T, let $Obs_{fail} = \{\sigma \in Obs_{(\mathrm{T,M})} \mid \ell(\sigma) = \textbf{fail}\}$ and $Obs_{pass} = \{\sigma \in Obs_{(\mathrm{T,M})} \mid \ell(\sigma) = \textbf{pass}\}$,

$$\begin{cases} \mathsf{M} \text{ passes } \mathsf{T} & \text{iff } Obs_{fail} = \emptyset \wedge Obs_{pass} = Pur(\mathsf{T}) \\ \mathsf{M} \text{ fails } \mathsf{T} & \text{otherwise.} \end{cases}$$

Given a test suite *TS*, we also denote that M passes *TS* iff for all T $\in$ *TS* M passes T, and M fails *TS* otherwise.

3.3 State Labelings of Test Cases

Given a specification S, a test case T should be "sound", that is, for any implementation M, if M and S are trace-equivalent, then M passes T.

In the context of trace equivalence, a conforming implementation should have the same traces as a given specification. Therefore each test case specifies certain sequences of actions, which are either valid or invalid traces of the specification. The purpose of a test case is to verify that an IUT has implemented the valid ones and not any of the invalid ones. Accordingly, we conclude that all test cases for trace equivalence must be of the following form [15]:

Definition 6 (*Test cases for trace equivalence*): Given an LTS specification S, a test case T is said to be a test case for S w.r.t. $\approx$, if, for all $\sigma \in Tr(t_0)$ and $\{t_i\} = t_0\textbf{-after-}\sigma$, the state labeling of T satisfies

$$\ell_{\approx}(t_i) = \begin{cases} \textbf{pass} & \text{if } \sigma \in Tr(s_0) \wedge init(t_i) \cap out(s_0\textbf{-after-}\sigma) = \emptyset \\ \textbf{fail} & \sigma \notin Tr(s_0) \\ \textbf{inconclusive} & \text{otherwise.} \end{cases}$$

A test suite for S w.r.t. $\approx$ is a set of test cases for S w.r.t. $\approx$.

From this definition, we have the following proposition [15]: Given a test case T for S w.r.t. $\approx$, for any LTS M, if M $\approx$ S, then M passes T.

Since in trace semantics test cases for S are represented as valid or invalid traces of S, given a sequence $\sigma \in \Sigma^*$, let $\sigma = a_1.a_2 \ldots .a_n$, a test case T for S w.r.t. $\approx$ can be obtained by constructing an LTS T $= t_0{-}a_1{\rightarrow} t_1 \ldots t_{n-1}{-}a_n{\rightarrow} t_n$ and then labeling T according to Definition 6. A sequence that is used to form a test case is also called a *test sequence.*

3.4 Fault Model and Fault Coverage

The goal of conformance testing is to gain confidence in the correct functioning of the implementation under test. Increased confidence is normally obtained through time and effort spent in testing the implementation, which, however, is limited by practical and economical considerations. In order to have a more precise measure of the effectiveness of testing, a fault model and fault coverage criteria [1] are introduced, which usually take the mutation approach [1], that is, a fault model is defined as a set of all faulty LTS implementations considered. Here we consider a particular fault model $\mathcal{F}(m)$ which consists of all LTS implementations over the alphabet of the specification S and with at most m multi-states, where m is a known integer. Based on $\mathcal{F}(m)$, a test suite with complete fault coverage for a given LTS specification with respect to the trace equivalence relation can be defined as follows.

Definition 7 (*Complete test suite*): Given an LTS specification S and the fault model $\mathcal{F}(m)$, a test suite *TS* for S w.r.t. $\approx$ is said to be complete, if for any M in $\mathcal{F}(m)$, M $\approx$ S iff M passes *TS*.

We also say that a test suite is *m-complete* for S if it is complete for S in respect to the fault model $\mathcal{F}(m)$. A complete test suite guarantees that for any implementation M in the context of $\mathcal{F}(m)$, if M passes all test cases, it must be a conforming implementation, and any faulty implementation in $\mathcal{F}(m)$ must be detected by failing at least one test case in the test suite.

4 STATE IDENTIFICATION IN SPECIFICATIONS

Similar to the case of FSMs, in order to identify states in a given LTS specification, at first the specification is required to have certain testability properties, two of which are the so-called reducibility and observability.

4.1 Trace Observable System

Definition 8 (*Trace observable system (TOS)*): Given an LTS S, a deterministic LTS $\overline{\mathsf{S}}$ is said to be the trace observable system corresponding to S, if $\overline{\mathsf{S}} \approx$ S and $\overline{\mathsf{S}}$ is reduced in trace semantics.

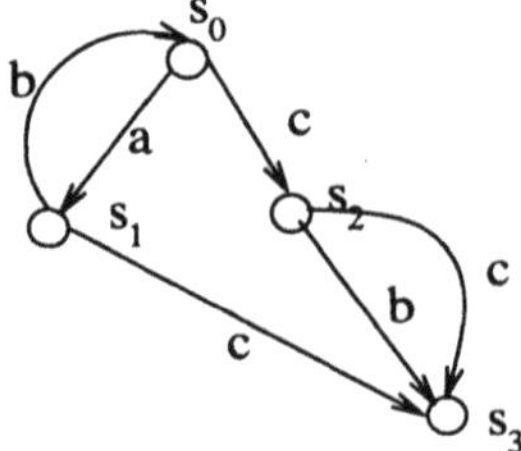

Figure 2 A corresponding trace observable system of Figure 1.

From the above definition, the TOS $\overline{S}$ of S is deterministic, reduced and trace-equivalent to S; moreover, the TOS $\overline{S}$ is unique for all LTSs trace-equivalent to S. There are the algorithms and tools that transform a given LTS into its TOS form [3]. For the LTS in Figure 1, the TOS is given in Figure 2.

In the context of trace semantics, for any LTS, the corresponding TOS models all its observable behavior. Therefore, for test generation, any LTS considered can be assumed to be in the TOS form.

4.2 State identification Facilities

There are the following facilities of state identification which can be adapted from the FSM model to the LTS model. Here we assume that the given LTS specification S is in the TOS form that has n states $s_0, s_1, \ldots s_{n-1}$, where s_0 is the initial state.

Distinguishing Sequence
Given an LTS S, we say that an observable sequence distinguishes two states if the sequence is a trace for one of the two states, but not for both. A *distinguishing sequence* for S is an observable sequence that distinguishes any two different states. Formally, $\sigma \in \Sigma^*$ is a distinguishing sequence of S if for all $s_i, s_j \in S, i \neq j$, there exists $\sigma' \in Pref(\sigma)$ such that $\sigma' \in Tr(s_i) \oplus Tr(s_j)$, where $A \oplus B = (A \backslash B) \cup (B \backslash A)$.

There are LTSs in the TOS form without any distinguishing sequence. As an example, the LTS in Figure 2 has no distinguishing sequence.

Unique Sequences
A *unique sequence* for a state is an observable sequence that distinguishes the given state from all others. Formally, $\sigma_i \in \Sigma^*$ is a unique sequence for $s_i \in S$, if, for all $s_j \in S, i \neq j$, there exists $\sigma_i' \in Pref(\sigma_i)$ such that $\sigma_i' \in Tr(s_i) \oplus Tr(s_j)$. Let S have n states, a tuple of unique sequences $< \sigma_0, \sigma_1, \ldots, \sigma_{n-1} >$ is said be set of unique sequences for S. If there exists $\sigma \in \Sigma^*$ such that $\sigma_i \in Pref(\sigma)$, for $0 \leq i \leq n-1$, then σ is a distinguishing sequence. The notion of unique sequences, also called unique event sequences in [3], corresponds to that of FSM-based UIO sequences [12].

For the LTS in Figure 2, we may choose $< a, b.a, b.a, c >$ as its unique sequences. Note that unique sequences do not always exist. For example, if the transition $s_2 - c \rightarrow s_3$ in Figure 2 is deleted, then no unique sequence exists for s_3 in the resulting LTS.

Characterization Set
If a set of observable sequences, instead of a unique distinguishing sequence, is used to distinguish all the states of S, we have a so-called *characterization set* for S. A characterization set for S is a set $W \subseteq \Sigma^*$ such that for all $s_i, s_j \in S, i \neq j$, there exists $\sigma_i \in Pref(W)$ such that $\sigma_i \in Tr(s_i) \oplus Tr(s_j)$.

There exists a characterization set W for any S in the TOS form. For the LTS in Figure 2, we may choose $W = \{a, b.a\}$.

Partial Characterization Set
A tuple of sets of observable sequences $< W_0, W_1, \ldots, W_{n-1} >$ is said to be *partial characterization sets*, if, for all $s_i \in S, 0 \leq i \leq n-1$, and for all $s_j \in S, i \neq j$, there exists $\sigma_i \in Pref(W_i)$ such that $\sigma_i \in Tr(s_i) \oplus Tr(s_j)$. The notion of partial characterization sets correspond to the notion of partial UIO sequences in [5].

Obviously, since the given S is in the TOS form, in other words, none of its two states are trace-equivalent, there exist partial characterization sets for S. We also note that the union of all partial characterization sets for S is a characterization set for S. For the LTS in Figure 2, we may choose $< \{a\}, \{b.a\}, \{b.a\}, \{a, b\} >$ as its partial characterization sets.

Harmonized State Identifiers
A tuple of sets of observable sequences $< H_0, H_1, \ldots, H_{n-1} >$ is said to be a set of *harmonized state identifiers* for S, if it is a tuple of partial characterization sets for S and for $i, j = 0, 1, \ldots, n-1, i \neq j$, there exists $\sigma \in Pref(H_i) \cap Pref(H_j)$. H_i also is said to be a harmonized identifier for $s_i \in S$. The harmonized identifier for s_i captures the following property: for any different state s_j, there exists a sequence σ_i in $Pref(H_i)$ that distinguishes s_i from s_j and σ_i is also in $Pref(H_j)$.

Harmonized state identifiers always exist, just as partial characterization sets do. As an example, for the LTS in Figure 2, we can choose the harmonized state identifiers $H_0 = \{a, b\}, H_1 = \{b.a\}, H_2 = \{b.a\}, H_3 = \{a, b\}$.

5 STATE IDENTIFICATION IN IMPLEMENTATIONS

Similar to FSM-based testing, we assume that the given implementation is an LTS M whose set of all possible actions is limited to the set of actions Σ of the specification S (the correct interface assumption [1]). We also have a reliable reset, such that the state entered when this implementation is started or after the reset is applied is the initial state (the reliable reset assumption). In the case of nondeterminism, it makes no sense to identify single states of

M, so M is also assumed to be a TOS, in which each multi-state consist of a single state. For this reason, we require that S is in the TOS form, so that a state identification facility can be developed from S and also can be used to identify the states of M.

In order to identify the states of the implementation M, the number of states of M is also assumed to be bound by a known integer m. Therefore, M is also a mutant according to the fault model $\mathcal{F}(m)$.

Similar to FSM-based testing [7], there are also the two phases for LTS-based testing. In the first phase, the used state identification facility is applied to M to check if it can also properly identify the states in M. Once M passes the first phase, we can in the second phase test whether each transition and its tail state are correctly implemented. We present the structure of tests for the two phases using harmonized state identifiers as an example. In order to perform the first testing phase, proper transfer sequences are needed to bring M from the initial state to those particular states in M to which H_i should be applied. Moreover, it should be guaranteed that all the sequences in H_i are applied to the same particular state in M. Since a reliable reset is assumed, we can guarantee this in a way similar to FSM based testing: after a sequence in H_i is applied, the implementation M is reset to the initial state, and brought into the same particular state by the same transfer sequence,and then another sequence in H_i is applied. This process is repeated until all the sequences are applied.

Accordingly, let Q be a *state cover for* S, i.e. for each state s_i of S, there exists exactly one input sequence σ in Q such that $s_0 - \sigma \rightarrow s_i$, similar to FSM based testing, we can use $< N_0, N_1, \ldots N_{n-1} >$ to cover all states of M (a *state cover for* M), where

$$N_i \doteq \{\sigma \in Q@(\Sigma^0 \cup \Sigma_1 \cup \ldots \cup \Sigma^{m-n}) \mid s_0 = \sigma \Rightarrow s_i\}$$

and construct a set of test sequences to be executed by M from the initial state in the first testing phase as follows:

$$TS_1 = \bigcup_{i=0}^{n} N_i @ H_i$$

Inituitively, sequences of the sets N_i are used to reach n required states, as well as all possible $(m-n)$ additional states in M. Harmonized state identifiers H_i are applied to identify all states in M. In order to execute a given sequence $\sigma = a_1.a_2 \ldots a_k$ from the initial state m_0, we can convert σ into an LTS $t_0 - a_1 \rightarrow t_2 \ldots - a_k \rightarrow t_k$ and then compose this LTS with M in parallel composition $t_0 \parallel m_0$. Due to nondeterminism, it is possible that this run ends before the final action of this sequence is executed. Several runs are needed to exercise all the possible paths of M that can be traversed by this sequence (the complete testing assumption).

Using TS_1, we can make test cases for LTS S for the first testing phase by transforming the sequences in TS_1 into the corresponding LTSs as above and then labeling the LTSs according to Definition 6. In the following, this transforming and labeling process is always implied if we say that a test suite is obtained from a given set of test sequences.

After TS_1 is successfully executed, all the states of M which execute all traces of H_k are grouped in the same group $f(s_k)$, where $0 \leq k \leq n-1$.

In the second phase of testing, for testing a given defined transition $s_i\text{-}a\!\rightarrow s_j$ in S, it is necessary to first bring M into each state $m_k \in f(s_i)$, then apply a at this state to see if a can be executed; moreover, let M be in m_l after a is executed, it is necessary to check that $m_l \in f(s_j)$ which should be verified by H_j. (Note that due to nondeterminism, m_k may really be a multi-state, the action that is expected to check may not be executed in a time, so the above process should be tried several times.) On the other hand, we should further check if any undefined transition out of s_i has been implemented in M, i.e. for each $b \in \Sigma$, if $s_i \not\text{-}b\!\rightarrow$, then check that $m_k = b \Rightarrow$ does not exist. Because if $m_k\text{-}b\!\rightarrow$ exists, M is surely an invalid implementation, so it is not necessary to verify the tail state after b is executed.

Obviously, N_i may be used to bring M to any state $m_k \in f(s_i)$. Using this state cover, we can obtain a *valid transition cover* $< E_0, E_1, \ldots E_{n-1} >$, where

$$E_i = \{\sigma \in \bigcup_{k=0}^{n-1}(N_k@\Sigma) \mid s_0 = \sigma \Rightarrow s_i\}$$

which covers all transitions that should be present in any conforming implementation, and an *invalid transition cover* $\overline{E}$,

$$\overline{E} = \{\sigma.a \in \bigcup_{k=0}^{n-1}(N_k@\Sigma) \mid \exists s_i \in S\ (s_0 = \sigma \Rightarrow s_i \neq a \Rightarrow)\}$$

which covers all transitions that should be absent in any conforming implementation.

Next, H_i is used to verify the tail states of reached after each sequence in E_i. Excluding the transitions that have already been tested in the first testing phase, we can construct the set of test sequences for the second testing phase as follows:

$$TS_2 = \overline{E} \cup (\bigcup_{i=0}^{n-1}(E_i \backslash N_i)@H_i)$$

We conclude that the set of test sequences is expressed as follows, by combining the two sets of test sequences for the first and second testing phases:

$$TS \;=\; TS_1 \cup TS_2 = \overline{E} \cup (\bigcup_{i=0}^{n-1} E_i@H_i)$$

We have seen that the above process is an analogue of the checking experiments for the FSM model, except that invalid transitions need to be tested although their tail states need not to be verified. Similarly, it is expected that a test suite which is derived from S based on the above process is complete with respect to trace equivalence for $\mathcal{F}(m)$. In the next section, we present the test generation methods, based on the facilities presented in Section 4.2.

6 TEST GENERATION

6.1 Methods

Based on the above state identification techniques, we have a number of methods for constructing a set TS of test sequences for a given LTS specification S and with certain fault coverage for $\mathcal{F}(m)$. Let S be given in the TOS form with n states. We can obtain the state cover for implementation $< N_0, N_1, \ldots N_{n-1} >$, the valid transition cover for implementation $< E_0, E_1, \ldots E_{n-1} >$ and the invalid transition cover for implementation $\overline{E}$ as presented in the above section. Let $E = \bigcup_{i=0}^{n-1} E_i$ and $N = \bigcup_{i=0}^{n-1} N_i$.

The DS-method

Similar to the FSM-based DS-method [8], we use a distinguishing sequence σ for S to form a test suite for S, as follows.

$$TS \quad = \quad E@\{\sigma\} \cup \overline{E} \tag{1}$$

Theorem 1 *Given an LTS specification* S *in the TOS form and a distinguishing sequence* σ *for* S*, the test suite obtained from TS as given in (1) is m-complete for* S *w.r.t.* $\approx$.

Unlike the traditional FSM-based DS-method, the LTS-based DS-method does not construct a single test sequence since a reliable reset exists. It seems that, in case of deadlock, the reset is the only way to continue test execution.

The US-method

Let $< \sigma_0, \sigma_1, \ldots, \sigma_{n-1} >$ be a set of unique sequences for S, then a test suite for S, which is an analogue of that derived by the FSM-based UIO-method [12], can be formed as

$$TS \quad = \quad (\bigcup_{i=0}^{n-1} E_i@\{\sigma_i\}) \cup \overline{E} \tag{2}$$

As a specific case, unique sequences might be prefixes of the same (distinguishing) sequence. For the same reason explained in relation with the DS-method, the US-method does not combine unique sequences using the rural Chinese postman tour algorithm to obtain an optimal single test sequence.

Since unique sequences do not always exist, partial characterization sets can be used instead of unique sequences. This corresponds to the improvement on the UIO-method in [5]. Although partial characterization sets exist for any LTS in the TOS form, like the US-method, the improvement can not guarantee that a derived test suite is m-complete.

A similar method borrowing the notion of UIO sequences in the FSM model is proposed in [3], in which unique sequences are called unique event sequences. This method does not check invalid transitions, so it may not cover a fault where an undefined transition has been implemented.

The Uv-method

In order to obtain an m-complete test suite, the US-method can be improved such that

$$TS = N@(\bigcup_{i=0}^{n-1} \sigma_i) \cup (\bigcup_{i=0}^{n-1} (E_i \backslash N_i)@\{\sigma_i\}) \cup \overline{E} \quad (3)$$

Theorem 2 *Given an LTS specification* S *in the TOS form and a set of unique sequences* $< \sigma_0, \sigma_1, \ldots, \sigma_{n-1} >$ *for* S, *the test suite obtained from TS as given in (3) is m-complete for* S *w.r.t.* $\approx$.

The Uv-method usually drives a test suite of length larger than the US-method. However, unlike the US-method, it guarantees full fault coverage. The Uv-method corresponds to the FSM-based UIOv-method [18].

The W-method

Given a characterization set W for S, we form a test suite for S by the following formula. This is an LTS-analogue of the FSM-based W-method [4].

$$TS = E@W \cup \overline{E} \quad (4)$$

Theorem 3 *Given an LTS specification* S *in the TOS form and a characterization set W for* S, *the test suite obtained from TS as given in (4) is m-complete for* S *w.r.t.* $\approx$.

We note that in the case that $|W| = 1$, the W-method is the DS-method.

The Wp-method

Let W be a characterization set for S and $< W_0, W_1, \ldots, W_{n-1} >$ be partial characterization sets for S, similar to the FSM-based Wp-method [7], the Wp-method uses the following test sequences to form a test suite for S

$$TS = N@W \cup (\bigcup_{i=0}^{n-1} (E_i \backslash N_i)@W_i) \cup \overline{E} \quad (5)$$

Theorem 4 *Given an LTS specification* S *in the TOS form, a characterization set W and partial characterization sets* $< W_0, W_1, \ldots, W_{n-1} >$ *for* S, *the test suite obtained from TS as given in (5) is m-complete for* S *w.r.t.* $\approx$.

Obviously, the Wp-method derives usually a test suite of length smaller than the W-method because $W_i \subseteq W$. We note that the Uv-method is a specific case of the Wp-method, in which the union $\bigcup_{i=0}^{n-1} \sigma_i$ is a characterization set and $< \{\sigma_0\}, \{\sigma_1\}, \ldots, \{\sigma_{n-1}\} >$ are partial characterization sets.

The HSI-method
Let $< H_0, H_1, \ldots, H_{n-1} >$ be harmonized state identifiers for S, similar to the FSM-based HSI-method [10, 9], The HSI-method follows completely the approach presented in the above section to form a test suite for S.

$$TS = (\bigcup_{i=0}^{n-1} E_i@H_i) \cup \overline{E} \quad (6)$$

Theorem 5 *Given an LTS specification* S *in the TOS form and harmonized state identifiers* $< H_0, H_1, \ldots, H_{n-1} >$ *for* S, *the test suite obtained from TS as given in (6) is m-complete for* S *w.r.t.* $\approx$.

Since the union $\bigcup_{i=0}^{n-1} H_i$ is a characterization set, the HSI-method usually derives a test suite of length smaller than the W-method.

6.2 Examples

Assuming that the specification is given in Figure 2, with the HSI-method, we can derive a 4-complete test suite, which checks trace equivalence for this specification, as well as to the specification in Figure 1, as follows.

	s_0	s_1	s_2	s_3
State Identifiers H_i	$a,\ b$	$b.a$	$b.a$	$a,\ b$
State Cover Q	ε	a	c	$a.c$
Valid Transition Cover E_i	$\varepsilon,\ a.b$	a	c	$a.c,\ c.b,\ c.c$
Invalid Transition Cover $\overline{E} = \{b,\ a.a,\ c.a,\ a.c.a,\ a.c.b,\ a.c.c\}$				

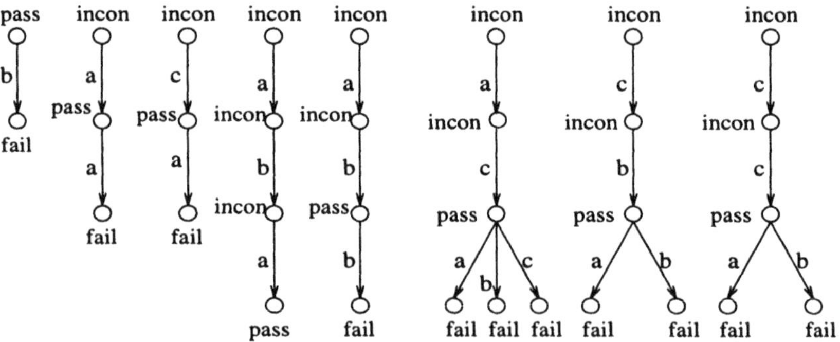

Figure 3 A complete test suite for the LTS specification in Figure 2.1.

$TS = \{b,\ a.a,\ c.a,\ a.b.b,\ a.b.a,\ a.c.a,\ a.c.b,\ a.c.c,\ c.b.a,\ c.b.b,\ c.c.a,\ c.c.b\}$. The corresponding test cases are shown in Figure 3.

7 CONCLUSION

In this paper, we have redefined, in the LTS model, the notions of state identification, which were originally defined in the formalism of input/output finite state machines (FSMs). Then we presented corresponding test derivation methods for specifications given in the LTS formalism that derive finite tests with fault coverage for trace equivalence. Note that the existing FSM-based methods are not directly applicable to LTSs, because LTSs assume rendezvous interactions making no distinction between inputs and outputs.

The notions of state identification in the LTS realm are distinguishing sequence, unique sequences, characterization set, partial characterization sets and harmonized state identifiers. The test generation methods based on these techniques are the DS-method, the US-method, the Uv-method, the W-method, the Wp-method and HSI-method. Among these methods, the DS-method, Uv-method, the W-method, the Wp-method and the HSI-method guarantee complete fault coverage.

8 REFERENCES

[1] G. v. Bochmann and A. Petrenko. Protocol testing: Review of methods and relevance for software testing. (1994) *ACM 1994 Intern. Symp. on Soft. Testing and Analysis*, 109–124.

[2] E. Brinksma. A theory for the derivation of tests. (1988) *PSTV VIII*, 63–74.

[3] A. R. Cavalli and S. U. Kim. Automated protocol conformance test generation based on formal methods for LOTOS specifications. (1992) *5th IWPTC*, 212–220.

[4] T. S. Chow. Testing software design modeled by finite-state machines. (1978) *IEEE Trans. on Soft. Engi.*, SE-4(3):178–187.

[5] W. Chun and P. D. Amer. Improvements on UIO sequence generation and partial UIO sequences. (1992) *PSTV XII*, 245–259.

[6] S. Fujiwara and G. v. Bochmann. Testing nonterministic finite state machine with fault coverage. (1991) *4th IWPTS*, 267–280.

[7] S. Fujiwara et al. Test selection based on finite state models. (1991) *IEEE Trans. on Soft. Engi.*, SE-17(6):591–603.

[8] F. C. Hennie. Fault-detecting experiments for sequential circuits. (1964) *5th Symp. on Switching Circuit Theory and Logical Design.*

[9] G. Luo, et al. Selecting test sequences for partially-specified nondeterministic finite machines. (1994) *7th IWPTS*, 91–106.

[10] A. Petrenko. Checking experiments with protocol machines. (1991) *4th IWPTS*, 83–94.

[11] D. H. Pitt and D. Freestone. The derivation of conformance tests from LOTOS specifications. (1990) *IEEE Trans. on Soft. Engi.*, SE-16(12):1337–1343.

[12] K. Sabnani and A. T. Dahbura. A protocol test generation procedure. (1988) *Comput. Net. and ISDN Syst.*, 15(4):285–297.

[13] Q. M. Tan, A. Petrenko, and G. v. Bochmann. Modeling basic LOTOS by FSMs for conformance testing. (1995) *PSTV XV*, 137–152.

[14] Q. M. Tan, et al. Testing trace equivalence for labeled transition systems. (1995) TR 976, Dept.of I.R.O., University of Montreal, canada.

[15] Q. M. Tan, A. Petrenko, and G. v. Bochmann. A framework for conformance testing of systems communicating through rendezvous. (1996) *IEEE 26th Intern. Symp. on Fault-Tolerant Computing*, 230–238.

[16] J. Tretmans. *A Formal Approach to Conformance Testing.* (1992) Ph.D. thesis, Hengelo, Netherlands.

[17] J. Tretmans. Testing labelled transition systems with inputs and outputs. (1995) *8th IWPTS*, 461–476.

[18] S. T. Vuong and et al. The UIOv-method for protocol test sequence generation. (1990) *2th IWPTS*, 203–225.

9 BIOGRAPHY

Qiang-Ming Tan received the B.S. degree and the M.S degree in computer science from Chongqing University, Chongqing, China, in 1982 and 1984, respectively. Since 1993, he has been with the Université de Montréal, Canada for the Ph.D. degree in conformance testing on communication protocols. From 1984 to 1992, he was a lecturer in the Department of Computer Science of Chongqing University. Now he is also working with Claremont Technology Inc. Canada.

Alexandre Petrenko received the Dipl. degree in electrical and computer engineering from Riga Polytechnic Institute and the Ph.D. in computer science from the Institute of Electronics and Computer Science, Riga, USSR. In 1996, he has joined CRIM, Centre de Recherche informatique de Montréal, Canada. He is also an adjunct professor of the Université de Montréal, where he was a visiting professor/researcher from 1992 to 1996. From 1982 to 1992, he was the head of a research department of the Institute of Electronics and Computer Science in Riga. From 1979 to 1982, he was with the Networking Task Force of the International Institute for Applied Systems Analysis (IIASA), Vienna, Austria. His current research interests include high-speed networks, communication software engineering, formal methods, conformance testing, and testability.

Gregor v. Bochmann (M'82-SM'85) received the Dipl. degree in physics from the University of Munich, Munich, West Germany, in 1968 and the Ph.D. degree from McGill University, Montréal, P.Q., Canada, in 1971. He has worked in the areas of programming languages, compiler design, communication protocols, and software engineering and has published many papers in these areas. He holds the Hewlett-Packard-NSERC-CITI chair of industrial research on communication protocols in Université de Montréal, Montréal. His present work is aimed at design methods for communication protocols and distributed systems. He has been actively involved in the standardization of formal description techniques for OSI. From 1977 to 1978 he was a Visiting Professor at the Ecole Polytechnique Fédérale, Lausanne, Switzerland. From 1979 to 1980 he was a Visiting Professor in the Computer Systems Laboratory, Stanford University, Stanford, CA. From 1986 to 1987 he was a Visiting Researcher at Siemens, Munich. He is presently one of the scientific directors of the Centre de Recherche Informatique de Montréal (CRIM).

12

An approach to dynamic protocol testing

Sangjo Yoo, Myungchul Kim, and Deukyoon Kang
Korea Telecom Research & Development Group
Sochogu Umyundong 17, Seoul, 137-792
{sjyoo, mckim, dykang}@sava.kotel.co.kr

Abstract

Protocol conformance testing aims at checking if a protocol implementation conforms to the standard (or specification) it is supposed to support. The results of testing can be classified into global verdict showing the tested system is either error-free or faulty, and local verdict indicating whether each element (e.g., a transition in the FSM) of the system is implemented correctly or not. In reality, the conventional protocol test procedure may give wrong local verdicts in the initial stages of testing because the procedure uses predetermined test sequence. In this paper, we propose a dynamic procedure for protocol testing using Test Sequence Tree (TST). The procedure allows us to get local verdicts more correctly than the conventional methods. The TST is reconfigured dynamically to obtain accurate verdicts for the untested elements by feedback of the local verdicts of the tested elements. The proposed technique was tested on the ITU-T Q.2931 signalling protocol. Our results showed that the fault coverage of our test procedure is better than the conventional methods. An extension of the proposed dynamic testing technique to the nondeterministic FSM is also discussed.

Keywords

Protocol testing, test sequence tree, test path selection

Testing of Communicating Systems, M. Kim, S. Kang & K. Hong (Eds)
Published by Chapman & Hall © 1997 IFIP

1 INTRODUCTION

Protocol conformance testing determines the conformance of a protocol implementation to its specification (or standard). Conformance testing of implementations with respect to their standards before their deployment to networks is important to vendors and network operators to promote interoperability and to ensure correct behaviour of the implementations.

There has been much work on automatic test sequence generation methods from Finite State Machine (FSM) (S. Naito, 1981)(G. Gonenc, 1970)(Krishan Sabnani, 1988)(Tsun S. Chow, 1978). Among them, Transition tour (T), Distinguishing Sequence (DS), Unique Input/Output (UIO), and characteristic set (W) methods are well known. DS, UIO, and W methods have better fault coverage than the T method (Deepinder P. Sidhu, 1989). The test sequences generated by DS, UIO, and W methods provide both global and local verdicts. Global verdicts indicate whether the tested system is error-free or faulty, whereas local verdicts show whether each element (e.g., a transition in FSM) of the system is implemented correctly or not. Local verdicts also provide information on error locations (indicating where the faulty transitions of the system are) and the degree of conformance that are not provided in global verdicts. However, the conventional protocol test procedure may give wrong local verdicts to the Implementation Under Test (IUT) having faulty elements because it uses fixed test sequences that are predetermined. The fixed test sequences are usually obtained by the conventional test sequence generation method (e.g., DS, UIO, or W methods).

In this paper, we propose a new dynamic procedure using the Test Sequence Tree (TST) for protocol testing which provides more accurate local verdicts than the conventional one. The TST is reconfigured dynamically during testing to get correct verdicts for untested elements by feedback of the local verdicts of the tested elements.

The rest of the paper is organized as follows; Section 2 surveys test sequence generation methods and test procedures, and points out the problem of the conventional test procedure. Section 3 describes the principles of dynamic protocol testing illustrated with a simple FSM. Our algorithm for dynamic protocol test procedure is proposed in Section 4. As a case study, the proposed model is applied to the IUT-T Q.2931 protocol (B-ISDN signalling procedure) in Section 5. In Section 6, extension of the test procedure to nondeterministic FSM is discussed. Finally, Section 7 concludes the paper.

2 PROBLEM STATEMENT

2.1 Overview of test sequence generation and test procedure

Protocol conformance testing includes a number of steps as shown in Figure 1, i.e., generating test sequence, applying the test sequence to the IUT, and analyzing the test results.

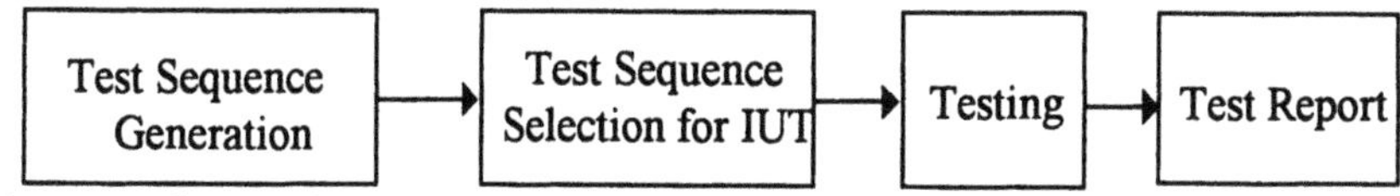

Figure 1 Conventional procedure for protocol conformance testing.

Test sequence generation methods based on FSM such as DS, UIO, and W methods are often used to produce the test sequences. In FSM, a transition consists of a head state, an input/output, and a tail state. The head state denotes the starting state of the transition and the tail state denotes the ending state of the transition. To test each transition of a machine, the following conventional procedure is usually applied to generate a sub test sequence for each transition:

1) Bring the IUT into a desired state starting from an initial state with the shortest path.
2) Apply the inputs to the IUT and observe the outputs from the IUT.
3) Verify that the IUT ends in the expected state.

Note that the well-known test sequence generation methods (e.g., DS, UIO, or W methods) are used in step 3). The test sequence for the entire machine is generated by concatenating the sub test sequences (Deepinder P. Sidhu, 1989) Myungchul Kim, 1995). In order to enhance efficiency, test sequence optimization using Chinese Postman algorithm (M. U. Uyar, 1986)(Alfred V. Aho, 1988), multiple UIOs (Shen Y. -N., F. Lombardi, 1989), and others techniques (Deepinder P. Sidhu, 1989)(Mon-Song Chen, 1990) have been proposed.

A paper (Samuel T. Chanson, 1992) which is closely related with our work makes use of the Abstract Test Case Relation Model (ATCRM) to derive test cases dynamically. In this paper, test purposes are obtained from the sequence of transitions from the initial state to all reachable states. There are some disadvantages to this approach:

1) Because the unit of test is a test purpose (i.e., a set of transitions) rather than focusing on one transition, it is harder to localize the errors.
2) The number of test purposes is large because the ATCRM covers all possible behaviors of the IUT and the test purposes consist of all possible paths from the initial state in ATCRM.

While Chanson and Li's work could be a solution to the problem stated, our method has the following advantages:

1) The number of test purposes is less with the same fault coverage
2) The ability of error localization is better because our model focuses on transitions not paths.

2.2 Problem of conventional protocol test procedure

The conventional protocol test procedure uses fixed test sequences. This gives rise to inefficiency. With reference to Figure 2-a, if transition A of IUT is implemented incorrectly, then the test sequence will give fail verdicts to all untested transitions following A (given by the bold arrows).

If the protocol has an alternative path consisting of transitions that are implemented correctly, then Figure 2-b shows that transitions B, C, D and E should not fail the test. Thus, the protocol test procedure using fixed test sequence may give wrong local verdicts to transitions that are implemented correctly because faulty transitions may affect subsequent transitions.

The conventional test procedure consists of deriving sub test sequences for each transition, optimizing the length of test sequence for machine, applying it to IUT, and then analyzing the test results. Table 1 shows the sub test sequences for each transition of the FSM in Figure 3 by getting the shortest path from the initial state to the head state of a transition to be tested, executing the transition, and verifying the tail state of the transition using the UIO method. The reset operation is added before each sub test sequence in order to ensure that sub test sequences always start at the initial state. Among the sub test sequences of Table 1, if the sub test sequence of transition i is included in the sub test sequence j, the testing for the transition i can be performed by the sub test sequence j as well. Thus the optimized test sequence can be derived as given in Table 2.

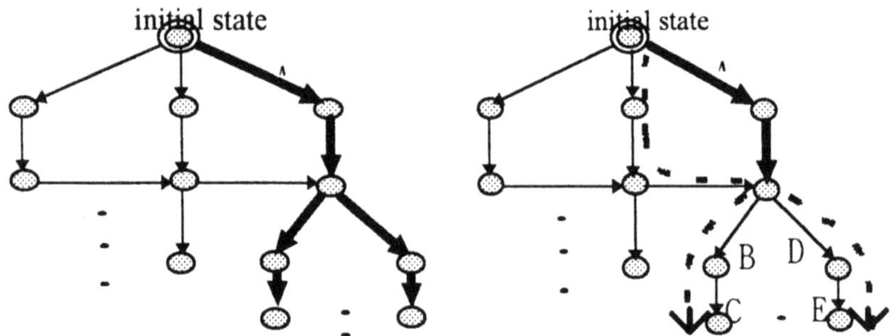

a) using shortest path from the initial state
b) using other correctly implemented path from the initial state
: state → : transition
➞ : transition assigned as "fail" because of the faulty transition A
- - › : path using correctly implemented transitions

Figure 2 Test result comparison by path selection.

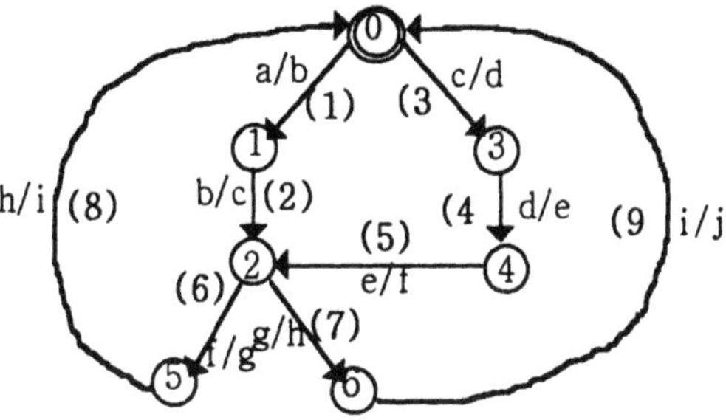

Figure 3 Finite State Machine M.

Table 1 Sub test sequences using UIO for the machine M in Figure 3

Transition	*Sub test sequence*	*Transition*	*Sub test sequence*
(1)	[reset/null, a/b, **b/c**]	(6)	[reset/null, a/b, b/c, f/g, **h/i**]
(2)	[reset/null, a/b, b/c, **f/g**]	(7)	[reset/null, a/b, b/c, g/h, **i/j**]
(3)	[reset/null, c/d, **d/e**]	(8)	[reset/null, a/b, b/c, f/g, h/i, **a/b**]
(4)	[reset/null, c/d, d/e, **e/f**]	(9)	[reset/null, a/b, b/c, g/h, i/j, **a/b**]
(5)	[reset/null, c/d, d/e, e/f, **g/h**]		

* The transition presented in **bold characters** is the UIO sequence for the transition.

Table 2 Optimized test sequence for the machine M in Figure 3

Test sequence for the machine M
[reset/null, a/b, b/c, f/g, h/i, a/b, reset/null, c/d, d/e, e/f, g/h, reset/null, a/b, b/c, g/h, i/j, a/b]

However, if the IUT has a faulty implementation of transition (1) (e.g., generating the output 'd' for the input 'a'), not only transition (1) is assigned a "fail" verdict, transitions (2), (6), (7), (8), and (9) will also be assigned the "fail" verdicts even though they are implemented correctly because transition (1) is part of the test sequence for testing those transitions. If transitions (3), (4), and (5) are implemented correctly, then we may adopt the path consisting of transitions (3), (4), and (5) as an alternative path for testing transitions (6), (7), (8), and (9). In this way, we can provide more accurate test results by isolating the effect of the faulty transition (1).

3 DYNAMIC SELECTION OF TEST PATH

We now propose a new test procedure for selecting an appropriate path dynamically from the initial state to the transition to be tested depending on the local errors in the IUT. The dynamic selection of test path makes it possible to get more accurate intermediate test results. Before proposing our dynamic path selection method, it is necessary to define some terms:

Definitions

- A **Set of Transitions (ST)** is the set of all transitions in a FSM M.

  $ST=\{t_1, t_2, \ldots, t_i, \ldots, t_n\}$ *where*

  t_i *=<a head state, an input/output, a tail state> and*

  n *= the total number of transitions of machine M.*

- A **Unique Path (UP_i)** is a path including transition t_i and a transition to verify the tail state of t_i, if there is only one possible path from the initial state to t_i,
- A **Set of Transitions in UP_i (STU_i)** is the set of all transitions in Up_i.

  $STU_i = \{t_1, \ldots, t_k\} \quad (0< k \leq n).$

- A **Path Test Sequence (PTS_i)** is the test sequence for transition t_i. PTS_i^q, the test sequence for t_i, is generated as follows: 1) apply the q-th path (if it exists) which brings the IUT from the initial state to the head state of t_i, 2) apply t_i, and 3) apply DS, UIO, or W methods to verify the tail state of t_i.

  $PTS_i^q = Path_i^q @\, t_i\, @Verification(\text{for the tail state of the } t_i)$

  where, @ *: concatenation of sequence and*

  $Path_i^q$ *= sequence of transitions of the q-th path from the initial state to* t_i.

- A **Test Sub-Sequence Tree ($TSST_i$)** is the set of all PTS_is for t_i.

  $TSST_i = \{PTS_i^1, \ldots, PTS_i^q, \ldots, PTS_i^j\}$

  where j *= the # of possible paths from the initial state to* t_i.

- A **Test Sequence Tree (TST)** for FSM M is

  $TST_M = \{TSST_1, \ldots, TSST_i, \ldots, TSST_n\}.$

Let us demonstrate how TST is set up initially and reconfigured dynamically during testing based on the results of local verdicts. For the Finite State Machine M in Figure 4, by using UIO sequence for tail state verification, the TST_M (Test Sequence Tree) for testing each transition in M is given in Figure 5.

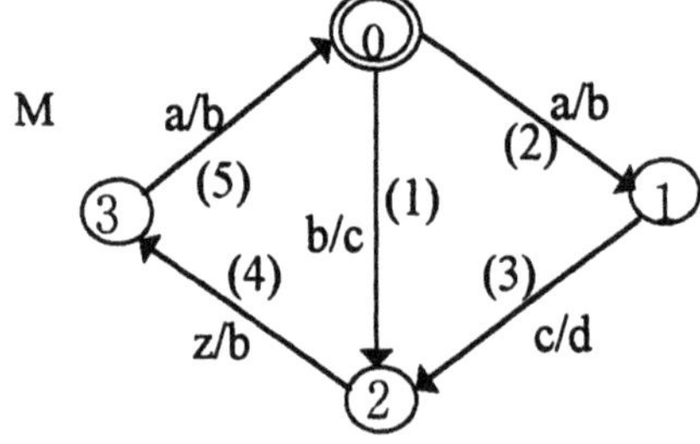

Figure 4 Finite state machine M.

Table 3 UIO sequences for machine

State	UIO sequence
0	[b/c]
1	[c/d]
2	[z/b]
3	[a/b, a/b]

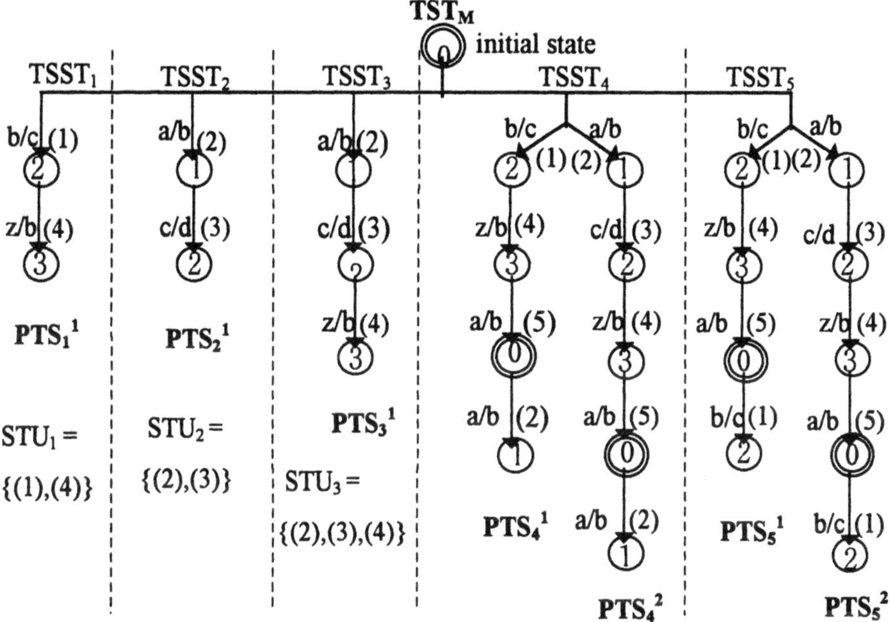

Figure 5 TST_M before testing.

For testing transition (1), there is only one path from the initial state to the transition. Therefore, $TSST_1$ for transition (1) is obtained by concatenating [b/c] in transition (1) and UIO sequence [z/b] for state 2 that is the tail state of transition (1).

For testing transition (4), there are two possible paths to bring the IUT to transition (4) from the initial state. The first one is [b/c] in transition (1) that is the shortest path from the initial state to transition (4), and the second one is [a/b, c/d] passing through transitions (2) and (3). (Notation '[a/b, c/d]' stands for concatenation of 'a/b' and 'c/d'.) Therefore, $TSST_4$ for testing transition (4) consists of PTS_4^1 = [b/c, z/b, a/b, a/b] and PTS_4^2 = [a/b, c/d, z/b, a/b, a/b]. As shown in Figure 5, note that $TSST_1$, $TSST_2$, and $TSST_3$ have STUs.

In using the TST in Figure 5, we start with testing transition (1) which is closest to the initial state. If a fault on the transition is detected; the test result of transition (1) is assigned a "fail" verdict; the transition is registered as a "faulty transition". To reconfigure TST_M as a result of the failure of transition (1), TST_M is searched to find a STU having transition (1) as its element. If a STU_i has transition (1) as an element of the set, the corresponding transition t_i is assigned a "fail" verdict automatically and does not have to be tested. In the case of TST_M in Figure 5, there is no transition having transition (1) as an element of its STU.

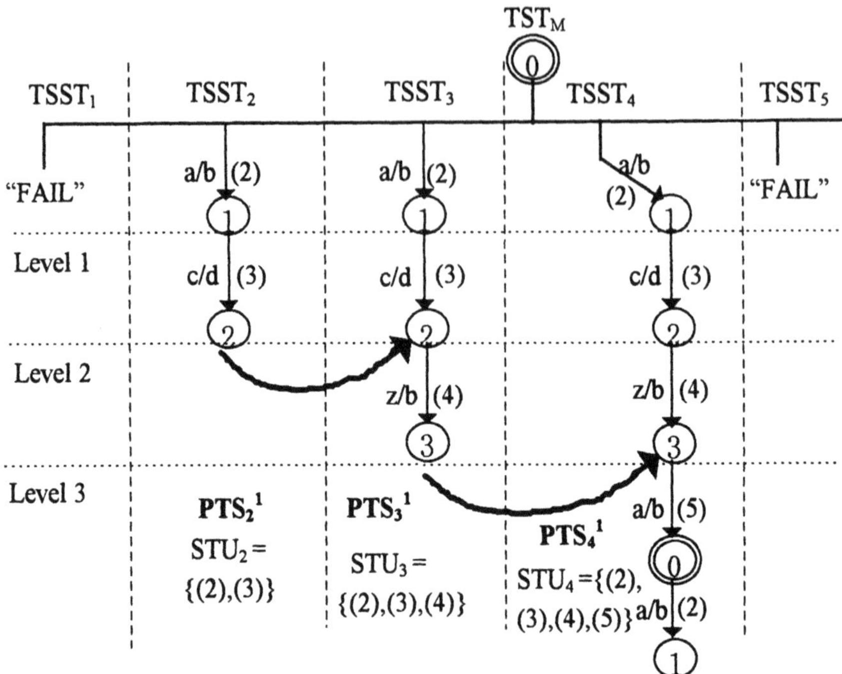

Figure 6 Reconfigured TST_M' right after detecting transition (1) is faulty.

The next step is to eliminate all PTSs which contain transition (1) from TST_M. In Figure 5, PTS_4^1 of $TSST_4$ for transition (4) and PTS_5^1 and PTS_5^2 of $TSST_5$ for transition (5) include transition (1). Therefore these three paths are eliminated from TST_M. As a result, we have only one PTS for transition (4) so that STU_4 must be created. In case of $TSST_5$ for transition (5), there is no remaining PTS for the transition so that the "fail" verdict is assigned automatically without testing. Figure 6 shows the reconfigured TST_M' after the testing of transition (1).

As shown in Figure 6, if the "pass" verdict is assigned to transition (2) as the result of observation [a/b, c/d], PTSs are searched to locate those having the same transition path (namely, [a/b, c/d]) up to level 2 from the initial state. As a result, PTS_3^1 of $TSST_3$ and PTS_4^1 of $TSST_4$ are selected. Of those, PTS_3^1 of $TSST_3$ having shorter length from the initial state is chosen for the next test. Since the transitions up to level 2 of PTS_3^1 are the same as those of PTS_2^1, only the test sequence corresponding to the transition [z/b] is applied for testing transition (3) of the IUT.

In the example given in Figure 4, if we do not use the proposed test procedure, transition (4) would be assigned a "fail" verdict because of the faulty implementation of transition (1) regardless of whether transition (4) is implemented correctly or not. The test procedure proposed can be depicted as in Figure 7.

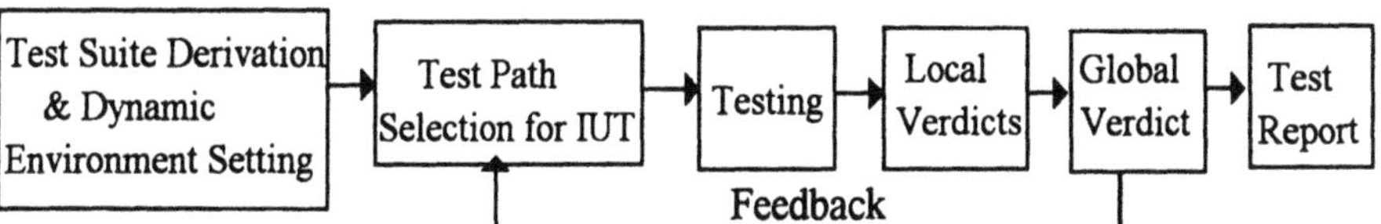

Figure 7 The proposed test procedure using dynamic selection of test path.

By using the result of every local verdict, TST is dynamically reconfigured to select an appropriate path test sequence.

The proposed dynamic test procedure has the following properties:

1) It uses an alternative path (if it exists) during testing when there is a problem (e.g., a faulty transition is detected) in the preamble path which brings the IUT to the transition to be tested.
2) If all possible paths from the initial state to a transition to be tested include faulty transitions, then the transition is automatically given a "fail" verdict without testing.
3) If a transition has passed the test, testing of further transitions starting from the current IUT state condition is performed without reset (i.e., restarting from the initial state) to minimize testing effort.

4 PROPOSED TEST PATH SELECTION ALGORITHM

In this section, we give an algorithm for the dynamic test path selection procedures proposed. It consists of two steps: initial construction of TST and Dynamic Test Path Selection.

(Step 1 : Initial Construction of TST before Testing)

begin
  construct TSST;
  setup TST;
  compute STU;
end

For each transition in ST={$t_1, \ldots, t_i, \ldots, t_n$}, $TSST_i$ is constructed using DS, UIO, or W method for tail state verification. For testing of t_i, all PTS_is are generated. The PTS_is are rearranged in increasing order of length from the initial state to t_i. As a result, we obtain $TSST_i = \{PTS_i^1, \ldots, PTS_i^q, \ldots, PTS_i^j\}$. Using TSST for each transition of FSM, the Test Sequence Tree (TST) is setup. Let the set of TSSTs be ordered according to the distance from the initial state. Therefore, we have $TST = \{TSST_1, \ldots, TSST_i, \ldots, TSST_n\}$. If there is only one path from the initial state to T_i, then compute STU_i for $TSST_i$.

(Step 2 : Testing and Dynamic Test Path Selection)

begin
  for t_1 **to** t_n **in** $t_i \in$ ST **do**
   begin

```
if "pass" or "fail" verdict is already assigned for t_i then
        continue;
    else
    begin
        for q:=1 to q:=j do   ——— (1)
        begin
            execute transitions in PTS_i^q ;
            if unexpected output is observed then
            begin
                if PTS_i^q is the last path of t_i then
                begin
                    assign "fail" verdict to t_i test;
                    if any TSST_k of TST has STU and t_i is an element
                        of STU_k then assign "fail" to t_k test of TST;
                    break;
                end
            end
            else
            begin
                assign "pass" verdict to t_i test;
                If PTS_k exists then——— (2)
                begin
                    jump into level p of PTS_k;
                    break;
                end
            end
        end
    end
end
end
```

For each transition in ST, we obtain a local verdict. In statement (1), j is the number of all possible paths from the initial state to t_i in the q-th PTS. As a result of executing transitions in PTS_i^q, if unexpected output is detected and PTS_i^q is the last path of t_i, then the verdict of t_i is "fail" and $TSST_k$ having t_i in its STU_k is also assigned "fail" without testing. If the output is correct, a "pass" verdict assigned to t_i. In statement (2), we try to find PTS_k with the shortest sequence matching the sequence from the initial state to level p that is the last position of the currently executed t_i. If the PTS_k exists, then jump to level p of PTS_k.

5 COMPARISON OF EXPERIMENTAL TEST RESULT FOR B-ISDN Q.2931 SIGNALLING PROTOCOL

In this section, we compare our new test procedure using dynamic test path selection method with the conventional one by applying both of them to real communication protocol testing. Figure 8 shows the simplified FSM for the call establishment and clearing procedure for the user side of ITU-T Q.2931 protocol. ITU-T Q.2931 is a recommendation for the User Network Interface (UNI) signalling protocol that is used in Asynchronous Transfer Mode network (ATM) and Broadband Integrated Digital Network (B-ISDN). The FSM of Figure 8 has 7 states and 15 input/output transitions.

Table 4 shows the UIO sequence for each state of the FSM in Figure 8. Table 5 and Table 6 list the test sequences for transitions using shortest path and Test Sequence Tree (TST) according to the proposed method, respectively. In the conventional test procedure, the test sequence for each transition presented in Table 5 is fixed by using only one path from the initial state and is not changed during testing. However in case of the proposed test procedure, multiple paths for transition (10), (11), (12), (13), (14), and (15) are allowed as shown in Table 6, and the path to be used is dynamically selected during testing.

To compare our test procedure with the conventional one for the FSM in Figure 8, a fault model is used. Generally, faults for FSM can be classified into three cases (Deepinder P. Sidhu and Ting-kau Leung, 1989) :

1) Produce an unexpected output for a given input and move to an expected state.
2) Produce an expected output for a given input and move to an unexpected state.
3) Produce an unexpected output for a given input and move to an unexpected state.

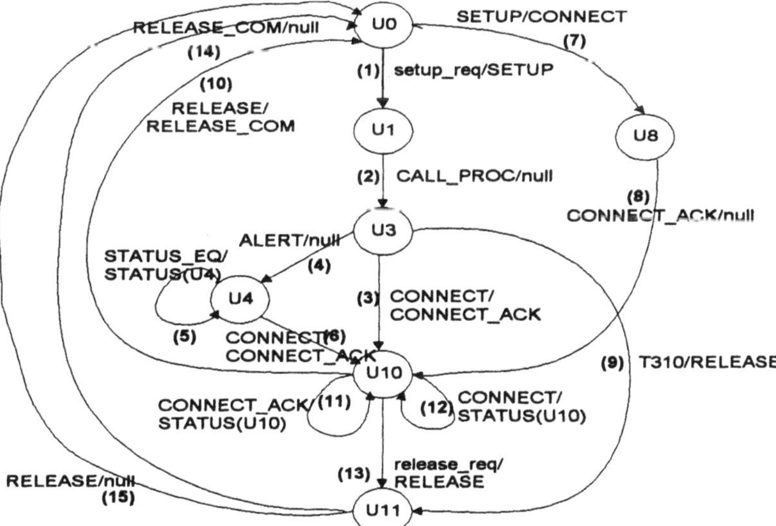

Figure 8 Simplified FSM of ITU-T Q.2931 signalling user side protocol for call establishment and clearing.

Table 4 UIO sequence for each state

State	*UIO sequence*	*State*	*UIO sequence*
U0	setup_req/SETUP	U8	CONNECT_ACK/null
U1	CALL_PROC/null	U10	release_req/RELEASE
U3	T310/RELEASE	U11	RELEASE_COM/null
U4	STATUS_EQ/STATUS(U4)		

Table 5 Test sequence for each transition using shortest path from the initial state

Transition	*Test sequence*	*Transition*	*Test sequencefor*
(1)	(1)-(2)	(9)	(1)-(2)-(9)-(14)
(2)	(1)-(2)-(9)	(10)	(7)-(8)-(10)-(1)
(3)	(1)-(2)-(3)-(13)	(11)	(7)-(8)-(11)-(13)
(4)	(1)-(2)-(4)-(5)	(12)	(7)-(8)-(12)-(13)
(5)	(1)-(2)-(4)-(5)-(5)	(13)	(7)-(8)-(13)-(14)
(6)	(1)-(2)-(4)-(6)-(13)	(14)	(7)-(8)-(13)-(14)-(1)
(7)	(7)-(8)	(15)	(7)-(8)-(13)-(15)-(1)
(8)	(7)-(8)-(13)		

Table 6 Test Sequence Tree (TST) structure using the proposed method

TSST	*PTS*	*Test sequence*	*TSST*	*PTS*	*Test sequence*
$TSST_1$	PTS_1^1	(1)-(2)	$TSST_{12}$	PTS_{12}^1	(7)-(8)-(12)-(13)
$TSST_2$	PTS_2^1	(1)-(2)-(9)		PTS_{12}^2	(1)-(2)-(3)-(12)-(13)
$TSST_3$	PTS_3^1	(1)-(2)-(3)-(13)		PTS_{12}^3	(1)-(2)-(4)-(6)-(12)-(13)
$TSST_4$	PTS_4^1	(1)-(2)-(4)-(5)	$TSST_{13}$	PTS_{13}^1	(7)-(8)-(13)-(14)
$TSST_5$	PTS_5^1	(1)-(2)-(4)-(5)-(5)		PTS_{13}^2	(1)-(2)-(3)-(13)-(14)
$TSST_6$	PTS_6^1	(1)-(2)-(4)-(6)-(13)		PTS_{13}^3	(1)-(2)-(4)-(6)-(13)-(14)
$TSST_7$	PTS_7^1	(7)-(8)	$TSST_{14}$	PTS_{14}^1	(7)-(8)-(13)-(14)-(1)
$TSST_8$	PTS_8^1	(7)-(8)-(13)		PTS_{14}^2	(1)-(2)-(9)-(14)-(1)
$TSST_9$	PTS_9^1	(1)-(2)-(9)-(14)		PTS_{14}^3	(1)-(2)-(3)-(13)-(14)-(1)
$TSST_{10}$	PTS_{10}^1	(7)-(8)-(10)-(1)		PTS_{14}^4	(1)-(2)-(4)-(6)-(13)-(14)-(1)
	PTS_{10}^2	(1)-(2)-(3)-(10)-(1)	$TSST_{15}$	PTS_{15}^1	(7)-(8)-(13)-(15)-(1)
	PTS_{10}^3	(1)-(2)-(4)-(6)-(10)-(1)		PTS_{15}^2	(1)-(2)-(9)-(15)-(1)
$TSST_{11}$	PTS_{11}^1	(7)-(8)-(11)-(13)		PTS_{15}^3	(1)-(2)-(3)-(13)-(15)-(1)
	PTS_{11}^2	(1)-(2)-(3)-(11)-(13)		PTS_{15}^4	(1)-(2)-(4)-(6)-(13)-(15)-(1)
	PTS_{11}^3	(1)-(2)-(4)-(6)-(11)-(13)			

For simplicity, we use only the fault model given by case 1) in this paper. For the FSM in Figure 8, we assume that all transitions of the FSM are possible faulty transitions. Also, for a given number of faulty transitions, we compute the possible faulty machines as shown in Table 7. For example, there are 105 different FSM implementations in case the FSM has two faulty transitions. Because there are 15 testing of transitions for each faulty FSM implementation, we have 1,575 test results in total. For the ideal tester that can identify perfectly all faulty and correct transitions, the number of "pass" transitions is 13 because there are two faulty transitions among the fifteen transitions. When we apply the test sequences of Table 5 generated by the conventional method, the average number of "pass" transitions is 8.219 for the implementation that has two faulty transitions since we have 863 "pass" results. On the other hand, our proposed test procedure using dynamic test path selection method get 9.514 average "pass" transitions because we obtain 999 "pass" results. Table 7 shows that the dynamic testing method produce more accurate test results for the individual transitions.

Table 7 Experimental test results using the fault model

The # of Faulty transitions	*Total # of different FSM implement-ations*	*The # of all possible test results*	***Conventional test method using fixed test sequence***		***Proposed test method using dynamic test path selection***		***Ideal tester***
			Total # of "pass" results	*The # of average "pass" transitions*	*Total # of "pass" results*	*The # of average "pass" transitions*	*The # of "pass" transitions*
1	15	225	168	11.2	182	12.133	14
2	105	1,575	863	8.219	999	9.514	13
3	455	6,825	2,692	5.916	3,171	6.969	12
4	1,365	20,475	5,690	4.17	7,111	5.21	11
5	3,003	45,045	8,610	2.867	10,834	3.608	10
6	5,005	75,075	9,606	1.919	11,923	2.382	9
7	6,435	96,525	8,016	1.2456	9,642	1.4984	8
8	6,435	96,525	5,019	0.78	5,778	0.898	7
9	5,005	75,075	2,340	0.4675	2,566	0.5127	6
10	3,003	45,045	795	0.2647	834	0.2777	5
11	1,365	20,475	188	0.1377	191	0.1399	4
12	455	6,825	28	0.0615	28	0.0615	3
13	105	1,575	2	0.019	2	0.019	2
14	15	225	0	0	0	0	1

As shown in Figure 9, the fault coverage of the proposed test method is closer to the ideal tester than that of the conventional method. This shows that the proposed test procedure using dynamic test path selection method can be used more efficiently and effectively in testing for product implementation or in acceptance testing for procurement. This is particularly useful when the proposed test procedure is used for debugging in the protocol implementation phase.

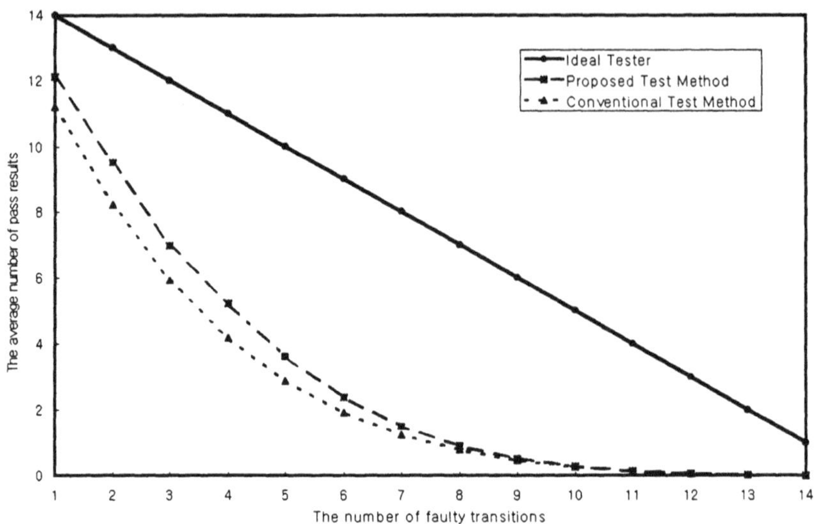

Figure 9 Test results comparison with ideal tester.

6 EXTENDING TO NONDETERMINISTIC FSM

In each state of the machine, if only one transition rule is executable, the machine moves to a new control state in a deterministic way. On the other hand, if more than one transition rules are executable for the same input, a nondeterministic choice is made to select a transition. The machine that can make such choices is called a nondeterministic machine. Most test sequence generation methods assume the FSM is deterministic. However many communication protocols exhibit nondeterminis. In this section, we present a method to extend our test procedure proposed in Section 3 to the observable nondeterministic FSM. The nondeterminism of ATM/B-ISDN may arise from the following reasons:

1) nondeterminism caused by options allowed in the specifications (e.g., in the ITU-T Q.2931 recommendation and ATM Forum UNI specification, sending a CALL-PROCEEDING message as the response of receiving a SETUP message is optional.).
2) nondeterminism caused by the messages which can be sent and received at any time for error report, check, or recovery (e.g., in the ITU-T Q.2931 recommendation and ATM Forum UNI specification, the STATUS-ENQUIRY message to ask for the state of peer entity can be sent in any state and at any time except the null state.).

For the nondeterministic FSM given in Figure 10, assume the next transition in state 5 is decided in a nondeterministic way. The output of either 'y' or 'z' can be observed as the response to the input 'x'. In this case, we say that transition (2) and transition (3) are in "companion relationship" in state 5 and state 5 is a "nondeterministic node". The TST of the FSM in Figure 10 is constructed in Figure 11. The transitions connected by the dashed lines are in companion relationship with each other. If the expected output 'y' for the input 'x' in transition (2) of PTS_m^1 for the *transition m* test is not observed, but the unexpected output 'z' in transition (3) in state 5 is observed instead, then, move to the next transition level (i.e., level j) of the path PTS_n^1 that uses the same path transition as PTS_m^1 up to level *i* and apply the remaining test sequence to the IUT. In case of testing *transition n*, if the output 'y' is observed instead of 'z' in transition (3) of PTS_n^1, move to level *j* of PTS_m^1 and continue the testing. By using the above procedure, nondeterministic FSMs can also be tested efficiently based on the proposed Test Sequence Tree and dynamic test path selection method.

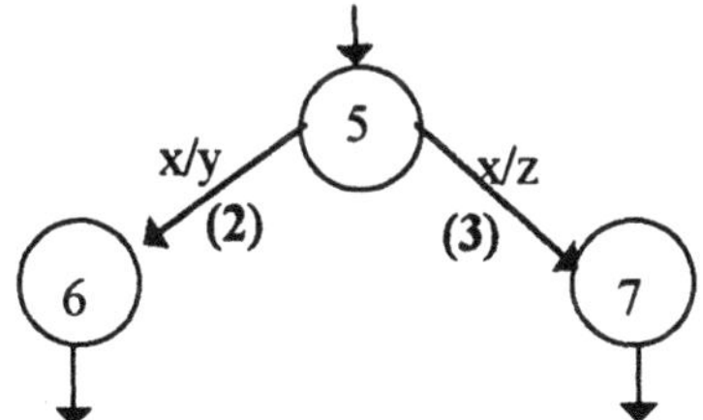

Figure 10 A part of nondeterministic FSM.

In addition, our approach avoids testing of duplicate paths at nondeterministic nodes as illustrated in the example above. During testing, if we get one of the outputs allowed against an input given at a nondeterministic node, the testing proceeds to the transition matching the output. The original transition is marked as "not tested yet". On the other hand, the transition in companion relationship with the original one is tested and it's verdict is given. This approach avoids duplicate testing on nondeterministic nodes, and thus provides more effective protocol testing to nondeterministic FSM.

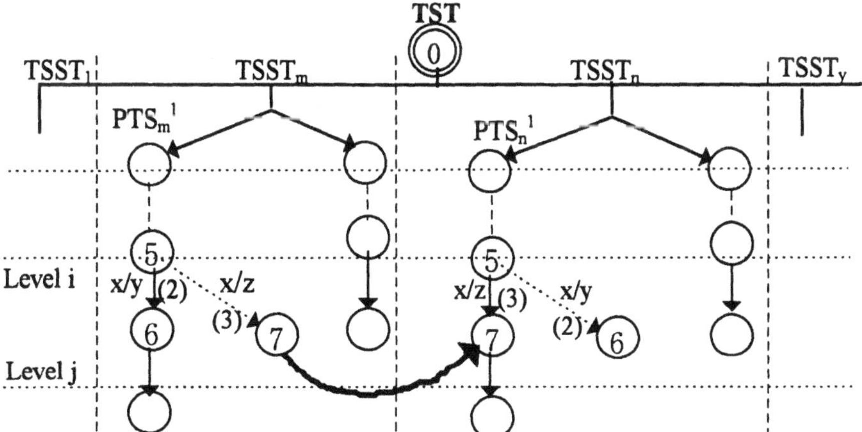

Figure 11 Dynamic test path selection method for the FSM of Figure 10.

7 CONCLUSION

In this paper, we have proposed a new dynamic protocol testing procedure that produces more correct local verdicts which helps to reduce the testing overhead. The Test Sequence Tree (TST) is the basic data structure used. The TST is reconfigured dynamically for the untested transitions during testing based on the results of local verdicts of already tested elements. We have applied our proposed dynamic test path selection algorithm to the FSM describing ATM/B-ISDN signalling protocol and compared it with the conventional method in terms of fault coverage. The results showed that the proposed test procedure generates more accurate verdicts and can be used more efficiently and effectively in testing. This method can be used for product implementation or in acceptance testing for procurement. Finally we have also presented some initial ideas in extending our proposed test procedure to deal with nondeterministic FSMs.

ACKNOWLEDGEMENTS
The authors would like to thank to Dr. Samuel T. Chanson and Dr. Sungwon Kang for their helpful comments on this paper.

8 REFERENCES

S. Naito and M. Tsuoyama (1981), "Fault detection for sequential machines by transition tours", IEEE 11th Fault Tolerant Comp. Symp., 238-43.

G. Gonenc (1970), "A method for the design of fault detection experiments", IEEE Trans. on Computer., vol C-19, pp. 551-558.

Krishan Sabnani and Anton Dahbura (1988), "A Protocol Test Generation Procedure", Computer Networks and ISDN Systems, vol.15, 285-97.

Tsun S. Chow (1978), "Testing Software Design Modeled by Finite-State Machines", IEEE Trans. on Software Eng., vol.SE-4, no.3, 178-87.

Deepinder P. Sidhu and Ting-kau Leung (1989), "Formal Methods for Protocol Testing: A Detailed Study", IEEE Trans. on Software Eng., vol.15, no.4, 413-26.

Myungchul Kim, S.T. Chanson, and Sangjo Yoo (1995), “Design for testability for protocols based on formal specifications”, The 8th Int’l Workshop on Protocol Test Systems, 257-69.

M. U. Uyar and A. T. Dahbura (1986), “Optimal Test Sequence Generation for Protocols: The Chinese Postman Algorithm Applied to Q.931”, IEEE Global Telecom. Conf.

Alfred V. Aho, Anton T. Dahbura, David Lee, and M. Umit Uyar (1988), “An Optimization Technique for Protocol Conformance Test Generation Based on UIO Sequences and Rural Chinese Postman Tours”, Protocol Specification, Verification, and Testing, VIII, 75-86.

Shen, Y.-N., F. Lombardi, and A. T. Dahura (1989), “Protocol Conformance

Testing Using Multiple UIO Sequences", Protocol Specification, Verification, and Testing, IX, 131-43.
Mon-Song Chen, Yanghee Choi, and Aaron Kershenbaum (1990), "Approaches Utilizing Segment Overlap to Minimize Test Sequences",Protocol Specification,Verification,and Testing, IX,85-98.
S. T. Chanson and Qin Li (1992), "On Static and Dynamic Test Case Selections in Protocol Conformance Testing", The 5th Int'l Workshop on Protocol Test Systems, 225-67

9 BIOGRAPHY

Sangjo Yoo received BA in electric communication engineering from Hanyang Univ. in 1988 and MS in electrical engineering from the Korea Advanced Institute of Science and Technology in 1990. Currently he is with the Korea Telecom R&D Group as a member of technical staff.
Myungchul Kim received BA in electronics engineering from Ajou Univ. in 1982, MS in computer science from the Korea Advanced Institute of Science and Technology in 1984, and Ph.D in computer science from the Univ. of British Columbia in 1992. Currently he is with the Korea Telecom Research and Development Group as a managing director, Chairman of Profile Test Specification-Special Interest Group of Asia-Oceania Workshop, and is the Co-Chair of the 10th IWTCS'97. His research interests include protocol engineering on telecommunications and multimedia.
Deukyoon Kang received BA in electronics engineering from Kumoh Nat'l Institute of Technology in 1993 and MS in computer science from Pohang Institute of Science and Technology in 1995. Currently he is with the Korea Telecom R&D Group as a member of technical staff.

13

Sensitivity analysis of the metric based test selection

Jadranka A. Curgus
Boeing ISS Research & Technology
Bellevue, WA, USA

Son T. Vuong, Jinsong Zhu
Department of Computer Science
University of British Columbia
Vancouver, B.C., Canada V6T 1Z4
Email: {vuong, jzhu}@cs.ubc.ca

Abstract

In our earlier work [7, 3], we introduced a novel concept of *test distance* and an effective multi-pass metric-based method of test selection for communication protocols. The main weight of the method rests on the concept of *test distance*, which forms the basis of a metric definition to guide the convergent test selection process. In this paper, we discuss a sensitivity analysis of this metric based test selection method. We present empirical results regarding the sensitivity of common metric definitions to various aspects of protocols such as recursion levels, multiple concurrent connections, transition or event patterns.

Keywords

Protocol testing, test selection, test coverage

Testing of Communicating Systems, M. Kim, S. Kang & K. Hong (Eds)

1 INTRODUCTION

Formal work on test coverage metrics for protocols had been long overdue when our metric based test selection method [7, 3] was first introduced; it provides a new, analytical way of assessing the quality of a test suite in terms of its coverage of the specification. Contrary to fault targeting models [1], where detection of a predefined fault classes (also called fault model [1]) is the test purpose, this metric based method seeks to achieve trace equivalence of the test suite with the specification, and therefore measures the coverage by examining how "closely" the test suite covers the specification.

The metric definition and the test selection method are interesting in that the former can be shown to lead to a compact metric space and the latter is tantamount to a convergent test selection process where the more test sequences are selected the closer the selected set tends to the orginal set, i.e. there are no relevant, peculiar test cases or groups of test cases that may be missed out in the selection process due to mere overlook, as are usually the cases in heuristic test selection. Furthermore, the metric defined is made general and flexible by a number of parameters which can be tuned according to the expert knowledge of the specific protocols and potential faults.

The definition of the metric, in terms of test distances, does not have to be related to fault detection capabilities of the test suite, since as long as the specification can be eventually covered, all faults that can occur should be discoverable during testing. However, the definition will certainly affect the effectiveness of the convergence process, since a "bad" definition of a test distance may make the process so slow that a large amount of test cases are needed to attain a desired coverage. The question so induced would be: how can one be sure that a metric is effective?

We looked at this problem, and believe that an effective metric should be able to capture important factors in a protocol specification, such as recursion levels, multiple concurrent connections, parallelism, and transition (or event) patterns, since they constitute major characteristics of a protocol. In other words, if a metric can incorporate these properties effectively, we can expect that it will effectively cover the specification with a reasonable amount of test cases. The difficulty, however, lies in analytically determining the effectiveness of a metric in handling them, since those properties may present themselves radically different in various protocols, and different metrics may be needed in different situations. We therefore resort to experimental methods that show the sensivity of common metrics to the properties.

In order to produce the results closer to real situations, we decide to use, in our experiment, a reasonably complex, real-life protocol, the Internet Stream Protocol [4] *, and a number of common metrics (to be defined later). Although the results may not be extrapolated to all other protocols and/or

*A recent revision of this protocol (ST2+, RFC 1819) is available within the internet community.

metrics, we believe it can give us important initial results that help to design further experiments in assessing more metrics and protocols. With more results and additional "representative" real-life protocols, it will definitely help understand protocol properties and how a metric can capture them effectively.

The rest of the paper is organized as follows. We first give a brief overview of the metric based method, followed by a description of our experiment settings in Section 3. Section 4 presents our serial experiments on the sensivity of a metric to various properties. We conclude by discussing the observations and further research work.

2 OVERVIEW OF THE METRIC BASED METHOD

As already stated, the purpose of the metric based method is to generate test cases that cover the specification. A specification can be considered as describing the behaviour space of a protocol, where *execution sequences* constitute the control part. The whole space can be infinite, either an execution sequence can be infinite, or there are infinite number of execution sequences. Therefore, in order to cover the space within the computer system and time resources limit, approximations have to be made.

The metric based method solves this problem by defining a metric space over the behaviour space made of execution sequences. A set of finite execution sequences (a *test suite*) as approximations of infinite sequences, can be selected based on the metric. Furthermore, a finite number of test suites, which approximates the infinite behaviour space, can be generated based on the restriction of test cost. The important property of this approximation process is that the series of test suites can converge to the original specification in the limit. Thus, we have a way to achieve coverage of the specification with an arbitrary degree of precision limited only by the test cost.

The metric is built on the concept of test distance between two execution sequences. The distance satisfies the requirement that the resulting space be a metric space, so that we have the nice property of finite covers of an infinite space [3]. It should also grasp the intuitive requirement of testing relationships between execution sequences, so that a concept of "closeness" of sequences can be understood. This closeness actually represents a notion of testing representativeness: the closer the two sequences, the more likely they'll yield the same test result.

Formally, we define a test distance as [3]:

Definition. Let s, t be two (finite or infinite) execution sequences in S and let $s = \{(a_k, \alpha_k)\}_{k=1}^{K}$, and $t = \{(b_k, \beta_k)\}_{k=1}^{L}$, $K, L \in \mathbf{N} \cup \{\infty\}$. The testing distance between two execution sequences s and t is defined as

$$dt(s,t) = \sum_{k=1}^{\max\{K,L\}} p_k \delta_k(s,t)$$

where

$$\delta_k(s,t) = \begin{cases} |r_{\alpha_k} - r_{\beta_k}| & \text{if } a_k = b_k \\ 1 & \text{if } a_k \neq b_k \end{cases}$$

If s and t are of different lengths then the shorter sequence is padded to match the length of the longer sequence, so that $\delta_k = 1$ for all k in the padded tail.

Note that in the above definition, p and r are functions satisfying the following properties:

P1 $\{p_k\}_{k=1}^{\infty}$ is a sequence of positive numbers such that $\sum_{k=1}^{\infty} p_k = p < \infty$.
P2 $\{r_k\}_{k=0}^{\infty}$ is an increasing sequence in $[0, 1]$ such that $\lim_{k\to\infty} r_k = 1$. Put $r_\infty = 1$.

This definition guarantees that the space (S, dt) is a metric space, and more importantly it is totally bounded and complete [3]. It ensures the existence of finite covers for infinite metric space (S, dt), which is the theoretical foundation for the approximation process and also the test selection algorithm described below.

The selection algorithm for the metric based method (MB selection method for short), is a multi-pass process in which each pass is realized by the selection function SELECT(T_0, G, ϵ, C), which returns a selected set T, being an ϵ-dense subset of the original set G of test cases generated by some test generator, such that the cost of T (which includes the initially selected set T_0) is less than some given threshold cost C.

The cost function of a test case can be defined to represent the resources (time and space) required to execute that test case, e.g. its length. The cost of a set of test cases can be defined simply as the sum of the cost of the individual test cases in the set.

Metric-based test selection algorithm

Step 1. Initially, the selected set T is empty, G is the given (generated) set of test cases, ϵ is the initial target distance, and C is the given cost threshold for the selected set.

Step 2. While Cost$(T) < C$ do T = SELECT$(T, G, \epsilon = \epsilon/k, C)$ for some scaling factor $k > 1$, applied at each iteration (that is, each pass).

Step 3. Stop. No further pass is possible because any other test case added to the set T of test cases selected so far violates the cost constraint.

The function SELECT(T, G, ϵ, C) (that is, each pass in Step 2 of the above general algorithm) can be specified as follows:

Step 2.1 Let $X = G \backslash T$, i.e. G excluding T.

Step 2.2 If X is empty return T and exit, else remove (randomly) some test case t from X.

Step 2.3 If $dt(t, T) < \epsilon$ goto Step 2.2.

Step 2.4 If $\text{Cost}(T \cup \{t\}) < C$ then add t to the selected set T.

Step 2.5 Goto Step 2.2.

As can be seen, the target distance ϵ decreases with each pass in the algorithm. The multipass algorithm generates Cauchy sequences, each being formed by test cases selected over successive passes. The Cauchy sequences converge to limits that are themselves in the original set of test cases, and any infinite test case in the original set can be approximated by some converging sequence of selected test cases. Thus, the algorithm performs test selection as a convergent approximation process, which ensures that no test cases or regions of test cases are overlooked.

3 EXPERIMENT SETTINGS

Having described the methodology of metric based test selection, we now look at the experiments we are going to do. The above algorithm selects test cases repeatedly until a certain test cost is reached. We can thus observe how test cases are selected with a certain test distance definition.

To get meaningful results, typical protocols and metric definitions should be used. However, there is generally no consensus on which protocols would be "typical". We therefore decide to use the Internet Stream Protocol, a protocol that is both practical (i.e., used in real applications) and interesting (i.e., relatively complex). We consider it as "typical" in the sense that it possesses interesting protocol properties as appeared in most real-life protocols. The results should therefore at least shed some light on similar protocols.

The Internet Stream Protocol has been used for years as the primary protocol in a voice and video conferencing system operating over the wideband and terrestrial wideband networks. This version of the specification was issued in 1990, and is known as ST-II [4] *. The ST Control Message Protocol (SCMP) is mainly used in our experiments.

The basic functionality of SCMP is to build routing trees between a sender, called *origin*, and one or more receiving applications, called *targets*. One such routing tree will typically include one or more additional nodes along each of its paths, called *intermediate agents*. ST (the data phase) then simply transmits data down the routing tree, from an origin to one or more targets. The data flow over the routing tree is referred to as a *stream*. Each of the communicating entities within a routing tree runs an ST-II protocol and is

*Resource Reservation Protocols (of which ST-II is one representative) are an important class of protocols that is currently receiving undivided attention of the internet community, and other communications forums concerned with real-time streaming (multimedia) protocols.

called an *ST-II agent*. One ST-II agent can simultaneously act as one or more origins, routers, and targets.

From the testing point of view, the ST-II protocol makes it interesting for the following reasons:

1. There are two different sources for the concurrency in ST-II: a) multiple protocol connections, and b) simultaneous execution of origin, target, and intermediate agent functional entities within one ST-II agent.
2. With up to 17 different types of messages for SCMP part of ST-II, ST-II becomes one of the more complex protocols, especially if the combined state space composed of one origin, one target, and one intermediate agent module is considered.
3. The protocol is non-symmetrial since individual connections are trees. The usual concept of a "peer" partner in communication, as a convenient symmetrical counterpart in testing, has to be abandoned.

We carried out the experimentation under the assumption that an ST-II protocol implementation is tested as an entity, composed of independent parallel executions of (several) origin, intermediate or target agents. Because of the non-symmetricity of the communication pattern, the temporal order of protocol primitives arriving from IUT as a response to different upstream or downstream agents (even in one-stream environments) is unpredictable and should be abstracted. Therefore, the test architecture applicable in this setting is the interoperability test architecture, where a set of lower PCO-s must be observable.

3.1 Test development system

The test system that we use to conduct the experiments is the TESTGEN+ system [8]. A functional diagram of the system is shown in Figure 1.

The TESTGEN Module

TESTGEN [8] is the test generator module in an experimental protocol TEST Generation ENvironment for conformance testing developed at the University of British Columbia. The generated test suites incorporate both control flow testing and data flow testing with parameter variation. Both types of testing are controlled by a set of user-defined constraints which allows a user to focus on the protocol as a whole or just on restricted areas of the protocol.

TESTGEN begins with a protocol description in an extended transition system formalism (a subset of Estelle [5]) and ASN.1 description of the input service primitives, output service primitives, and protocol data units. Once all constraints are defined, TESTGEN identifies subtours within the specified

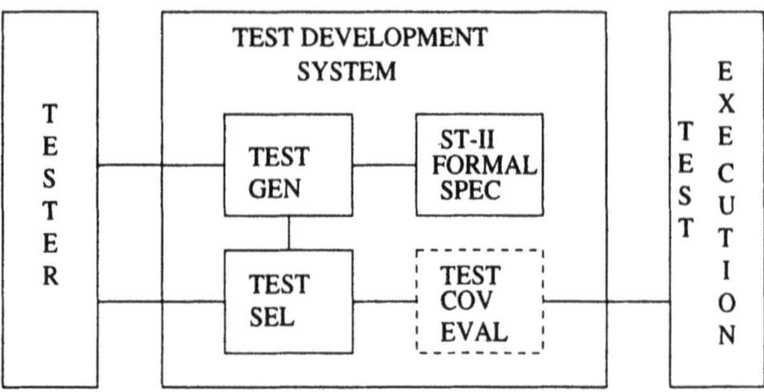

Figure 1 The Functional Structure of the Test Development System.

protocol which satisfy the minimum and maximum usage constraints. Each subtour then undergoes the parameter variation defined by the types of service primitives in the subtour and the parameter variation constraints.

The TESTSEL Module

The TESTSEL module (which optionally includes an independent test coverage evaluator TESTCOV EVAL module), is an implementation of test case selection (as presented in in Section 2) and test case coverage evaluation based on the distance and coverage metrics within the metric-based method. Since the test coverage metric we are using is guaranteed to be convergent when Cauchy sequences are followed, and the algorithm produces such sequences, a set of test cases which converges to the initial test suite will be iteratively yielded.

The two modules, TESTGEN and TESTSEL have been integrated into a test development environment similar to the one already used for the development of the InRes protocol test sequences [6]. The seed files for test suites used in experimentation are generated by TESTGEN. These can be fed into the TESTSEL module directly, or first passed through an independent interleaving algorithm. This algorithm produces random independent interleavings and random recursive levels of a specified number of streams and target agents, in order to obtain test suites exercising multiple simultaneous connections, concurrency of different agents, and higher levels of recursion. The output of the TESTSEL module and the random interleaving algorithm are then compared and analysed.

4 EXPERIMENT RESULTS

We first developed an Estelle specification of the ST-II origin, intermediate, and target agents (refer to Appendix E of [2] for details), and an ASN.1 description of the data part. The specifications are then fed to our TESTGEN system to generate the sets of subtours. These sets are named *originseeds*, *intermediateseeds*, and *targetseeds*, corresponding to the origin, intermediate, and target agents. The purpose of these sets is to serve as the starting sets for the random interleaving and recursion generation process that we implemented specifically for the use in the experiments. They should preferably provide a thorough coverage of the relevant tours through each individual agent's state space in their simplest form (i.e., without concurrency or recursion).

The random independent interleaving module generates test suites with concurrency of: a) multiple streams (simultaneous connections) within the same ST-II agent; b) multiple agent functional entities within the same ST-II agent; and c) multiple targets within the same stream. Concurrency due to the interleaving of multiple intermediate agents within the same stream has been omitted. In the experiments, we examine the effects of the recursion and concurrency of the ST Control Message Protocol only, since the data phase of this protocol is trivial.

4.1 Stability and granularity of the test selection process

The next two experiments investigate the granularity and stability of the test selection process guided by a typical metric function.

The metric function p_k takes values $p_k = 2^{10-k}$, for $k = 1, 2, \ldots$. The function r_k takes values $r_k = 1 - \frac{1}{2^k}$ for all $k = 1, 2, \ldots$. 20 sets of 100 test cases were generated, representing random interleavings of 1 through 10 simultaneous network connections, and 10 through 80 ST-II target agents. Test sets are labelled ts$c.t$, indicating the number of simultaneous connections c and targets t that were allowed in their generation.

Experiment 1.1

The first series of 10 test sets contained test cases with individual events labelled by event names only. (This would be suitable for testing event variations at global levels, e.g. does an implementation correctly react to an event e after some event d.) Figure 2 shows the number of test cases selected versus refinements in the test density. x axis represents the density range of 0 through 1024.00, the diameter of the metric space generated by this metric. For this metric, results for ts10.35 and ts10.50 coincide. Also, the line styles of ts10.30 and ts10.35 are reused for ts1.80 and ts2.70 - the lower lines belong to ts1.80 and ts2.70.

Experiment 1.2

The second series of 10 test sets contained test cases with individual events labelled by event names, transition tour they belong to and the stream identifier. (This is a much finer identification which would allow detailed testing of event variations within individual connections and with respect to event parameter data variations). Figure 3 shows the granularity of the selection process for this case.

In both cases, i.e. given both very fine (many distinct event labels), and much coarser (relatively few distinct event labels) event space, there is a steady increase in the number of test cases selected as the density approaches 0. Figures at the low end of density spectrum indicate, that even at very fine granularity levels of density, this metric function is still capable of distinguishing still finer test suites. More test cases are selected for test suites in the second series of tests, for the same levels of density, number of simultaneous connections and recursion, since more distinct event labels can be identified. In both graphs, ts2.60 and ts2.70 occupied the middle portion of the graph, with the suites ts1.t (ts5.t and ts10.t) almost entirely below (above, resp). Given same density levels and event labelling, fewer test cases are almost always selected for test suites involving fewer simultaneous connections, indicating a poorer event pattern. At the same time, given the same number of simultaneous connections, the effect of higher vertical recursion (more targets) on the number of test cases selected is not very noticeable, given the low values of the r_k series for moderate values of k in this example.

The stability and granularity of the MB selection which are observed in these two experiments, are confirmed throughout the test selection and coverage analysis experiments that follow.

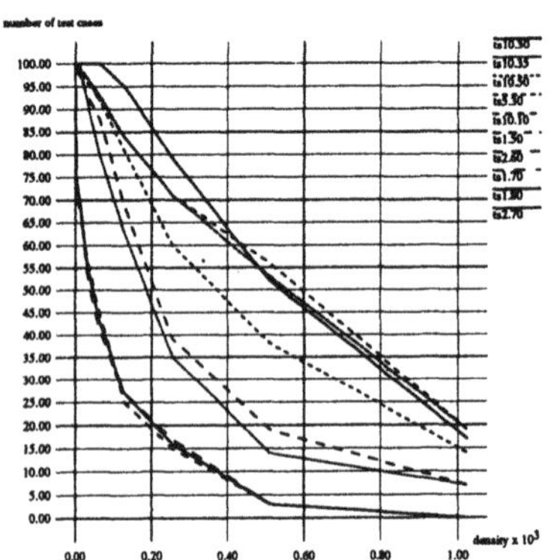

Figure 2 Granularity of MB test selection with event name identification.

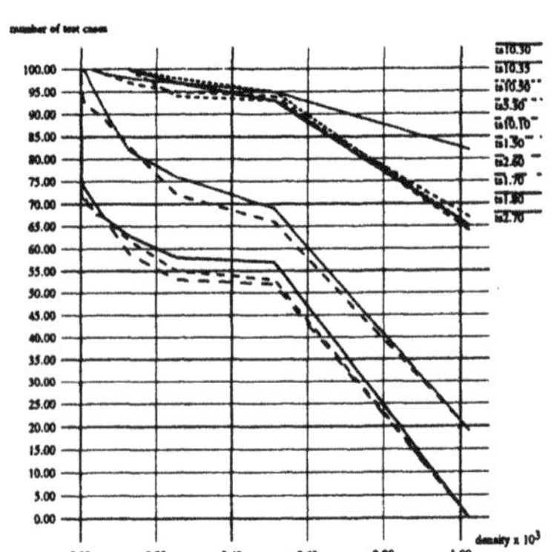

Figure 3 Granularity of MB test selection with event name, gate and data parameters identification.

4.2 Identifying transition tours

Experiment 2

This experiment investigates the efficiency of applying metric based method for transition tour selection, for single connection environments, with multiple agent functionality and with moderate vertical recursion. We allow three different Finite State Machines to represent individual connections.

We use the seed files, i.e., originseeds, intermediate seeds, and targetseeds, for each of the three agents of the ST-II protocol machine. We hypothesize that they identify all transitions of each agent. We did not have a tool to generate or verify the transition tours, consequently the value of the test is in the efficiency estimate rather than the exactness of the transition tour identification.

The metric used is $p_k = 2^{9-k}$. Also, the recursion was given a minimal weight through the r function as in experiments 1.1 and 1.2.

Let N_T be the minimum level of the reachability tree corresponding to the specification by which all transitions have been traversed by a subtour. We observe that the largest N_T is for the intermediate agent, where $N_T = 9$, for ACK packets to DISCONNECT or REFUSE PDU-s. We generated 8 test suites, containing randomly selected one-connection environments with the upper bound on vertical recursion (number of targets) equal to 30. All transition tours from seed files were added to these sets in order to simulate test suites which have transition coverage. Generated test suites had 177, 277, 377, 577, 677, 777, 877 and 977 test cases, and are designated ts.n, where n is the number of test cases they contained. Figure 4 (enlarged portion in Figure 5) shows the number of test cases selected by the progressive passes of the test selection algorithm, for each of the test sets, as the density covers the range from 511 to 0.998047. At 0.998047 the algorithm is guaranteed to have identified transition cover of the three agents, when they work in isolation, provided the experiment's hypothesis is satisfied.

The greater variety in event patterns and recursion is expected, as the randomly generated test sets grow larger. This accounts for more test cases selected, at equal density levels, for larger test sets.

The results show that, even with larger test suites, the algorithm is not overly wasteful, when used for identifying test suites with fault coverage at least equal to that of the transition tour method. Since the metric used in the experiment yields a complete metric space, the experiment simulates a test selection environment where, first, a test suite of transition tours equivalent coverage is selected, after which the process further proceeds by identifying still finer subsets of the initial set, with respect to variations in patterns of recursion and event sequencing. Completeness guarantees complete coverage of the initial set in the limit.

4.3 Sensitivity to vertical recursion

Experiment 3

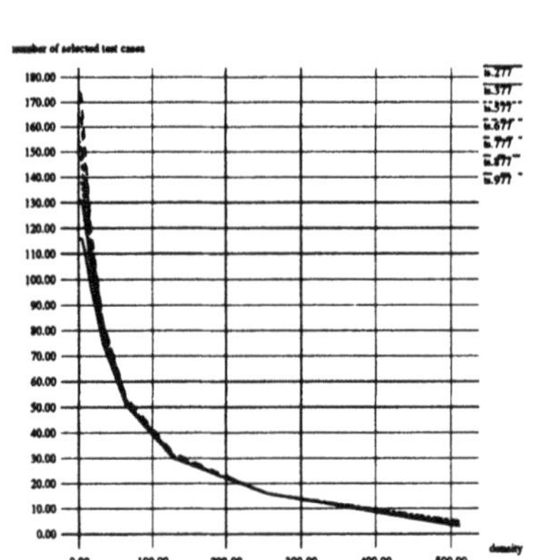

Figure 4 Selecting Transition Tours with moderate vertical recursion.

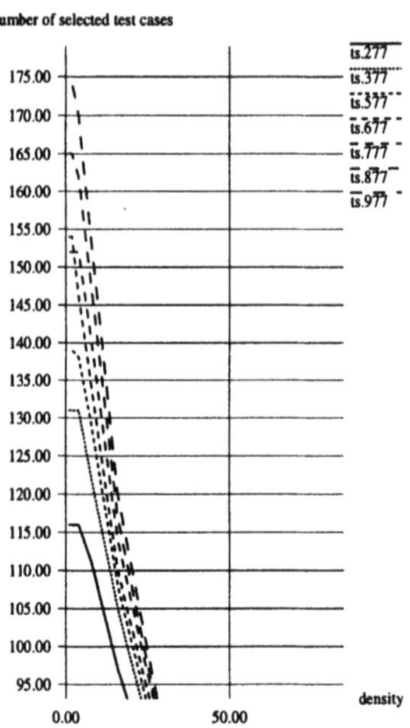

Figure 5 Selecting Transition Tours with moderate vertical recursion.

In this experiment we investigate the effect of vertical recursion on the mutual density of test sets. The metric function used is the same as in the previous experiment. We use test suites of 100 test cases each, randomly generated, with low numbers of simultaneous connections (1 - 3), and moderate vertical recursion (10 - 30) targets. The experiment shows that sets generated with a certain number of simultaneous connections and a certain event pattern are less dense in the sets with the same characteristics, if they contain less recursion than those sets.

Figure 6 shows this effect of vertical recursion on mutual coverage of two series of 6 test suites each. Sets $r_1, \ldots, r_6$ are randomly generated sets of test cases, and $s_1, \ldots, s_6$ are their corresponding sets (same number of simultaneous connections and event patterns), but with a certain reduction in recursion when calculated with respect to the corresponding set (i.e. the same subscript) from the r series. The test suite generation specifics and the amount of reduction in recursion (in percentages) are shown in Table 1. No special effort was taken to fairly space the recursion levels in either r or s series of test sets.

The effect is plotted by representing the mutual density for each pair of sets (r_k, s_k), $k = 1, \ldots, 6$ running along the x-axis. Connecting the points of the scatter graph allows for easier viewing. The density points of sets $s_1, \ldots, s_6$

Table 1 Experiment 3 - Characteristics of test sets.

set	connections	targets	rec. reduction for sets s_k
r1	1	10	30
r2	2	10	50
r3	3	30	100
r4	1	30	25
r5	2	30	100
r6	3	20	70

in sets $r_1, \ldots, r_6$ are connected by the r/s line, and the density points of sets $r_1, \ldots, r_6$ in sets $s_1, \ldots, s_6$ are connected by the s/r line.

We generally found that with random generation, sets of 100 test cases were sufficient, under given protocol characteristics, for the mutual coverage figures to show sensitivity towards vertical recursion at reduction levels in the range of 25-100 percent. This is due to the fact that, given a sufficient number of test cases in a test suite and a test generation algorithm that does not necessarily cluster recursive events in few clusters with poor distribution, lower bounds of recursion are more likely to generate less variety in event cluster recursion. Consequently, it will be easier for the set which has higher bounds on recursion to approximate event patterns and recursive levels of poorly recursive sets than vice versa, under such conditions.

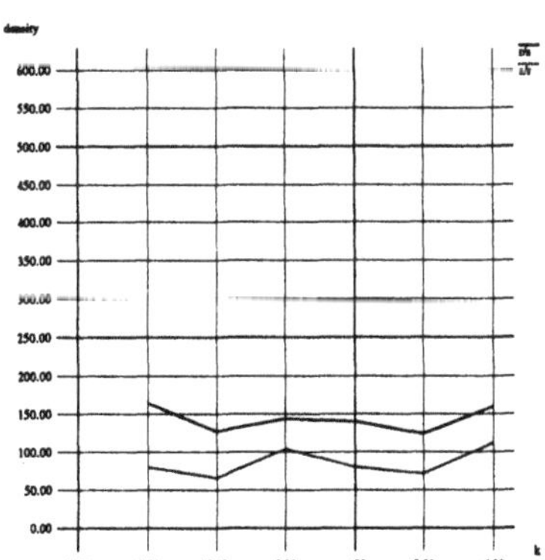

Figure 6 Sensitivity to vertical recursion.

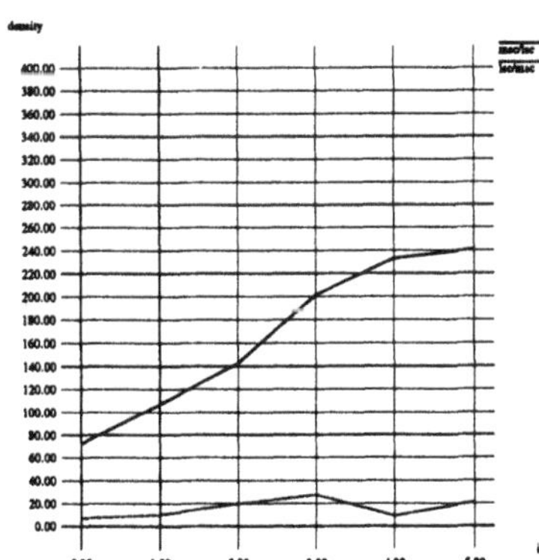

Figure 7 Mutual densities of some sets with moderate and low number of simultaneous connections.

In a related experiment (not shown here), pairs of randomly generated sets, with moderate numbers of simultaneous connections, one with low and one with high recursion, were compared for mutual coverage. Under the same metric function as in this experiment, the results were inconclusive. We identified the cause by examining sequences which contributed to high maximum distances. We concluded that a new event pattern in a test sequence, likely to appear early in a sequence with many simultaneous connections, could easily offset the effect of better recursion distribution under this metric.

Joint conclusion from both experiments is that the quality of test suite expressed through metric based coverage functions is sensitive towards vertical recursion, however a metric with adequate sensitivity to recursion should be used if it is to be a decisive factor of the quality of test suite.

4.4 Sensitivity to parallelism

Experiment 4

In this experiment we investigate the impact of testing with low and moderate network connection parallelism on the mutual coverage of test suites that differ with respect to the level of parallelism. The experiments were conducted under quite general circumstances. All test sets, 6 with low and 6 with moderate parallelism of network connections, were randomly generated, containing 100 test cases each. Therefore a certain potential existed for inconclusive results due to choosing a particularly rare combination of events in either case. However, the mutual coverage results were consistent in 3 different metric spaces and throughout 36 different comparisons. We therefore concluded that the results were a good indication of the MB coverage function to identify suites with greater simultaneity of network connections, in a general case. The results are reported in Figures 7, 8, 9.

The characteristics of individual test suites, with respect to the number of simultaneous connections and the number of targets are given in Table 2. Suites involved moderate vertical recursion (10-30 targets), and were put into "low simultaneous connection" (lsc*id*) set if involving 1 to 3 simultaneous connections. Likewise, "moderate simultaneous connection" sets (msc*id*) involved 10 - 35 simultaneous connections. Suites with equal *id*-s were compared.

The comparisons were carried out in metric spaces generated by metrics with

1. $p_k = 1$, for $k < 400$, and $p_k = \frac{1}{2^{k-400}}$, for $k \geq 400$.
2. $p_k = 1$, for $k < 25$, and $p_k = \frac{1}{2^{k-25}}$, for $k \geq 25$.
3. $p_k = 1$, for $k < 15$, and $p_k = \frac{1}{2^{k-15}}$, for $k \geq 15$.

and the series r_k were defined as in Experiment 1.2.

The plots for each of these metric spaces are in Figures 7, 8, 9, respectively. X-axis is labelled with set *id*-s, from 0 through 6. The plots show the density of sets $lsc0, \ldots, lsc5$, in sets $msc0, \ldots, msc5$, resp. These scatter points are

Table 2 Experiment 6 - Characteristics of randomly generated test sets.

test set id	connections	targets	cases generated
msc0	10	30	100
lsc0	1	30	100
msc1	15	30	100
lsc1	1	30	100
msc2	20	30	100
lsc2	2	30	100
msc3	25	10	100
lsc3	3	30	100
msc4	30	10	100
lsc4	1	10	100
msc5	35	10	100
lsc5	2	10	100

connected by a line labelled msc/lsc. Similarly, the line labelled lsc/msc connects points that show densities of sets $msc0, \ldots, msc5$ in sets $lsc0, \ldots, lsc5$, resp.

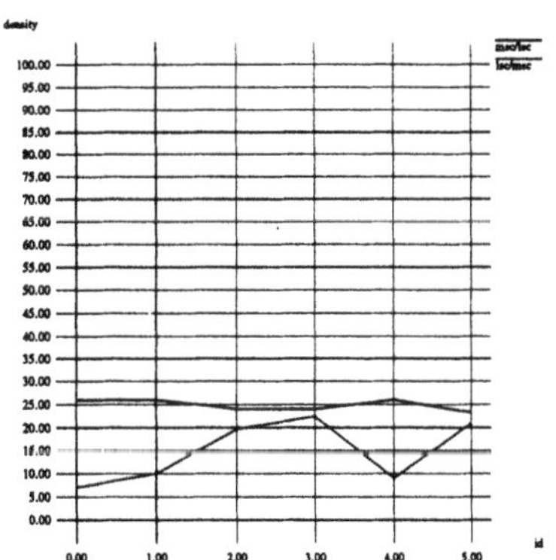

Figure 8 Mutual densities of some sets with moderate and low number of simultaneous connections.

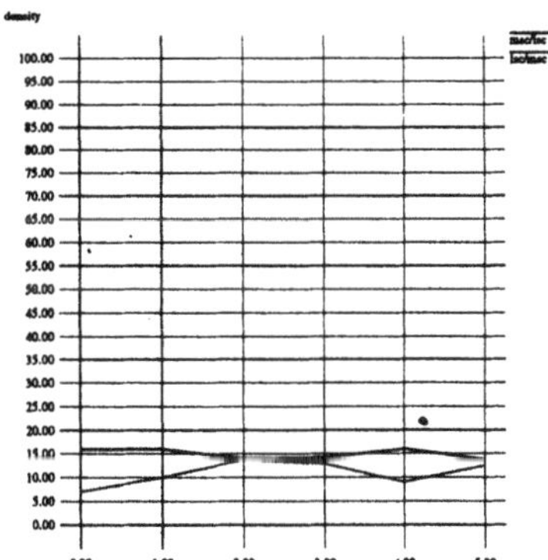

Figure 9 Mutual densities of some sets with moderate and low number of simultaneous connections.

Although the test sets with moderate parallelism are consistently more dense in low-parallelism sets in all three metric spaces considered, significant difference exists in the magnitude of the mutual density variation. Metric 1 evaluates all patterns of very long sequences with equal value, up to a large $k = 400$. Therefore, the density variation in the space generated by this metric can be attributed to the much larger length of sequences involving many connections. This, however, is a definite indication of the number of connections exercised, given fixed vertical recursion, but not necessarily of the simultaneity of such connections. Therefore, the next two metric spaces favoured the average and upper end case length of sets ls*cid*, which was found to be in the range of 11 - 36. (Lengths of ms*cid* test cases were in the range of 60 - 145.) Consequently, they really evaluated the greater variation in pattern of test sequences generated by many simultaneous connections, especially in view of the fact that all sequences longer than 15 (25) would contribute at most 1 to the final distance. This is an extreme case which definitely did not give *a priori* advantage to longer test sequences of sets ms*cid*. It is therefore conceivable that a more moderate metric would yield results consistent with this experiment's results.

4.5 Sensitivity to a missing event or event combination

The purpose of the next two experiments is to investigate the sensitivity of mutual densities of two test suites, when one of the test suites does not contain an event or event combination, or does so only to a certain lesser extent. All test sets contain 100 test cases. Metric is the same as in Experiment 1.1, unless otherwise noted.

Experiment 5.1

First, the sensitivity analysis was carried out for the environments involving no parallelism and moderate vertical recursion. In this experiment, 20 randomly generated sets of 100 test cases each were compared for mutual coverage. Test sets s_1 through s_{10} are random interleavings of one connection, up to 30 ST-II targets. Test sets o_1 through o_{10} are random interleavings of one connection, up to 30 ST-II targets, where ST-II agent origin does not include sequences with a PDU HAO (HID APPROVE) as a response to a CN (CONNECT) PDU. Figure 10 shows the density deviation for the densities of sets o_k in randomly generated test sets s_k, connected by the line s/o. For comparison, also shown are plots of all other density figures available in the experiments: a) densities of sets s_k in sets o_k, connected by a line o/s, b) densities of sets s_k in sets s_{k+1}, connected by a line s/s, and c) densities an oracle set originoracle in sets o_k, line o/originoracle.

Experiment 5.2

In this experiment we investigated environments with moderate parallelism and moderate vertical recursion. This experiment shows the effects of a missing event (SP or PDU) in a more complex test suite. 3 groups of test sets, 100 random test cases each, were generated to randomly contain between 5 to 10 simultaneous connections, and 10 to 30 targets. The set s_1 through s_6

was generated taking into account all available SPs and PDUs for all ST-II agents. Sets r_1 through r_6 were generated with DSI (DISCONNECT PDU from the direction of origin) missing from intermediate seed sets. Likewise, sets t_1 through t_6 were generated with HCT PDU missing from target agent seed sets.

Figure 11 shows that applying general metric to mutual coverage of such sets yields inconclusive results. Lines s/r, r/s, s/t and t/s connect scatter points of densities of sets r, s, t, and s in sets s, r, s and t respectively, where densities are calculated only between sets with equal subscripts.

We did observe, however, that at these levels of parallelism and recursion, the mutual coverage of randomly generated test sets with all events included into seed files, consistently showed better density figures.

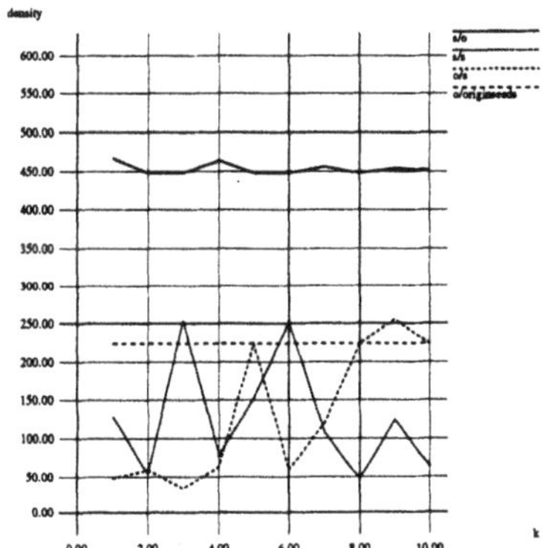

Figure 10 Sensitivity to missing events - single connection with recursion.

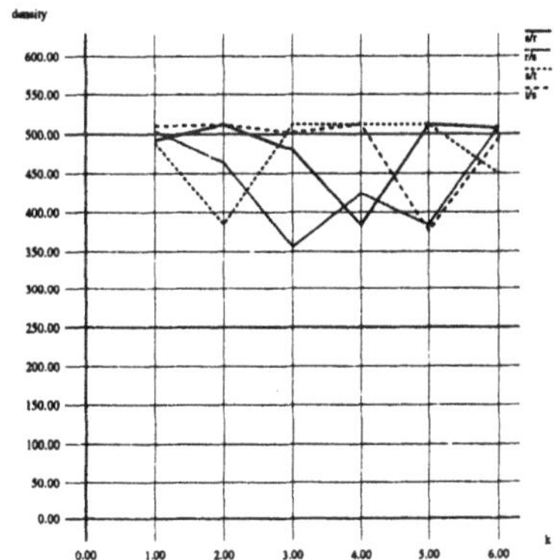

Figure 11 Sensitivity to missing events - general metric function, multiple connections with recursion.

In a related experiment, we improved sets t from the previous experiment, by adding 10 percent and 70-80 percent of sequences, involving the HCT PDU. Figure 12 shows the range of densities of sets $t_1, \ldots, t_6$ (all HCT event sequences deleted), $t_{1_{aehct}}, \ldots, t_{6_{aehct}}$ with almost all HCT events missing (10 percent sequences with it), and $s_1, \ldots, s_6$ with 70-80 percent sequences with HCTs (which is the typical content in a randomly generated test set of these characteristics), in randomly generated test sets $rs_1, \ldots, rs_6$, applying a general metric function (from Experiment 1.1).

Although the results explain the densities in Figure 11 to a certain extent, it still would be necessary to have a knowledge of the expected densities at particular levels of vertical recursion and parallelism, in order to use general metric functions for missing event identification.

In the following experiments, a special metric function was designed to identify the coverage of a particular event (HCT PDU in this case). Its definition is the same as in Experiment 1.1, except for the cases were neither of the events at a position k in two sequences is HCT: if this is the case, the summand $p_k \delta_k$ for this position k is zero. This metric function measures the coverage of the event HCT both with respect to its position in sequences and the level of recursion.

Figure 13 shows the application of the metric function to specially identify the coverage of HCT event, and its improvement with improving t-test sets with sequences containing event HCT, up to its average 70-80 percent representativeness as observed in randomly generated test sets. These are calculated as densities of t sets in randomly generated s sets.

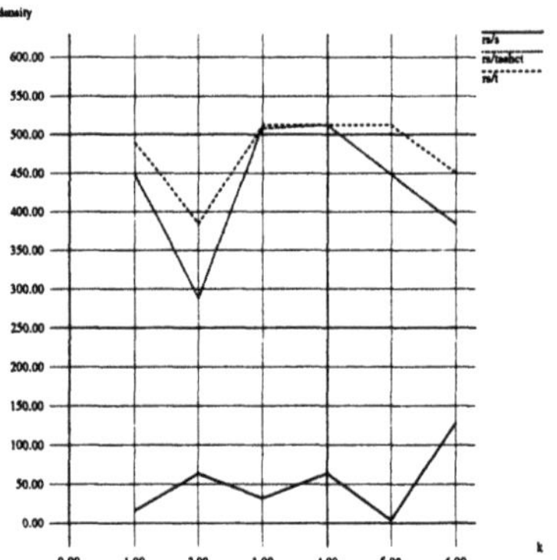

Figure 12 Sensitivity to missing events - general metric function, moderate parallelism and recursion.

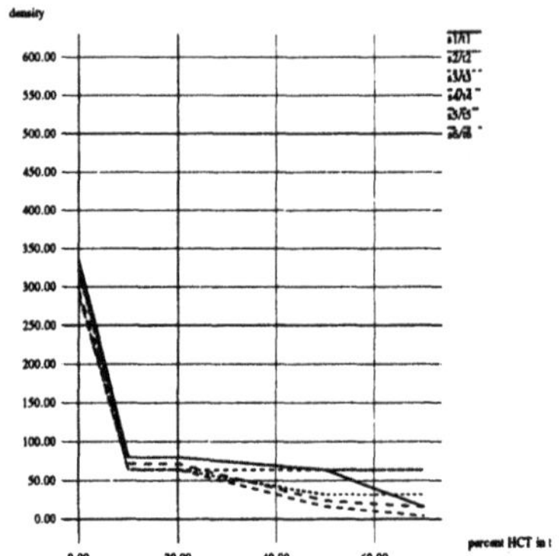

Figure 13 Sensitivity to missing events - special metric, moderate parallelism and recursion.

Figure 14 shows the effect of improving sets t_1 to t_6 with 10, 20,50 and 70-80 percent of sequences which contain HCT in case of general metric. These are calculated as densities of t sets in randomly generated s sets. These do not improve as fast as in the previous metric.

All subsequent examples use the special metric designed to identify missing event content. The following figures show the sensitivity of this metric in various situations.

Figure 15 shows the gain in the quality of t by increasing its HCT content, by plotting the density of randomly generated test suites, in sets tid, whose special event content (HCT content) is improved by adding 10 - 70 percent test sequences involving the event considered.

The next 2 figures show how the plots for 2 combinations of randomly generated test sets s and sets t from this same example, meet in the range of 16-64, as the coverage of the event HCT increases. (These are excerpts for fixed combinations of test sets from Figures 6.17 and 6.20.)

Figure 18 shows mutual coverages of the randomly generated sets and set

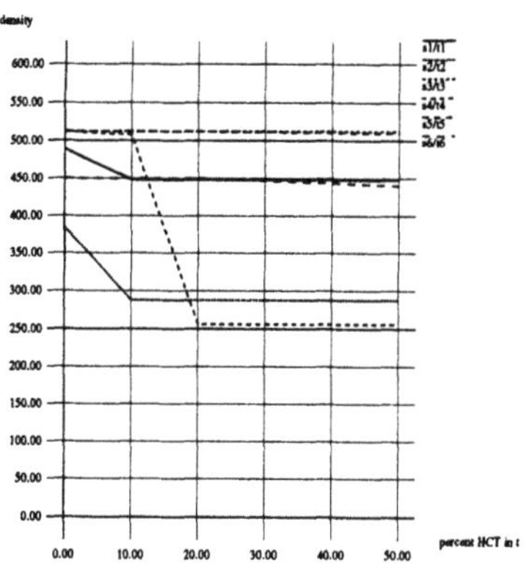

Figure 14 Sensitivity to missing events - general metric function, moderate parallelism and recursion.

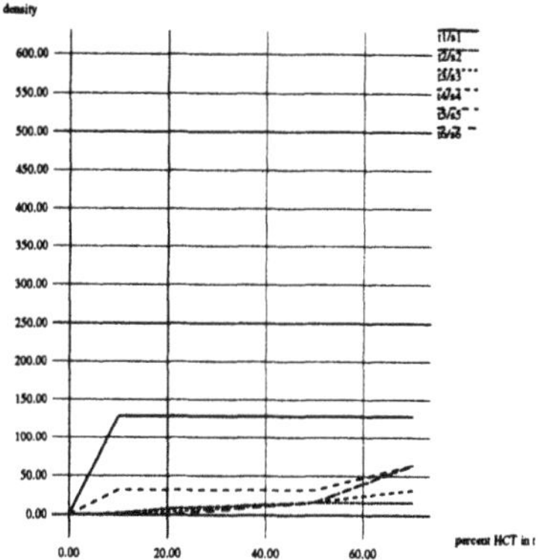

Figure 15 Sensitivity to missing events - special metric function, moderate parallelism and recursion.

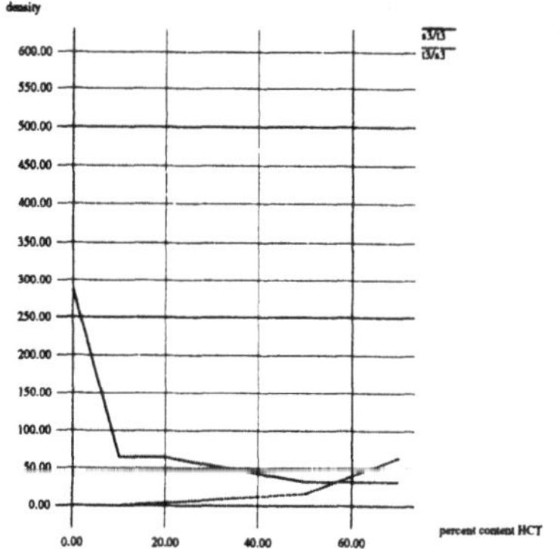

Figure 16 Sensitivity to missing events - special metric, moderate parallelism and recursion.

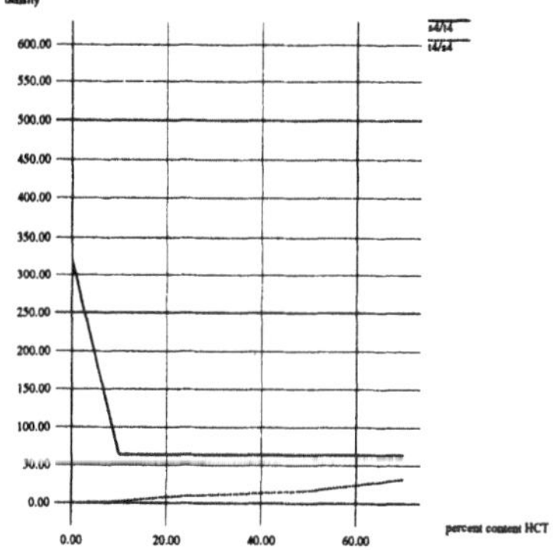

Figure 17 Sensitivity to missing events - special metric, moderate parallelism and recursion.

combinations in the same metric. Sets typically include 70-80 percent tests with HCT. Random sets were generated using 5-10 connections, 10-30 targets. Experiments show that in such cases, mutual coverages of randomly generated sets also tend to settle in this same range.

5 CONCLUSIONS

We have performed a series of experiments to explore the metric based method's ability to identify or select test suites with certain levels of vertical recursion,

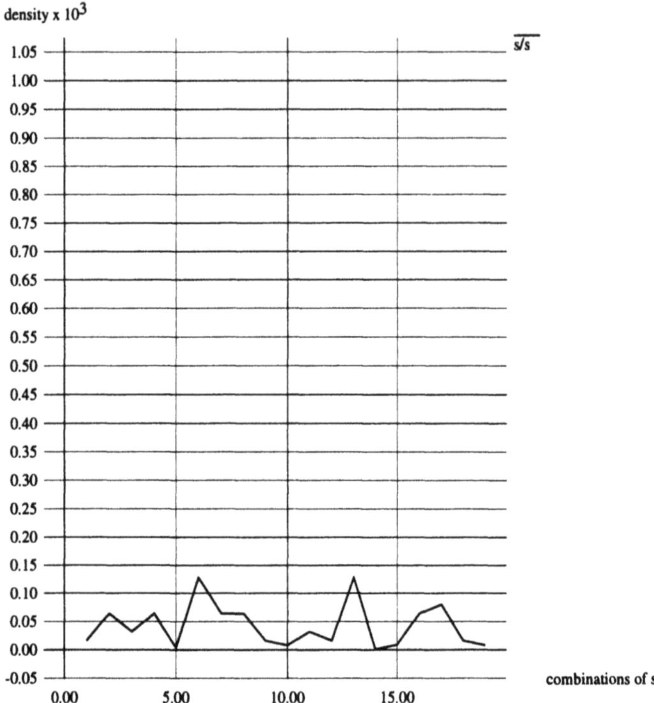

Figure 18 Sensitivity to missing events - special metric on random sets.

multiple connection capabilities, and transition or event patterns. The results show that the metric functions were stable and able to handle fine granularity of protocol behavior. The metric functions are shown to be sensitive to both recursion and parallelism, which are often primary sources of complexity in protocol specifications. The effect of missing events or event patterns with different metric functions and moderate parallelism and recursion were also observed.

In this empirical study, we have endeavored to provide a reasonably complete assessment of the sensitivity of the method by applying it to a fairly complex, real-life protocol, the ST-II protocol, which belongs to a class of protocols currently in the center of interest for multimedia streaming communications and the networking communities in general. It also exhibits most sources of protocol complexities, including recursions and multiple concurrent connections. We find the test selection method to be empiritically attractive in general. However, for a more thorough investigation, further experiments can be conducted for other "representative" complex protocols, with other metrics and even larger sets of test suites. Obviously, a fair amount of work would be demanded for each set of sensitivity assessment experiments performed on each additional protocol.

Further work may also include more experiments from which heuristics could be extracted to guide the choice of metric functions. Last, but not least, we may also study the sensitivity of metric functions to fault detection capability of the test suite generated. This is intuitively intriguing since by covering the specification sufficiently, faults should have a good chance to get detected.

6 REFERENCES

[1] G.v. Bochmann and et al. Fault models in testing. In J. Kroon, R.J. Heijink, and E. Brinksma, editors, *Protocol Test Systems, V*, Leidschendam, The Netherlands, October 1991.

[2] J.A. Curgus. *A metric based theory of test selection and coverage for communication protocols.* PhD thesis, Dept. of Computer Science, Univ. of British Columbia, June 1993.

[3] J.A. Curgus and S.T. Vuong. A metric based theory of test selection and coverage. In *Proc. IFIP 13th Symp. Protocol Specification, Testing, and Verification*, May 1993.

[4] C. Topolcic (ed). Experimental internet stream protocol, version 2 (ST-II); RFC-1190. *Internet Requests for Comments*, (1190), October 1990.

[5] Y. Lu. On TESTGEN, An Environment for Protocol Test Sequence Generation, and Its Application to the FDDI MAC Protocol. Master's thesis, The University of British Columbia, 1991.

[6] M. McAllister, S. T. Vuong, and J. Curgus. Automated test case selection based on test coverage metrics. In *Protocol Test Systems V.* Elsevier Science Publishers B.V. (North Holland), 1992.

[7] S.T. Vuong and J.A. Curgus. On test coverage metrics for communication protocols. In *Proc. 4th Int. Workshop on Protocol Testing System*, 1991.

[8] S.T. Vuong, H. Janssen, Y.Lu, C. Mathieson, and B. Do. TESTGEN: An environment for protocol test suite generation and selection. *Computer Communications*, 17(4):257–270, April 1994.

Analyzing performance bottlenecks in protocols based on finite state specifications

Sijian Zhang
P.O.Box 30004, 8602 Granville St., Vancouver, BC, Canada V6P 5A0
(email: <sijian@unixg.ubc.ca>)

Samuel T. Chanson
Dept. of Computer Science, HongKong Univ. of Science and Technology
Clear Water, Hong Kong (email: <chanson@cs.ust.hk>)

Abstract

This paper studies the problem of identifying performance bottlenecks in communication protocols. The model used is a Finite State Machine extended with time and transition probabilities known as PEFSM. A definition of PEFSM is given and the bottleneck identification methods proposed are based on this performance model. Informally, a bottleneck with respect to a performance metric is defined as the transition among all the transitions in a PEFSM which would produce the largest marginal improvement of the performance metric if the time of the transitions were reduced by the same small amount. We present two methods to locate the bottleneck transitions with respect to two of the most important performance metrics, i.e., *throughput* and *queue wait time*. These methods are partially validated by simulation.

Keywords

Finite state, specification, performance, bottleneck

Testing of Communicating Systems, M. Kim, S. Kang & K. Hong (Eds)
Published by Chapman & Hall © 1997 IFIP

1 INTRODUCTION

Performance bottlenecks exist in almost all computer systems in various forms. System designers, managers, analysts and users have worked on identifying the performance bottlenecks in a computer system for a long time. A bottleneck can be a service center of a system [Leung88, Allen90] or, at a more abstract level, a system parameter. For instance, in [ZiEt92], the *sensitivities* of the parameters in the throughput expression are used to determine the throughput bottleneck.

There exist many definitions for performance bottlenecks and most of them are defined with respect to only *throughput* or *utilization* [Ferr78, Lock84, Leung88, Yang89, Allen90, ZiEt92]. Nevertheless, all definitions of the performance bottlenecks have a common characteristic: a bottleneck identifies the component[1] in the system which has the most significant impact on system performance. A small improvement to the bottleneck component can greatly improve system response time, throughput or utilization.

This paper is concerned with finding performance bottlenecks in communication protocols. We note that it is common to specify communication protocols as interacting Finite State Machines (FSMs) or Extended FSMs (EFSMs) which are FSMs extended with variables. Many standardized protocols are directly given as FSMs or EFSMs. Examples can be found in [Tane88, ISO2576, ISO7776]. At least two internationally standardized formal description techniques exist (Estelle [ISO8807] and SDL [SDL]) which provide a way of specifying protocols and distributed systems based on FSMs or EFSMs. Therefore, it is reasonable to define a performance model based on FSM or EFSM for use in performance analyses as well as bottleneck identification. In thc following section, we shall call such a performance model as *performance extended FSM* (PEFSM).

The PEFSM is essentially an FSM enhanced with time and transition probabilities. In the FSM of a PEFSM, *state* and *transition* are the two main constructs. *States* are conceptual while *transitions* have direct correspondence in the implementation of the protocol specified by the FSM. Thc execution time of a transition directly affects performance. Therefore, it is natural to transform the bottleneck detection problem to the one of identifying the *bottleneck transition* in a PEFSM. This is useful because once the bottleneck transition is identified, we know where we should focus our efforts in improving system performance.

The execution of a transition takes non-zero time. In our model, each transition is associated with a class of incoming messages (i.e., message type). Futhermore, because of causal relationship, the transition service time affects the subsequent messages. Reducing the transition time in a PEFSM will im-

---

[1] A "component" can be a hardware device, a software module or a system parameter as mentioned earlier.

prove the overall performance of the PEFSM. For example, reducing a transition time will increase the throughput of each class of outgoing messages since the recurrence times of each state are decreased. However, the degree of improvement to a performance metric depends on the selected transitions. The one which results in the most improvement with respect to a performance metric is called the *bottleneck* of this performance metric.

We use the concept of *weight* to indicate the relation between the reduction of transition time and the improvement of a performance metric. A weight with respect to a performance metric is computed for each transition in a PEFSM. The higher the weight of a transition with respect to a performance metric, the more the performance metric can be improved by reducing the service (or execution) time of this transition. As such, the transition with the *greatest weight* with respect to a performance metric is the *bottleneck transition* (with respect to the performance metric). For instance, if the weight of transition ¡i,j¿ (from state i to state j) in a PEFSM with respect to the *mean queue wait time* of a message class is greater than that of any other transition, then transition ¡i,j¿ is the bottleneck transition with respect to the mean queue wait time. In other words, if each transition time is independently reduced by the same amount, the mean queue wait time will decrease the most in the case of a reduction in transition ¡i,j¿.

In this paper, we focus on two of the most important performance metrics of a PEFSM. They are the *throughput rate* of a class of outgoing messages and the *mean queue wait time* of a class of incoming messages. The methods to compute the *weights* of transitions with respect to each performance metric are also discussed.

The first method is to use *partial derivatives*. In general, if a performance metric $\mathcal{K}$ can be expressed as a function of a set of parameters, t_1, t_2, $\cdots$, t_n,

$$\mathcal{K} = \mathcal{F}(t_1, t_2, \cdots, t_n),$$

and the derivatives of $\mathcal{K}$ with respect to t_1, t_2, $\cdots$, t_n exist, then the partial derivatives

$$\frac{\partial \mathcal{K}}{\partial t_1}, \frac{\partial \mathcal{K}}{\partial t_2}, \cdots, \frac{\partial \mathcal{K}}{\partial t_n},$$

indicate the relative impact of the change of each parameter on the performance metric. Therefore, the partial derivatives can be used as the weights of the parameters. In our studies, we compute the partial derivatives of the throughput of outgoing messages of a specific class with respect to each transition time. These derivatives are taken as the weights of the transitions with respect to throughput.

The second method is to use an approximation technique to compute the weights of transitions with respect to the mean queue wait time of a specific incoming message class. This method is useful in the case where the computation of partial derivatives is difficult.

The rest of the paper is organized as follows. Section 2 gives the definition of the performance model PEFSM. Section 3 presents a method to locate the bottleneck transition of the throughput rate of a class of outgoing messages. Section 4 presents a different method to compute the weights for each transition with respect to the mean queue wait time of a class of incoming messages. The method is partially validated by simulation. Section 5 discusses related work on defining and locating performance bottlenecks, and Section 6 summarizes this paper.

2 PERFORMANCE MODEL

The detailed definition of PEFSM can be found in [Zhang95]. Due to space limitation, only a brief description of PEFSM is given in this paper. Since a PEFSM contains an embedded FSM, we start with the definition of the FSM.

2.1 FSM

A finite state machine (FSM) which describes a protocol entity is formally defined as a six-tuple

$$M = (Q, I, O, \delta, \xi, q_0) \tag{1}$$

where Q – a finite set denoting *states*;
I – a finite set denoting *incoming message classes*;
O – a finite set denoting *outgoing message classes*;
δ – a function denoting *transitions*, i.e., $\delta : Q \times I \rightarrow Q$;
ξ – a function denoting *transition outputs*, i.e., $\xi : Q \times I \rightarrow O$;
q_0 – an initial state.

Note that an FSM of a communication protocol does not necessarily have any final state. This is because a protocol (such as that in the telephone system) can execute forever without termination.

2.2 PEFSM

During execution, the FSM of a protocol changes from state to state. The state changing process is a stochastic process. Our performance model is a model that describes this stochastic process.

We enhance the FSM with *time* and *probability* to define the performance model which is called the *performance extended FSM* (PEFSM). Each transition in the PEFSM is associated with a *transition time* and a *single-step transition probablity.* The transition time from state i to j is denoted as $^{\psi}\tau_{ij}$ which is the time period from the start to the completion of the transition. The single-step probability of the transition from state i to j is denoted as p_{ij} which is the probability that transition ¡i,j¿ will be executed when the PEFSM is in state i.

Formally, a **PEFSM**, denoted as Ψ, is defined as a pair

$$\Psi = (M, \mathcal{P}) \tag{2}$$

where $M = (Q, I, O, \delta, \xi, q_0)$ is the kernel which is an FSM whose formal definition is given in (1), and $\mathcal{P} = (\mathbf{P}, {}^{\psi}\mathbf{H})$ is the running environment expressed in terms of time and probability:

$\mathbf{P} = [p_{ij}]$ – a matrix of the single-step transition probabilities;
${}^{\psi}\mathbf{H} = [{}^{\psi}h_{ij}(t)]$ – a matrix of the probability density functions (p.d.f.s) of the transition times.

In a PEFSM, M, $\mathbf{P}$ and ${}^{\psi}\mathbf{H}$ are primitive data. They are assumed to be provided directly by the performance evaluator.

Let $\{X(t), t \geq 0\}$ be the state changing process of the PEFSM;
$t_0, t_1, ..., t_n, ...$ be the sequence of the epochs right before the processing of a transition is completed;
$X_0, X_1, ..., X_n, ...$ be the sequence of the states in the PEFSM corresponding to the time sequence $t_0, t_1, t_2, ..., t_n, ...$, respectively.

The components of a PEFSM and their relationships are formally defined in the following:

1. $X(t) \in Q$ for all $t \geq 0$.
2. $X(0) = X_0 = q_0$.
3. $X_n = X(t_n^-)$ [2] and $X_{n+1} = X(t_n)$.
4. $Pr\{X_{n+1} = j | X_n = i\} = p_{ij}$.
5. If $X_n = i$ and $X_{n+1} = j$, then $t_n - t_{n-1} = {}^{\psi}\tau_{ij}$.
6. ${}^{\psi}h_{ij}(t)$ is the p.d.f. of ${}^{\psi}\tau_{ij}$, i.e. $Pr\{{}^{\psi}\tau_{ij} = t | X_n = i,\ X_{n+1} = j\} = {}^{\psi}h_{ij}(t)$.

From the above definitions, it can be seen that the trajectory of the state variable X of a PEFSM is governed by M, $\mathbf{P}$ and ${}^{\psi}\mathbf{H}$. M determines the state space of X and the possible next value of X statically. The other parameters govern the dynamic control of X.

We assume that the transition probability matrix $\mathbf{P}$ is known. When the stochastic process of state changing is ergodic[3] the steady-state state probability vector, π, can be computed from $\mathbf{P}$ by solving the matrix equation (see [Zhang95] for details):

$$\pi \mathbf{P} = \pi.$$

[2] t_n^- denotes the epoch right before the time instant t_n. $t_n^- < t_n$ but t_n^- is infinitely close to t_n.

[3] A stochastic process is *ergodic* when it is recurrent non-null and irreducible [Allen90].

3 THROUGHPUT BOTTLENECK

In a PEFSM, an outgoing message is generated when a transition is processed. Therefore, the throughput rate of the outgoing messages of a class is equal to the recurrence rate of the transition associated with the class.

Let ${}^{\psi}\bar{e}_i$ and ${}^{\psi}\bar{e}_{ij}$ be the mean recurrence rates of state i and transition ¡i,j¿ of a PEFSM, respectively. Their relationship can be expressed as

$${}^{\psi}\bar{e}_{ij} = {}^{\psi}\bar{e}_i \cdot p_{ij}$$

where p_{ij} is the probability of transition ¡i,j¿.

When the PEFSM is in equilibrium, we have (see [Howa71](p.725))

$${}^{\psi}\bar{e}_i = \frac{\pi_i}{{}^{\psi}\bar{T}}.$$

Therefore,

$${}^{\psi}\bar{e}_{ij} = \frac{\pi_i}{{}^{\psi}\bar{T}} \cdot p_{ij} = \frac{\pi_i p_{ij}}{\sum_{u,v \in Q} \pi_v p_{uv} \, {}^{\psi}\bar{\tau}_{uv}}.$$

Let $\bar{\eta}_{ij}$ be the throughput rate of the class ij outgoing messages. Since class ij outgoing messages are associated with transition ¡i,j¿,

$$\bar{\eta}_{ij} = {}^{\psi}\bar{e}_{ij} = \frac{\pi_i p_{ij}}{\sum_{u,v \in Q} \pi_u p_{uv} \, {}^{\psi}\bar{\tau}_{uv}}.$$

The above equation gives the relationship of the throughput rate $\bar{\eta}_{ij}$ and the transition times ${}^{\psi}\bar{\tau}_{uv}$ $(u, v \in Q)$. For a given PEFSM, $\pi_i p_{ij}$ $(i, j \in Q)$ is fixed, so the change of $\bar{\eta}_{ij}$ varies inversely with the change in the value of the denominator of the above equation.

Using the derivatives, one can determine that the coefficient $\pi_v p_{uv}$ of ${}^{\psi}\bar{\tau}_{uv}$ in the denominator indicates the relative degree of the improvement on $\bar{\eta}_{ij}$ by transition uv $(u, v \in Q)$ compared to the other transitions. $\pi_v p_{uv}$ in fact is the steady-state probability of transition uv. Among the transitions, the one with the largest steady-state transition probability has the greatest impact on increasing the throughput rate $\bar{\eta}_{ij}$.

Therefore we define $\pi_v p_{uv}$ as the *weight* of the transition uv with respect to $\bar{\eta}_{ij}$. The bottleneck transition of $\bar{\eta}_{ij}$ is the transition which has the largest $\pi_v p_{uv}$ $(u, v \in Q)$.

Since $\pi_v p_{uv}$ $(u, v \in Q)$ is not related to $\bar{\eta}_{ij}$, we can further conclude that the bottleneck transition is the same for the throughput rates of all classes of outgoing messages.

4 QUEUE WAIT TIME BOTTLENECK

We say that an incoming message is a *firable* message of state i if it is associated with a transition starting from state i. If the PEFSM is not in state i

when a firable message of state i arrives, the message will be stored in a queue. We shall assume that there is a *first in first out* (FIFO) queue for each class of incoming messages. We are interested in identifying the bottleneck transition with respect to the mean queue wait time of a specific class of incoming messages.

It has been proved that [Zhang95] the mean queue wait times of different classes of firable messages of a state are the same, i.e.,

$$\overline{W}_{ij} = \overline{W}_i \quad (j \in Q)$$

if class ij (for all $j \in Q$) of messages arrive independently and in a Poisson pattern. So it is necessary only to compute the *overall* mean queue wait time of all classes of firable messages of a state in this case.

To compute the overall mean queue wait time $\overline{W}_i$, we construct the *virtual jobs* of state i and treat the queuing system of the PEFSM as an M/G/1 system. The virtual jobs of state i are the sequences of transitions where each sequence forms a *first passage* from state i to i in the PEFSM. The mean queue wait time can then be computed by applying the well-known solution technique for M/G/1:

$$\overline{W}_i = \frac{\lambda_i \overline{\zeta_i^2}}{2(1-\rho_i)} = \frac{\sum_{j \in Q} \lambda_{ij} \overline{\zeta_{ij}^2}}{2(1-\rho_i)}.$$

In the above equation, $\overline{\zeta_{ij}^2}$ is the second moment of the service time of class ij virtual jobs. A set of equations has to be solved in order to compute $\overline{\zeta_{ij}^2}$ $(i, j \in Q)$ (see [Zhang95] for details). The closed-form solution of $\overline{\zeta_{ij}^2}$ is difficult to obtain. So it is generally infeasible to compute the partial derivatives of $\overline{W}_i$ with respect to each transition time ${}^{\psi}\bar{\tau}_{uv}$ $(u, v \in Q)$ for use as the *weights* to identify the bottleneck transition of $\overline{W}_i$. Therefore, the following approximate solution is proposed instead.

4.1 An approximation approach

As mentioned earlier, the FSM of a PEFSM contains information on the service order of incoming messages. This ordering affects the performance of the PEFSM and should be taken into consideration to obtain more accurate results.

In general, we can assume the queuing system of a PEFSM to consist of a single server with a single queue. Figure 1 shows a queuing system which serves a PEFSM. The service order of incoming messages is controlled by the FSM of the PEFSM and the incoming messages.

An asynchronous incoming message to a PEFSM may arrive before the PEFSM is ready to process this message. When this happens, the message will have to wait in a queue. The *queue wait time* of this message is the

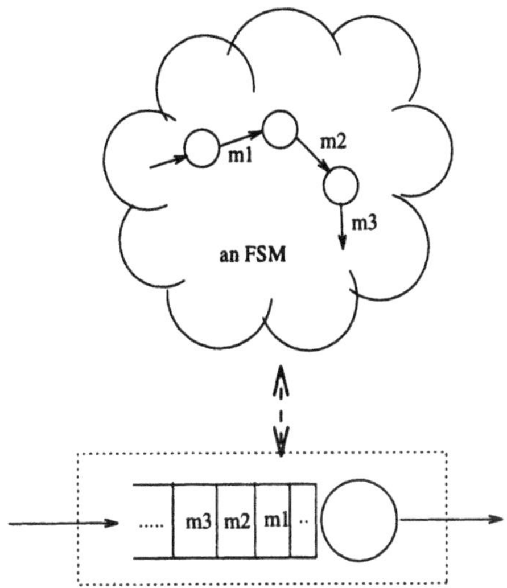

Figure 1: Service order implied by the FSM of a PEFSM.

elapsed time between the moment it arrives and the moment it is processed. From Figure 1, it is not difficult to see that the waiting period of this message includes not only the processing times of all the messages of the same class which arrived earlier that are still in the system, but also the processing times of the transitions associated with the messages of the other classes. These transitions bring the PEFSM to the right state so that the target message can be processed. For example, in Figure 1, message $m3$ has to wait for service until the transitions associated with $m1$ and $m2$ are processed.

Before we show how the incoming messages of a class in a PEFSM wait statistically when they arrive early, the definitions of *transition path* and *transition subpath* are first given.

Definition 4.1 (transition path) *A* transition path *of a PEFSM is a sequence of consecutive transitions in the PEFSM.*

Definition 4.2 (transition subpath ij) *A transition subpath ij of a PEFSM is a finite number of consecutive transitions in the PEFSM starting from state i and ending in state j. The first transition of the subpath is called the head of the subpath; the last transition is called the tail of the subpath.*

Figure 2 shows a state-transition tree of a PEFSM. Each state in the tree is a state in the FSM of the PEFSM, and each transition is a transition in the FSM. This tree includes all the possible transition subpaths to transition ¡i,j¿.

Suppose γ is an incoming message of class ij of the PEFSM and W_{ij} ($W_{ij} > 0$) is the queue wait time of γ. Furthermore, suppose transition subpath 1 in Figure 2 includes the transitions which must be processed before message γ.

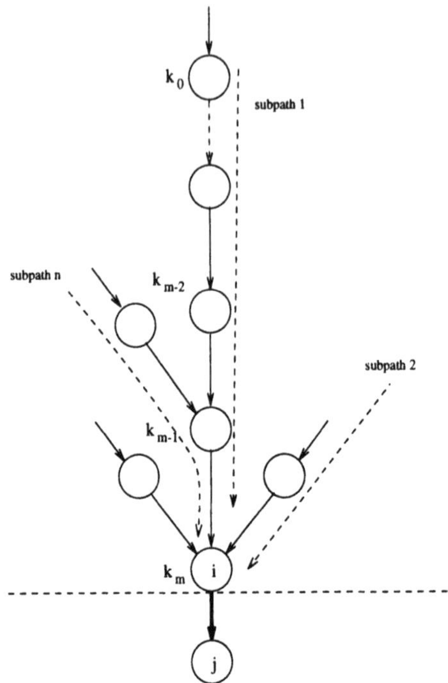

Figure 2: The subpaths to transition ¡*i*,*j*¿ in a PEFSM.

Assume the PEFSM is in state k_0 when message γ arrives. Then, the transition subpath $k_0 i$ includes the transitions which are seen by γ and will be processed before γ. Let these transitions be transitions $k_0 k_1$, $k_1 k_2$, $\cdots$, $k_{m-1} k_m$ ($k_m = i$), and D_1, D_2, D_3, $\cdots$, D_m be their transition times, respectively.

By definition, we have

$$\sum_{n=2}^{m} D_n < W_{ij} \leq \sum_{n=1}^{m} D_n. \tag{3}$$

Transition $k_0 k_1$ may already be in progress at the time γ arrives, in this case $W_{ij} \leq \sum_{n=1}^{m} D_n$.

γ will have to wait until the processing of all of these transitions is finished whether or not the incoming messages associated with them have already arrived. The decrease of the transition time of any of these transitions will reduce the queue wait time of γ, W_{ij}. Those transitions that appear in the transition subpath $k_0 i$ more than once will have a higher impact in reducing W_{ij}.

However, different messages of class ij may see different transition subpaths when they arrive. Furthermore, a transition may appear in more than one transition subpath. So the relative frequency of each transition subpath seen by γ should be taken into account in computing the *weights* for all the transitions with respect to W_{ij}. The relative frequency of a subpath can be computed using the *transition subpath probability* defined below.

Definition 4.3 (transition subpath probability) *The* probability *of a transition subpath $k_0 k_m$ is defined as:*

$$Pr\{subpath\ k_0 k_m\} = \prod_{n=1}^{m} p'_{k_{n-1}k_n}$$

where transition $k_{n-1}k_n$ $(n = 1, 2, ..., m)$ are the transitions in the subpath and $p'_{k_{n-1}k_n} = \pi_{k_{n-1}} p_{k_{n-1}k_n}$ is the steady-state probability of transition $k_{n-1}k_n$, and $p_{k_{n-1}k_n}$ is the single-step probability of transition $k_{n-1}k_n$.

Given the probabilities of transition subpaths, we can compute the relative frequency of a transition seen by a specific class of incoming messages. The frequency is simply the sum of the probabilities that this transition appears in all the possible transition subpaths which satisfy Inequality (3). It is useful to compute the relative frequency of a transition because decreasing the time of the transition with the highest frequency by the same amount will reduce the mean queue wait time of the specific class the most. Therefore, in this case, the frequency can be used as the *weight* in identifying the bottleneck transition of the mean queue wait time of the incoming messages of the specific class.

Let w_{uv} be the weight of transition uv. From the discussion above, we can define

$$w_{uv} = \sum_{l \in subpaths} (Pr\{\text{subpath } l\} \cdot Pr\{\text{transition } uv \text{ appears in the subpath } l\}).$$

Next, we present an algorithm to compute the weights given the mean queue wait time of a class of incoming messages.

4.2 Computation of weights

Assume that the single-step transition probabilities in $\mathbf{P}$ of a PEFSM are given, and the steady-state state probabilities π as well as the transition times of each transition have been computed.

Suppose $\overline{W}_{ij}$ is known either by computation or measurement. An algorithm to compute the weights with respect to $\overline{W}_{ij}$ is given in Figure 3.

Procedure 1 of the algorithm initializes all the weights to zero before calling Procedure 2. Procedure 2 computes the weights of all the transitions in the PEFSM. Using a recursive procedure, it traverses all the transition subpaths which end in state i and satisfy Inequality (3). The subpath starts backwards from state i. A transition is added to the current head of the subpath in each iteration. This transition becomes the new *head transition* of the subpath. The transition subpath grows until the sum of the mean transition times of the transitions along the subpath is larger than the given mean queue wait time,

Procedure 1 : *compute weights given mean queue wait time*
Inputs : $\overline{W}_{ij}$ – the mean queue wait time of class ij incoming messages;
Outputs : the weights of all the transitions in the PEFSM, w_{uv} $(u, v \in Q)$;
Steps :
1. initialize the weights of all the transitions to zero, i.e., $w_{ij} = 0$ for $i, j \in Q$;
2. call *Procedure 2* with arguments $(1, ij, \overline{W}_{ij})$.

Procedure 2 : *recursively backtrack to add the subpath probabilities to the transitions which are the heads of the subpaths*
Inputs : 1) the current subpath probability p;
2) the reference of the current transition uv;
3) the remaining waiting time R_w;
Outputs : weights of the transitions;
Steps :
1. if $R_w \leq 0$, return;
2. for (each transition which is immediately before the current transition uv in the FSM of the PEFSM, say transition ku) do :
   1) let $p = p * p'_{ku}$ where p'_{ku} is the steady-state transition probability of transition ku;
   2) let $w_{ku} = w_{ku} + p$;
   3) call *Procedure 2* with arguments $(p,\ ku,\ (R_w\ -\ {}^{\psi}\bar{\tau}_{uv}))$;
   endfor.

Figure 3: Algorithm for weight computation.

$\overline{W}_{ij}$. At each step, the current *subpath probability* is added to the weight of the head transition.

When Procedure 2 terminates, the weights of all the transitions in the given PEFSM with respect to the mean queue wait time, $\overline{W}_{ij}$, are computed. These weights reflect the relative frequency of the transitions seen by class ij incoming messages. If the transition time of each transition is reduced by the same amount one at a time, the one which has the largest weight will cause the largest improvement in the mean queue wait time of class ij messages. Therefore, the transition with the largest weight is the *bottleneck transition* with respect to the mean queue wait time.

4.3 Simulation results

Simulations have been conducted to validate the accuracy of the bottleneck identification method. The architecture of the simulation experiment is shown in Figure 4.

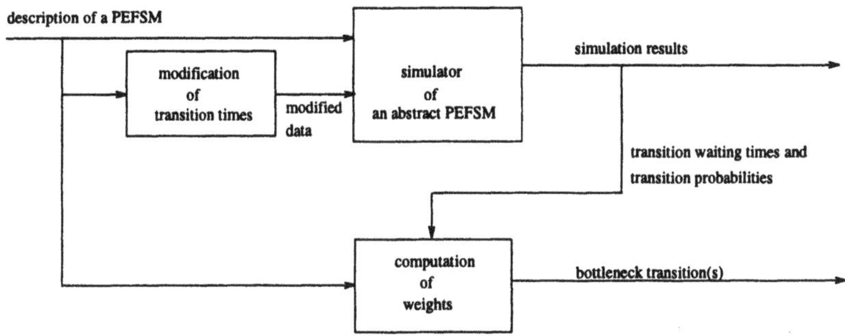

Figure 4: A simulation architecture.

The *simulator* module accepts a model description of the PEFSM and simulates the execution of transitions in the FSM of the PEFSM. The module contains an *incoming message generator* which generates the incoming messages to the PEFSM based on the given arrival model of the PEFSM.

The simulation results are fed to the *weight computations* module. This computation module also stores the description data of the PEFSM. The algorithms in Figure 3 are used to compute the weights of all the transitions with respect to the mean queue wait time of a specific transition. The transition with the largest weight is the *bottleneck transition.*

The *modification* module reduces the service time of each transition of the PEFSM by the same small amount one at a time. This module resends the modified data of the PEFSM to the simulation module and the simulation is rerun. All the mean queue wait times of class ij incoming messages in each run are recorded so as to verify if the bottleneck transition in fact causes the largest reduction in the mean queue wait time.

Several protocols were used in our experiments which showed that the proposed technique for bottleneck transition identification works in practice. We report the result of the *alternating bit protocol* in the following.

The FSM of the alternating bit protocol is given in Figure 5. The input data of the PEFSM are given in Columns 2, 3 and 4 of Table 1. The incoming data packets to be transmitted arrives in a Poisson pattern with a mean rate of 200.0 packets/second.

Columns 4, 5 and 6 are the simulation results. The steady-state transition probabilities were recorded in Column 4. These results agree with the results from computation of $p'_{ij} = \pi_i p_{ij}$ where π_i is the steady-state probability of state i, and p_{ij} is the single-step probability of transition ¡i,j¿. The weights were computed with respect to the mean queue wait time of a class of incoming messages and recorded in Column 5. Then, in each of the subsequent runs, we selected one of the transitions and reduced its service time by a small amount (0.002 second). The simulation was re-run with the modified data.

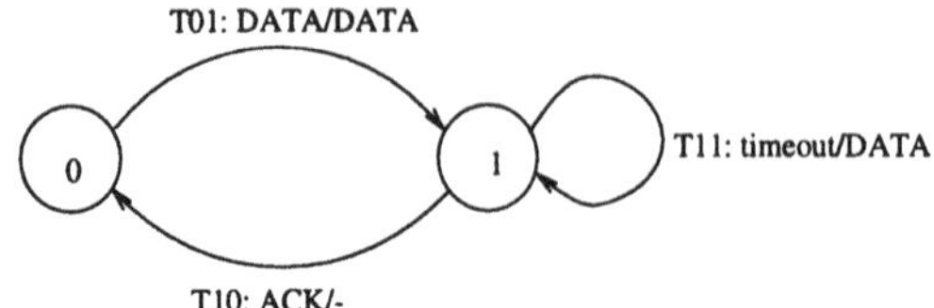

STATES :
state 0 : idle
state 1 : waiting for ACK
(acknowledgement)

TRANSITIONS :
T01 : sending a DATA packet
T10 : receiving an ACK
T11 : timeout retransmission

Figure 5: An FSM of the alternating bit protocol.

Table 1: A simulation result of bottleneck identification.

transition identifier	single-step tran. prob.	mean transition time[a]	transition probability[b]	transition weight[c]	queue wait time reduction[d]
T01	1.0	0.001	0.4738	0.3102	0.0069
T10	0.9	0.002	0.4738	**0.6207**	**0.0090**
T11	0.1	0.002	0.0525	0.0343	0.0010

[a] All times are in seconds.

[b] Steady-state transition probability.

[c] The weight is computed with respect to the mean queue wait time of the data packets.

[d] The new average queue wait time is measured by decreasing the mean service time of the corresponding transition by 0.002 second. The reduction of the queue wait time is equal to the original value minus the new value.

The reduction in mean queue wait time was recorded in Column 6. This procedure is repeated for all the transitions. From the table, we can see that reducing the service time of the transition with the largest weight causes the largest reduction in the mean queue wait time of that class of incoming messages. This result confirms the analysis given in this section.

More than 20 experiments with different work load parameters had been performed for several protocols. In most cases, the results from simulation agreed with the analytic results. Only 3 exceptions were found. However, even then, the reduction of the mean queue wait time by reducing the service time of the bottleneck transition was very close (within 15%) to the largest reduction. The reason why the proposed procedure occasionally does not correctly identify the bottleneck transition is that both the queue wait times and the transition times have variance and we use only the mean value to compute the weights for simpilicity.

5 RELATED WORK

Performance bottleneck detection and removal have received less attention than performance prediction. This is not only because the problem itself is hard but also because there is a lack of adequate formal definitions and effective analysis methods. Only a few methods have been proposed to locate the performance bottleneck of a software system.

The concept of *critical path* was first introduced in the context of *project management* and used to manage the progress of projects [Lock84]. It was later adopted for parallel processing systems [Yang89] where there are both parallel events and synchronization points. The critical path of a parallel program is the path in the program activity graph[4] which determines the program performance (e.g. shortest execution time). The possible domain of the potential bottleneck is that of the critical path. Other techniques are used for locating the bottleneck within the critical path.

Lockyer's *critical path analysis* [Lock84] is often used to identify bottlenecks in parallel or distributed systems which are modeled as acyclic directed graphs [Yang89, Wagn93]. However, only one transition in a PEFSM can be executed at a time. The execution of the transitions in a PEFSM are sequential and follows a certain order. There is no synchronization with other transitions in a single PEFSM. Therefore, the method of critical path analysis can not be directly applied to PEFSMs in identifying bottlenecks.

Although intuitively we all know what a bottleneck is, historically, the term *bottleneck* has had various definitions. They can be classified into the following two categories according to usage :

- analytical definitions
- measurement based definitions

Using derivatives is a common analytical approach to identifying a performance bottleneck. For example, in [Ferr78], the *derivatives* of the mean throughput rates with respect to the service rates of the constituent servers of the system are used to define performance bottlenecks analytically. If

$$\frac{\partial \overline{T}}{\partial \mu_i} > \frac{\partial \overline{T}}{\partial \mu_j} \; (j = 1, 2, ..., s; j \neq i)$$

then server Σ_i is the performance bottleneck of a system with s servers, where $\overline{T}$ is the mean throughput rate of the object system; μ_k is the service rate of server Σ_i (k=1,2,...,s).

However, this definition can not be used if $\overline{T}$ is not differentiable.

[4] A program activity graph is a directed graph which depicts the synchronization points of the whole system.

Utilization based techniques constitute another analytical way for determining *performance bottlenecks* [Leung88, Allen90]. Among the servers in a queuing network model, the one with the highest utilization or the one which first achieves 100% utilization with increasing workload on the system is considered to be the bottleneck of the system. However, this approach is not appropriate for a PEFSM because it is assumed to have only one service center.

Generally, the analytical definition is applied to a model of the system. When an implementation of the system already exists, *analyses* of data from *measurement* can be used to identify the bottleneck. In [ZiEt92], the bottleneck is defined as the performance parameter which is most sensitive to performance. The sensitivity of a parameter is defined as

$$sensitivity \stackrel{\text{def}}{=} \frac{\%ChangeInPerformance}{\%ChangeInParameter}$$

Intuitively, the sensitivity is similar to the *weight* to a certain extent. Both can be used in analytical approaches and measurement approaches.

6 CONCLUSION

We have proposed a methodology to identify performance bottlenecks based on a performance extended FSM model PEFSM. *Weights* are used to measure the impact of the reduction of each transition time on the improvement of a specific performance metric. The bottleneck with respect to a performance metric is defined to be the transition in the PEFSM with the maximum *weight*.

The methods to compute the *weights* of the transitions in a PEFSM with respect to two performance metrics are presented. The first method makes use of the closed-form expression of a performance metric such as *throughput*. This depends on the existence of both the closed-form expression of a performance metric and the partial derivatives of the performance metric with respect to each transition time. The second method uses an approximate recursive algorithm to compute the weights with respect to a performance metric. This method is used when no closed-form expression of the performance metric or derivatives exists.

The second method was used to identify the bottleneck transition with respect to the mean queue wait time of a specific class of incoming messages. It is more general than the method of derivatives. This second method can be applied to the PEFSM in which the arrivals of asynchronous messages are not Poisson, and the mean queue wait time may be obtained either by measurement or computation.

The mean transition time of the bottleneck transition can be reduced in two ways: reducing the *mean transition wait time* or the *mean transition service time*. To reduce the mean transition wait time, one may increase the arrival rate of the incoming messages associated with that transition. For example, we can increase the throughput rate of messages or decrease the queue wait time of messages in a specific workstation by shortening the *token* turnaround time for this workstation in a token ring network. To reduce the transition service time, one may try to improve the software implementation of that transition or use faster hardware to process the transitions.

7 REFERENCES

[Allen90] Arnold O. Allen. *Probability, Statistics, and Queueing Theory with Computer Science Applications.* Academic Press, Inc., second edition edition, 1990.

[Ferr78] Domenicao Ferrari. *Computer Systems Performance Evaluation.* Prentice-Hall, INC., Englewood Cliffs, New Jersey, 1978.

[Howa71] Ronald A. Howard. *Dynamic Probabilistic Systems*, volume Volume 1: Markov Models; Volume 2: SemiMarkov and Decision Processes. John Wiley & Sons, Inc., 1971.

[ISO2576] ISO TC97/SC16 N2576. *Formal specification of Transport protocol in Estelle.* ISO, 1986.

[ISO7776] ISO/TC 97, Information processing systems. *Information processing systems - Data communication - High-level data link control procedures - Discription of the of the X.25 LAPB-compatible DTE data link procedures.* ISO 7776 - 1986. ISO, first edition, 12 1986.

[ISO8807] ISO TC 97/SC 21. *Information processing systems - Open systems Interconnection - LOTOS - A formal description technique based on the temporal ordering of observational behaviour.* ISO 8807. Interational Organization for Standardization, 1989.

[Leung88] Clement H.C. Leung. *Quantitative analysis of computer systems.* Chichester; New York : Wiley, 1988.

[Lock84] K.G. Lockyer. *Critical path analysis and other project network techniques.* Pitman Publishing, 1984.

[SDL] CCITT. *Specification and Description Language - Recommendation Z.100.* Geneva, Switzerland: CCITT press, 1986.

[Tane88] Andrew S. Tanenbaum. *Computer Networks.* Pretice-Hall, Inc., 1988.

[Wagn93] A. Wagner, J. Jiang, and S. Chanson. TMON - A Transputer Performance Monitor. *Concurrency: Practice and Experience*, 5(6):511–526, 1993.

[Yang89] Cui-Qing Yang and Barton P. Miller. Performance Measurement for Parallel and Distributed Programs: A Structured and Automatic Approach. *IEEE Transactions on Software Engineering*, 15(12), 12 1989.

[Zhang95] Sijian Zhang. *Performance Modelling and Evaluation of Protocols based on Formal Specifications.* PhD thesis, The Univ. of British Columbia, 1995.

[ZiEt92] John A. Zinky and Joshua Etkin. Troubleshooting throughput bottlenecks using executable models. *Computer Networks and ISDN Systems*, 24(1), 3 1992.

8 BIOGRAPHY

Dr. Sijian Zhang

Dr. Sijian Zhang graduated from Beijing Univ. with B.Sc and M.Sc. He received his Ph.D. degree from Univ. of British Columbia, Canada in 1995. He is currently working for Hughes Aircraft of Canada. Dr. Zhang's research interests include protocols, conformance and performance testing, formal specifications and distributed computing.

Dr. Samuel Chanson

Dr. Samuel Chanson received his Ph.D. degree from the University of California, Berkeley in 1975. He was a faculty member at Purdue University for two years before joining the University of British Columbia where he became a full professor and director of its Distributed Systems Research Group. In 1993 Professor Chanson joined the Hong Kong University of Science & Technology as Professor and Associate Head of the Computer Science Department. He has chaired and served on the program committees of many international conferences on distributed systems and computer communications, including IEEE ICDCS, IFIP PSTV and IWTCS.

Dr. Chanson's research interests include protocols, high speed networks, multimedia communication and load balancing in workstation clusters. He has published more than 100 technical papers in the above areas.

PART SEVEN

Test Generation for Communicating State Machine

A conformance testing for communication protocols modeled as a set of DFSMs with common inputs

Atsushi Fukada†, *Tadashi Kaji*†*, *Teruo Higashino*†,
Kenichi Taniguchi† *and Masaaki Mori*††
†*Dept. Information and Computer Sciences,*
Osaka University, Toyonaka, Osaka 560, Japan
Tel : +81-6-850-6607 Fax : +81-6-850-6609
E-mail : {a-fukada, higashino, taniguchi}@ics.es.osaka-u.ac.jp
**Currently Hitachi Co. Ltd. E-mail : t-kaji@sdl.hitachi.co.jp*
††*Dept. of Information Processing and Management,*
Shiga University, Hikone, Shiga 522, Japan
E-mail : mori@biwako.shiga-u.ac.jp

Abstract

In this paper, we propose an effective conformance testing method for a subclass of protocols modeled as a set of DFSMs. The number of test cases in the proposed method is only proportional to the *sum* of those of states and transitions in a given set of DFSMs. In our method, we find a characterization set for each DFSM, which is used to test the DFSM alone in Wp-method, and the union of the characterization sets is used as a characterization set for the total system. For a set of DFSMs with common inputs, there may exist two or more tuples of states that have correct responses against a given characterization set. So, in order to identify each state *s* in a DFSM, we find a characterization set with some specific properties. Then we select a suitable

Testing of Communicating Systems, M. Kim, S. Kang & K. Hong (Eds)
Published by Chapman & Hall © 1997 IFIP

tuple of states containing the state s, and identify the state s by checking their response to the characterization set.

Keywords
Verification, protocol testing, test case selection and test coverage

1 INTRODUCTION

Conformance testing for communication protocols is highly effective to develop reliable communication systems. There are many research efforts on generating conformance test cases mechanically, which are well known as the TT-method [7], W-method [1], DS-method [2], UIO-method [9], and so on. Furthermore, there are some research papers focusing on more effective methods for generating test sequences . These efforts were mainly made for communication protocols modeled as a single deterministic finite state machine (DFSM). Recently similar research efforts have been done on non-deterministic or concurrent models [4, 5, 6, 11, 12].

According to the progress of computer networks, many kinds of protocols, which use several channels in parallel, are proposed. For such a protocol, it is quite natural that a protocol with several channels is considered as a set of DFSMs, each of which controls one channel and competes with other DFSMs for taking common inputs. Since a common input is taken by some DFSMs, the whole behavior of a given set of DFSMs is non-deterministic.

As a conformance testing method for such a non-deterministic FSM (NFSM) model, there is the GWp-method [6]. Such a method can be applied to a set of DFSMs mentioned above. However, since, in general, all reachable tuples of states of the DFSMs are considered as the states of the total system, the number of states of the total system is proportional to the product of those of the DFSMs. So, the number of generated test cases is also proportional to the product of those of the DFSMs. One of the methods free from such a drawback is to carry out the conformance testing for each DFSM independently. For example, the testing method in [5] is based on this idea and it treats protocols with communications among FSMs and internal actions. Since this method is based on the TT-method, it cannot identify the state after each transition is executed.

In this paper, for a specification modeled as a set of DFSMs with common inputs, we assume that the implementation under test (IUT, for short) is also modeled as a set of DFSMs (sub-IUTs) where the number of states of each sub-IUT does not exceed that of the corresponding DFSM in the specification. Under this assumption, we propose a testing method based on the GWp-method where the number of test cases is only proportional to the *sum* of those of states and transitions in a set of DFSMs. It identifies all states of DFSMs and confirms all transitions even if the IUT has any number of faults.

The proposed method uses the union of characterization sets W_i for all DFSMs A_i as a characterization set for the total system. Here, we assume that we can generate a characterization set which can identify all states in each DFSM even if the common inputs in the characterization set are received by several DFSMs non-deterministically. In order to identify a state s of a DFSM, we select a suitable tuple of states containing the state s, and identify the state s by checking its response to the characterization set.

The paper is structured as follows. In Section 2, we explain the model used in this paper, and then we define its fault model. In Section 3, the outline of the GWp-method for testing NFSMs is explained. In Section 4, we propose a testing method. In Section 5, the correctness of the proposed testing method is described. An example is given in Section 6. We conclude the paper in Section 7.

2 A SET OF DFSMS WITH COMMON INPUTS

2.1 Specification and Its Implementation

Definition 1 (Finite State Machine) A finite state machine (FSM) is defined as the following 6-tuple,

$$A = (S, X, Y, \delta, \lambda, s_0)$$

Here, S, X and Y are a finite set of states, a finite set of inputs and a finite set of outputs, respectively. δ is a transition function ($S \times X \rightarrow S$), and λ is an output function ($S \times X \rightarrow Y$). s_0 is the initial state of A. □

For two states s and t, we say that s is *equivalent* to t if $\lambda(s, \sigma^i) = \lambda(t, \sigma^i)$ holds for any input sequence σ^i. We say that a FSM M_1 is *equivalent* to a FSM M_2 if the initial state of M_1 is equivalent to that of M_2. A FSM M is said to be *minimal* if, for any two different states s and t of M, s is not equivalent to t. We say that a FSM M is *completely specified* (or *complete*) if both transition function and output function are defined for any pair of a state and an input. In this paper, if a FSM is not completely specified, we make the FSM complete as follows : For each pair of a state s and an input x whose transition and output functions are undefined, we add a new transition from s to itself whose output is empty and make the FSM complete. Here, we denote such an empty output by "ε". For such a new transition x/ε from a state s, we say that the FSM *ignores* input x at state s.

A FSM is said to be *initially connected* if there exists a transition sequence from the initial state to any state of the FSM, where the transition sequence may be a null sequence.

A FSM is said to be deterministic if, for any pair of a state s and an input x, $\delta(s, x)$ and $\lambda(s, x)$ are uniquely defined. Such a FSM is called a deterministic

FSM (DFSM). A FSM that is not a DFSM is called a non-deterministic FSM (NFSM). If two non-deterministic transitions from a state have the same outputs b for an input a, then we say that the non-deterministic transitions (a/b and a/b) are non-observable. Otherwise (for example, a/b and a/c), we say that they are observable non-deterministic transitions. We say that a NFSM is observable NFSM (ONFSM) if all the non-deterministic transitions in the NFSM are observable [10].

Next, we define a set of DFSMs used in this paper. We call this model as *Coupled DFSMs.*

Definition 2 (Coupled DFSMs) *Coupled DFSMs* are a k-tuple,

$$A = (A_1, A_2, \ldots A_k)$$

where $A_1, A_2, \ldots, A_k$ are DFSMs, respectively. Also, each $A_i (1 \leq i \leq k)$

$$A_i = (S_i, X_i, Y_i, \delta_i, \lambda_i, s_{i0})$$

must be a complete, initially connected and minimal DFSM. Furthermore, we suppose there is a reset operation so that the whole Coupled DFSMs are reset to their initial states at a time. □

Here, an input such as $x \in X_i \cap X_j$ is called a *common input*. If a common input x is given to Coupled DFSMs from the external environment, one of DFSMs takes the input x non-deterministically and the chosen DFSM returns a response (output). Here, it is assumed that A_i is not chosen whenever A_i ignores the input x and A_j does not.

Definition 3 (Specifications of Communication Protocols) A specification of a communication protocol dealt with in this paper is given as Coupled DFSMs

$$A = (A_1, A_2, \ldots A_k)$$

consisting of k DFSMs. We also suppose that this specification does not contain any non-observable and non-deterministic transitions as a whole. □

$A_i (1 \leq i \leq k)$ is said to be a *sub-specification* of A. Since we assume that a specification does not contain any non-observable and non-deterministic transitions, if there is a transition a/b in A_i, there may exist a transition a/c in A_j. However, there does not have to exist the same transition a/b in A_j.

Each implementation under test (IUT) I is given as follows.

Definition 4 (Implementation Under Test (IUT)) An implementation under test (IUT) I of a communication protocol is given as Coupled DFSMs consisting of k DFSMs

$$I = (I_1, I_2, \ldots, I_k)$$

where each I_j must satisfy the following properties: I_j has the same set X_j

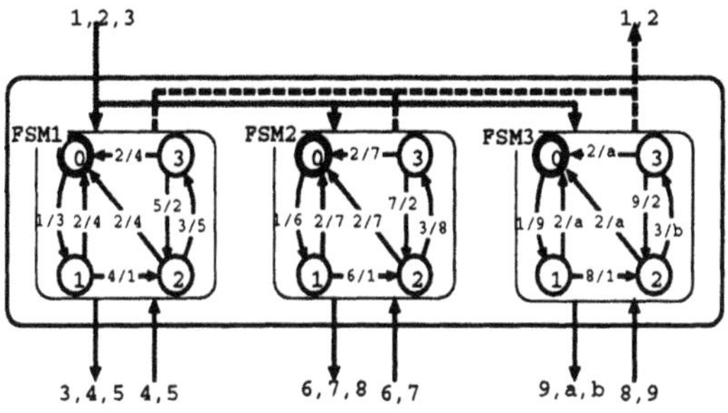

Figure 1 Example of Multi-Link Protocol.

Table 1 Inputs

	Inputs
1	link inc. req.
2	link dec. req.
3	data trans. req.
4,6,8	connect confirm
5,7,9	data trans. ack

Table 2 Outputs

	Outputs
1	connect confirm
2	data trans. ack
3,6,9	link inc. indication
4,7,a	link dec. indication
5,8,b	data trans. indication

Table 3 States

	States
0	disconnected
1	wait(connection)
2	connected
3	wait(data)

of input symbols as A_j. A set of output symbols in I_j is equal to the set $Y = Y_1 \cup Y_2 \cup \ldots \cup Y_k$ that is the set of all output symbols used in the protocol. That is, there may exist a fault that I_j gives an output y such as $y \notin Y_j \wedge y \in Y_i \ (i \neq j)$. The number of states of each I_j does not exceed that $(|S_j|)$ of A_j.

We also suppose that there exists a reliable reset operation so that the whole Coupled DFSMs can be reset to their initial states at a time. □

$I_j(1 \leq j \leq k)$ is said to be a *sub-IUT* of I. As stated above, we suppose that both a specification A and an IUT I are modeled by the same number k of DFSMs, however, the internal states of I are not observable. That is, the IUT I is considered to be a black-box and will be tested. On the other hand, the specification A must be ONFSM as a whole, but the IUT I could have non-observable and non-deterministic transitions.

Definition 5 (Conformance of Communication Protocol) For a specification $A = (A_1, A_2, \ldots, A_k)$ and an IUT $I = (I_1, I_2, \ldots, I_k)$ of communication protocols stated above, we define that I is a correct implementation of A if each sub-IUT I_j of I is equivalent to the corresponding sub-specification A_j of A. □

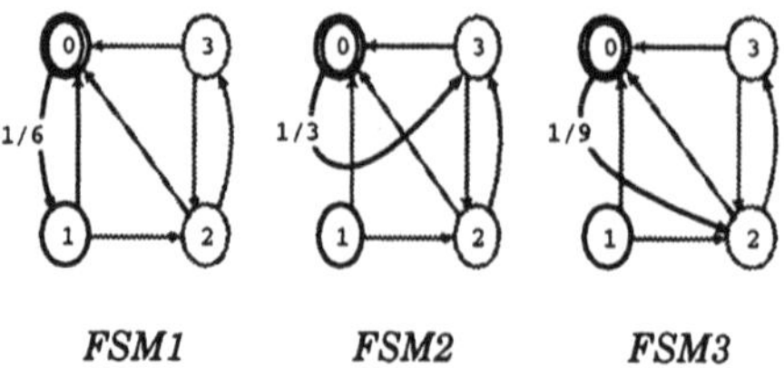

Figure 2 A Faulty IUT for Fig. 1.

2.2 Example Protocol in Our Model

We consider the protocol shown in Fig. 1 as an example. Fig. 1 represents a specification of a protocol which can dynamically vary the number of links between the lower and upper layers. On this specification, at most three links are set up by the orders from the upper layer, where the finite state control for each link is modeled as a DFSM. Table 1,2,3 represents the contents of inputs, outputs and states.

Whenever a "Link Increase request" is issued to this protocol by the upper layer, a link is newly set up by one of DFSMs which has not set up any link to the lower layer. In the case of "Link Decrease request" by the upper layer, one of DFSMs which has been holding a link to the lower layer cuts the link. Only executable DFSMs can compete with each other for these Link Increase/Decrease requests. That is, when an input 1 ("Link Increase request") is issued, one of DFSMs with states where the input 1 is not ignored is chosen by non-deterministically and responds to the input 1. For example, in case that FSM2 is in the state where the input 1 is ignored (such as state 1 in Fig. 1), FSM2 is never chosen. If all DFSMs are in the states where the input 1 is ignored, we consider that one of them ignores the input 1. In this specification, we can easily check which DFSM responds to each Link Increase/Decrease request from its output. That is, the specification is modeled as an ONFSM as a whole. Fig. 2 is an example of a faulty implementation of the protocol shown in Fig. 1. Hereafter, we assume that the italic-faced *FSMi* denotes a sub-IUT corresponding to the sub-specification FSMi. At the tuple of states $(0,0,0)$ of the Coupled DFSMs, their faults cannot be detected easily, since as a whole the outputs for the input 1 are the same as the specification (the set of $3, 6$ and 9).

3 OUTLINE OF GWP-METHOD

The GWp-method[6] is an extended version of the Wp-method so that ONFSMs can be dealt with. In the GWp-method, testing consists of two parts : one is state identification whether there exist all states of the specification in a given IUT, and the other is confirmation of transitions whether all transitions of the specification are correctly implemented in a given IUT. Both of testing are carried out by checking whether the output sequences obtained by applying test sequences to the IUT are equal to those from the specification. Since an ONFSM has non-deterministic actions, in general, a number of variety of output sequences may be produced as the response for an input sequence. So, in the GWp-method, we give each test sequence to the IUT for several times. If the obtained set of the output sequences is not equal to that from the specification, we regard that the IUT has a fault. If the sets are the same, we regard that the IUT returns the correct response for the test sequence and continue the test using other test sequences. If the IUT returns the correct response for all test sequences, then we regard that the IUT is a correct implementation of a given specification.

The set of test sequences in the GWp-methods is given by constructing two kinds of sets of sequences : the *characterization set* (W set) and *state cover* (V set).

Definition 6 (Characterization Set) The *characterization set* W for a FSM is given as a set of input sequences called *characterization sequences.* Each state in a given FSM must be uniquely identified by observing the set of output sequences obtained by applying all input sequences in W. □

Definition 7 (Transfer Sequences to Goal States) A *transfer sequence* to a state s in a FSM is an input sequence which makes the FSM move from the initial state to the goal state s. The set V of transfer sequences to all states in the FSM is called a *state cover.* □

In NFSM models, there may exist several reachable states for a given transfer sequence because of non-deterministic behavior. Therefore, for an ONFSM, if the ONFSM does not produce the expected output sequence for a given transfer sequence, we decide that the ONFSM does not reach the expected state because of non-deterministic behavior, and try the test again.

Definition 8 (Test Suite) The test suite for state identification is the concatenation $V.W$ of V and W. The test suite for confirmation of transitions is defined as follows using V, W and X (the set of input symbols) : $V.X \oplus W = \{\sigma.w | \sigma \in V.X, s_0 \xrightarrow{\sigma} s_i, w \in W\}$ □

4 PROPOSED TESTING METHOD

In the proposed method, for a specification $A = (A_1, A_2, \ldots, A_k)$ and an implementation $I = (I_1, I_2, \ldots, I_k)$ modeled as Coupled DFSMs, first, we give some conditions for the characterization set used in the testing. Then, using the characterization set satisfying the conditions, we will carry out the test whose cost is proportional to the sum of the numbers of states and transitions in the DFSMs. Like the GWp-method, the proposed testing method is divided into the following two parts: (1) state identification and (2) confirmation of transitions.

4.1 Construction of Characterization Set

In the proposed method, for state identification, we assume that we can construct a characterization set W_i for each sub-specification A_i which satisfies the following two conditions(Def.9 and 10).

Definition 9 (Condition for common inputs) For each characterization set W_i containing common inputs, W_i must be able to identify each state in A_i even if A_i ignores the common inputs in W_i. □

Then, we construct the following set W.

$$W = W_1 \cup W_2 \cup \cdots \cup W_k$$

For each input sequence σ in W, let $\overline{\sigma_i}$ denote the input sequence obtained from σ by deleting all the input symbols which the sub-specification A_i cannot respond to. And let W_i' denote the set of $\overline{\sigma_i}$ for all input sequences σ in W. Here, we also treat each W_i' as a characterization set for A_i.

As an example, let's consider the protocol in Fig. 1. Suppose that the following characterization sets are constructed.

$$W_1 = \{35, 4, 5\}, W_2 = \{16, 7, 6\}, W_3 = \{18, 9, 8\}$$

Then, we obtain the characterization set W for the total system as follows:

$$W = \{35, 4, 5, 16, 7, 6, 18, 9, 8\}$$

Also, the characterization sets

$$W_1' = \{35, 4, 5, 1\}, W_2' = \{16, 7, 6, 3\}, W_3' = \{18, 9, 8, 3\}$$

are obtained.

Here, we must take the above condition into consideration for the construction of W_1. In Table 4, even if we make a smaller characterization set $W_1'' = \{35, 4\}$, it can identify four states of FSM1. But the common input 3 in W_1'' may be taken by one of other DFSMs, that is, in Table 4, the corresponding output 5 with underline may be changed to ε. In this case, we cannot distinguish state 2 from state 3. So it is necessary to construct the characterization set W_1 which can identify each state in FSM1 even for such a case. That is the reason why we add another sequence 5 to W_1. By adding the sequence 5 to W_1, we can distinguish state 2 from state 3 even if the common input 3 is taken by another DFSM. Also in Tables 5 and 6, we add sequences 6 and 8 to W_2 and W_3, respectively, from the same reason.

Table 4 W_1

State	35	4	5
0	$\varepsilon\varepsilon$	ε	ε
1	$\varepsilon\varepsilon$	1	ε
2	$\underline{52}$	ε	ε
3	$\varepsilon 2$	ε	2

Table 5 W_2

State	16	7	6
0	$\underline{61}$	ε	ε
1	$\varepsilon 1$	ε	1
2	$\varepsilon\varepsilon$	ε	ε
3	$\varepsilon\varepsilon$	2	ε

Table 6 W_3

State	18	9	8
0	$\underline{91}$	ε	ε
1	$\varepsilon 1$	ε	1
2	$\varepsilon\varepsilon$	ε	ε
3	$\varepsilon\varepsilon$	2	ε

Definition 10 (Condition for response of other DFSMs) Let C_i denote the set of common inputs contained in W_i'. We assume that we can construct each W_i' such that every $A_j (j \neq i)$ has a state which ignores all common inputs in C_i. □

For state identification, we only assume the above two conditions. However, for confirmation of transitions, we need a further assumption. If a state s in a sub-specification A_i ignores a common input x, and if another sub-specification A_j produces an output y at every state for the common input x. In this case, we cannot confirm logically that $s \stackrel{x/\varepsilon}{\rightarrow} s$ in A_i is correctly implemented. The IUT produces the same outputs as the total system even if I_i produces the output y for the input x at state s, since at any tuple of states containing state s, I_j produces the output y.

Therefore, we also give the following assumption for each sub-specification A_i.

Definition 11 (Assumption for each DFSMs) Let D denote the set of common inputs. We assume that every A_i has a state which ignores each common input in D. □

In general, even if there exists a characterization set for each sub-specification, there may not exist a characterization set which satisfies the above three conditions. However, for most cases that there are not so many common input symbols in each characterization set, we believe that we can construct characterization sets which satisfy the above three conditions (for example, see an example in Section 6).

4.2 State Identification

We identify all the states in DFSMs as follows.

- Selection of tuples of states
  Let s_j^i denote a state in $A_j (j \neq i)$ which can ignore all inputs in C_i. For each state s_p of A_i, we treat $ss_p^i = (s_1^i, s_2^i, \cdots, s_k^i)$ $(s_i^i = s_p)$ as the tuple of

states to identify the state s_p. Note that when we give any input in C_i to the tuple of states ss^i_p, only the state s_p of A_i can make a response.

- Giving test sequences
  We give a test suite $V.W'_i$ to the tuples of states constructed above for several times. Here, each v^i_p in V is the sequence obtained by concatenating all transfer sequences $v^i_1, v^i_2, \cdots, v^i_k$ for the states $s^i_1, s^i_2, \cdots, s^i_k$ in ss^i_p. And we use the same v^i_p repeatedly while testing for a state. If we get an unexpected output sequence as the result for the transfer sequence v^i_p, we consider that non-deterministic behavior of Coupled DFSMs makes the IUT move to a tuple of states except ss^i_p and we try to give the transfer sequence v^i_p again. If we get the correct output sequence for v^i_p, then we observe a response of ss^i_p for each input of W'_i. If the obtained set of output sequences is not equal to that from the specification, we conclude that the IUT is faulty.

We apply the above method to all states in each sub-specification $A_i(1 \leq i \leq k)$. If we cannot find faulty states, we conclude that we have identified all the states in the IUT.

As an example, we try to identify states of FSM1 in Fig. 1.

- Selection of tuples of states
  The characterization set W'_1 has common inputs $C_1 = \{1, 3\}$. For example, both state 1 of FSM2 and state 1 of FSM3 ignore these common inputs 1 and 3. So, we select a tuple of states $(0, 1, 1)$ for identifying state 0 of FSM1. We select $(1, 1, 1)$ for state 1, $(2, 1, 1)$ for state 2 and $(3, 1, 1)$ for state 3, respectively.
- Giving test sequences
  We give the test suite $V.W'_1$ to the IUT for several times and observe a response from the IUT. This V is a set of transfer sequences to the chosen four tuples, for example, $V = \{11, 111, 1411, 14511\}$. For testing a tuples of states $(2, 1, 1)$, if the output sequence for a transfer sequence 1411 is not 3169, we decide that we couldn't transfer the IUT to a tuple of states $(2, 1, 1)$ because of non-deterministic behavior. Then we try the test again.

4.3 Confirmation of Transitions

For a transition $s_p \xrightarrow{x/y} s_q(x \in X, y \in Y \cup \{\varepsilon\})$ in each sub-specification A_i, we confirm the transition by dividing the following two cases.

If the input x is not a common input, we give a test suite $v.x.W'_i$ where v is the same transfer sequence used to identify the state s_p.

If the input x is a common input, we give a test suite $v.W \cup v.x.W'_i$. Here, the transfer sequence v in this test suite is constructed as follows. We find a tuple of states $ss^i_p = (s^i_1, s^i_2, \cdots, s^i_k)$ $(s^i_i = s_p)$ where each state s^i_j of $A_j(j \neq i)$

ignores the common input x. Then, v may be any transfer sequence to ss_p^i. Here, the test suite $v.W$ is used to identify all states in the tuple of states ss_p^i (note that W is the union of all characterization sets W_i' for identifying each state s_j^i). The test suite $v.x.W_i'$ is used for confirming the transition and identifying the state which the sub-IUT I_i reaches after the transition is executed.

We apply the above method to all transitions. If the IUT is not faulty, we conclude that the implementation of transitions for the specification is correct.

Note that the numbers of test sequences used for identifying states and confirming transitions are proportional to the sum of those of states and transitions in DFSMs, respectively.

5 CORRECTNESS OF TESTING METHOD

5.1 Correctness of State Identification

For a specification modeled as Coupled DFSMs, we must take the following two notices. One is the existence of the common inputs. Assume that for an identification of a state s in A_i, we can observe the expected output 1 after providing an input x. At this time, the IUT may be implemented badly, since it is possible that A_i doesn't give the output 1 and the other A_j gives the output. This is the reason why we cannot identify a state even if we can observe the expected response for the IUT. The other is our assumption that the IUT may have non-observable and non-deterministic transitions. On this case, for a given transfer sequence, even if we can observe the expected output sequence, we cannot guarantee that the IUT is led to the tuple of states which we want to lead. The IUT may be led to several tuples of states.

For example, the response of a tuple of states $(2,1,1)$ for W_1' is,

$$35/5\varepsilon, 4/\varepsilon, 5/\varepsilon, 1/\varepsilon$$

Here, a sub-IUT $FSM1$ may ignore the common inputs 1 and 3. Then the sub-IUT $FSM1$ has at least one of states which return the response like the following (there may exist both states) :

$$(35/5\varepsilon, 4/\varepsilon, 5/\varepsilon, 1/\varepsilon), (35/\varepsilon\varepsilon, 4/\varepsilon, 5/\varepsilon, 1/\varepsilon)$$

In order to consider all possibilities stated above, we introduce the state variables in Table 7 where the value of each variable is true if and only if there exists the corresponding state in the sub-IUT $FSM1$. The above condition can be expressed as the following logical formulas.

- For the tuple of states $(0,1,1)$, $\varphi 11 \vee \varphi 12$
- For the tuple of states $(1,1,1)$, $\varphi 13$
- For the tuple of states $(2,1,1)$, $\varphi 14 \vee \varphi 15$

Table 7 Var(FSM1)

Var	*Response for* W_1'
$\varphi 11$	$35/\varepsilon\varepsilon, 4/\varepsilon, 5/\varepsilon, 1/3$
$\varphi 12$	$35/\varepsilon\varepsilon, 4/\varepsilon, 5/\varepsilon, 1/\varepsilon$
$\varphi 13$	$35/\varepsilon\varepsilon, 4/1, 5/\varepsilon, 1/\varepsilon$
$\varphi 14$	$35/52, 4/\varepsilon, 5/\varepsilon, 1/\varepsilon$
$\varphi 15$	$35/\varepsilon 2, 4/\varepsilon, 5/\varepsilon, 1/\varepsilon$
$\varphi 16$	$35/\varepsilon 2, 4/\varepsilon, 5/2, 1/\varepsilon$

Table 8 Var(FSM2)

Var	*Response for* W_2'
$\varphi 21$	$16/61, 7/\varepsilon, 6/\varepsilon, 3/\varepsilon$
$\varphi 22$	$16/\varepsilon 1, 7/\varepsilon, 6/\varepsilon, 3/\varepsilon$
$\varphi 23$	$16/\varepsilon 1, 7/\varepsilon, 6/1, 3/\varepsilon$
$\varphi 24$	$16/\varepsilon\varepsilon, 7/\varepsilon, 6/\varepsilon, 3/8$
$\varphi 25$	$16/\varepsilon\varepsilon, 7/\varepsilon, 6/\varepsilon, 3/\varepsilon$
$\varphi 26$	$16/\varepsilon\varepsilon, 7/2, 6/\varepsilon, 3/\varepsilon$

Table 9 Var(FSM3)

Var	*Response for* W_3'
$\varphi 31$	$18/91, 9/\varepsilon, 8/\varepsilon, 3/\varepsilon$
$\varphi 32$	$18/\varepsilon 1, 9/\varepsilon, 8/\varepsilon, 3/\varepsilon$
$\varphi 33$	$18/\varepsilon 1, 9/\varepsilon, 8/1, 3/\varepsilon$
$\varphi 34$	$18/\varepsilon\varepsilon, 9/\varepsilon, 8/\varepsilon, 3/b$
$\varphi 35$	$18/\varepsilon\varepsilon, 9/\varepsilon, 8/\varepsilon, 3/\varepsilon$
$\varphi 36$	$18/\varepsilon\varepsilon, 9/2, 8/\varepsilon, 3/\varepsilon$

- For the tuple of states $(3, 1, 1)$, $\varphi 16$

In our proposed method, we have selected a state in A_j $(j \neq 1)$ which ignores all the common inputs in W_1'. So, each formula includes a state which returns the same response as the specification (e.g. $\varphi 11$ for the tuple of states $(0, 1, 1)$), and it may include a state whose response for a common input is ε (e.g. $\varphi 12$ for the tuple of states $(0, 1, 1)$).

For the sub-IUT $FSM2$, we can get the following logical product of formulas using state variables in Table 8.

$$(\varphi 21 \vee \varphi 22) \wedge \varphi 23 \wedge (\varphi 24 \vee \varphi 25) \wedge \varphi 26$$

For the sub-IUT $FSM3$, we can get the following logical product of formulas using state variables in Table 9.

$$(\varphi 31 \vee \varphi 32) \wedge \varphi 33 \wedge (\varphi 34 \vee \varphi 35) \wedge \varphi 36$$

On all of the cases, the formula for each tuple of states always has one state whose response is expected on the specification, and it may have another state whose response is equal to the expected response except that the output for the common input is changed to ε. And also, the set of formulas for one sub-IUT has no same state variables, that is, all state variables in the set of formulas are different. Since we assume that the number of states of each sub-IUT I_i does not exceed the number of states N_i of the corresponding sub-specification A_i, we must select at most N_i state variables to be true in order to make all of N_i formulas be true. So, we cannot select one state variable to be true so that two formulas can be true together. Then we must select only one of state variables in each formula to be true so that we make all of N_i formulas be true.

If we select the state variable whose response is not the expected output y but the empty output ε, we can consider that the output y, which we observed when we gave the characterization set to the IUT, was obtained from an other sub-IUT I_j. However, the formulas for the other sub-IUT I_j have only state variables whose response is expected on the specification or equal to the expected response except that the output for the common input is changed to ε. So, each state in the other sub-IUT I_j cannot produce the output y. Also, because of the limitation on the number of state variables to be true, we cannot select so much number of state variables.

From the above, all we can do is to select the state variables whose response is equal to the expected response on the specification in order to make all formulas be true. That is the reason why our method is correct for state identification.

We can identify the following four states for $FSM1$.

$$\varphi 11, \varphi 13, \varphi 14, \varphi 16$$

We can also identify all states in $FSM2$ and $FSM3$.

Here, we can say something about non-deterministic behavior of the IUT. Until now, when we give the IUT a transfer sequence v which can make A_i move to s_p, we cannot guarantee that the state of the sub-IUT I_i is s_p truly. However, now we can say that (1) each sub-IUT I_i has a state corresponding to each state in the sub-specification A_i, and (2) we can exactly lead the state of a sub-IUT to any state by the transfer sequence used to identify the state.

5.2 Correctness of Confirmation of Transitions

From now, we explain why by our proposal method (by giving test suite $v.x.W_i$ or $v.W \cup v.x.W_i'$), we can check whether each transition $s_p \overset{x/y}{\rightarrow} s_q (x \in X, y \in Y \cup \{\varepsilon\})$ in the sub-specification A_i is correctly implemented on the corresponding sub-IUT I_i.

(Case 1)

If the input x is not a common input, we give the IUT the test suite $v.x.W_i'$ where v is the transfer sequence used to identify the state s_p. We can guarantee that the starting state of the transition x/y is s_p truly since v is the sequence used to identify the state s_p. And by observing outputs from the IUT for $x.W_i'$, we can identify the state after x/y is executed.

(Case 2)

If the input x is a common input, we give the IUT the test suite $v.W \cup v.x.W_i'$. In general, when testing of state identification has not been finished, each sub-IUT may have a wrong state whose response is not equal to that on the sub-specification even if the response of the sub-IUT for all of W_i' is correct. But if testing of state identification has been finished, and if we can get the expected response, it shows the existence of the state of I_i corresponding to a state of the sub-specification A_i (because there are not any other possibilities). Then if we can observe the expected outputs when we give W for the tuple of states which the transfer sequence v leads ($ss_p^i = (s_1^i, s_2^i, \cdots, s_k^i)$ $(s_i^i = s_p)$), we can guarantee that we have led the IUT to ss_p^i.

Next we explain why we can confirm transitions by the test suite $v.x.W_i'$ by dividing into three cases.

(Case 2.1)

We first consider the case that the state is changed by execution of the transition $s_p \xrightarrow{x/y} s_q (x \in X, y \in Y \cup \{\varepsilon\})$. Because we have constructed the tuple of states so that a state s_j^i can ignore the common input x, the output must be only y when the input x is given at ss_p^i. At this time, by observing the response for W_i', we can check whether the common input x is taken into other DFSMs. If another DFSM takes the input x and produces the output y, and if the state s_p of the sub-IUT I_i is badly implemented as it can ignore the input x, then the response for W_i' is not for s_q but s_p, and then we can find the error.

(Case 2.2)
Secondly we consider the case that the state is not changed by execution of the transition $s_p \xrightarrow{x/y} s_q (x \in X, y \in Y \cup \{\varepsilon\})$, and the output y is ε. Since we have constructed the tuple of states ss_p^i so that every state in ss_p^i ignores the common input x, we can find the error if we cannot observe the output ε after giving x. We can also find the error that s_p and s_q are not the same if the response for W_i' is different from s_p's response.

(Case 2.3)
Lastly, we consider the case that the state is not changed by execution of the transition $s_p \xrightarrow{x/y} s_q (x \in X, y \in Y \cup \{\varepsilon\})$, and the output y is not ε. On this case we can test in a similar way. We have constructed the tuple of states ss_p^i so that every state in I_j $(j \neq i)$ ignores the common input x. On this case that the output y is not ε, however, we should consider that the IUT may be implemented as that state s_p of I_i can ignore the common input x and a state of another sub-IUT I_j can give the output y without changing its state. On the condition of $(s_p = s_q)$ and $(y \neq \varepsilon)$, we cannot deny such a possibility only by confirmation of the transitions of I_i. However the possible error is that the output y may be changed to ε like the testing for state identification. We can guarantee that there is not such a possibility by confirmation of all transitions in each sub-IUT I_j. That is, by confirming of the transitions at all states of each sub-IUT I_j, we can assure that each sub-IUT I_j can give either an output z which can be used only by A_j or only ε. So we can also assure that I_j cannot give the output y which can be used only by A_i, and finally we can guarantee that I_i gives the output y truly.

From the above, our proposal method guarantees that the implementation of transitions is correct.

6 EXAMPLE

We have applied our testing method to the abracadabra protocol. We have simplified an Estelle specification of the abracadabra protocol in [3] and modified it so that three sending processes run in parallel. Fig. 3 is its specification in our model where I/O symbols are represented symbolically. From this spec-

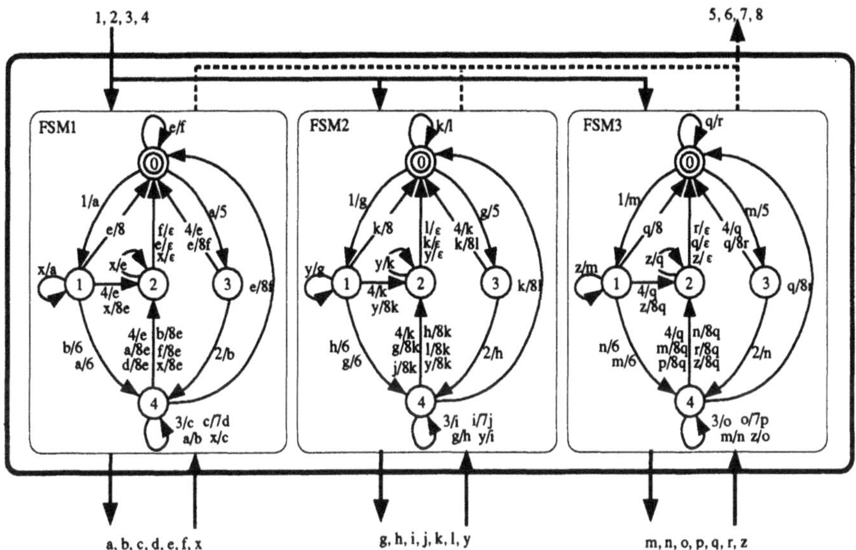

Figure 3 Example Spec. for Abracadabra Protocol.

ification, we have constructed characterization sets satisfying the conditions mentioned in Section 4.1.

7 CONCLUSION

In this paper, we propose a conformance testing method for communication protocols modeled as a set of DFSMs with common inputs. In the proposed method, we check each DFSM independently. So, the cost is only proportional to the sum of the numbers of states and transitions of DFSMs. Although we assume the existence of characterization sets which satisfy some conditions, we believe many communication protocols satisfy the conditions.

One of the future work is to extend the class of DFSMs so that we can treat the communications between DFSMs and internal actions in DFSMs.

8 REFERENCES

[1] T. S. Chow : "Testing Software Design Modeled by Finite-State Machines", IEEE Trans. on Software Engineering, Vol. 4, No. 3, pp.178–186, 1978.

[2] G. Gonenc : "A Method for the Design of Fault-Detection Experiments", IEEE Trans. on Computers, Vol. C-19, No. 6, pp. 551–558, 1970.

[3] ISO : "Information Technology – Open Systems Interconnection – Guidelines for the application of Estelle, LOTOS and SDL", ISO/IEC/TR 10167, 1991.

[4] H. Kloosterman : "Test Derivation from Nondeterministic Finite State Machines", Proceedings of the IFIP TC6 Fifth International Workshop on Protocol Test Systems, pp.297–308, 1992.
[5] D. Lee, K. K. Sabnani, D. M. Kristol and S. Paul : "Conformance Testing of Protocols Specified as Communicating FSMs", Proceedings of IEEE INFOCOM'93, pp.115–127, 1993.
[6] G. Luo, G. v. Bochmann and A. Petrenko : "Test Selection Based on Communicating Nondeterministic Finite State Machines using a Generalized Wp-method", IEEE Trans. on Software Engineering, Vol. 20, No. 2, pp.149–162, 1994.
[7] S. Naito and M. Tsunoyama : "Fault Detection for Sequential Machines by Transition Tours", Proceedings of the 11th IEEE International Symposium on Fault Tolerant Computing, pp.238–243, 1981.
[8] D. P. Sidhu and T.-K. Leung : "Formal Methods for Protocol Testing : A Detailed Study", IEEE Trans. on Software Engineering, Vol. 15, No. 4, pp.413–426, 1989.
[9] K. K. Sabnani and A. T. Dahbura : "A Protocol Testing Procedure", Computer Networks and ISDN Systems, Vol. 15, No. 4, pp.285–297, 1988.
[10] P. H. Starke : "Abstract Automata", North-Holland/American Elsevier, 1972.
[11] P. Tripathy and K. Naik : "Generation of Adaptive Test Cases from Nondeterministic Finite State Models", Proceedings of the IFIP TC6 Fifth International Workshop on Protocol Test Systems, pp.309–320, 1992.
[12] A. Ulrich and S. T. Chanson : "An Approach to Testing Distributed Software Systems", Proceedings of the IFIP WG6.1 15th International Symposium on Protocol Specification, Testing and Verification, pp.121–136, 1995.
[13] T. Kaji, T. Higashino and K. Taniguchi : "A Conformance Testing for Communication Protocols Modeled as Coupled DFSMs", Information Processing Society of Japan (IPSJ) SIG Notes 96–DPS–74, Vol. 96, No. 12, pp.19–24, 1996 (in Japanese).

9 BIOGRAPHY

Atsushi Fukada is a Master student of the Department of Information and Computer Sciences, Osaka University, Japan. His research interests include testing of communication protocols. Tadashi Kaji finished his Master course at the same department in 1996 and is currently working in Hitachi Co. Ltd.

Teruo Higashino and Kenichi Taniguchi are an Associate Professor and a Professor in the same department, respectively. Masaaki Mori is a Professor in Shiga University, Japan. Their research interests include communication protocols, formal description techniques and software engineering.

16

On test case generation from asynchronously communicating state machines

O. Henniger
GMD–German National Research Center for Information Technology
Rheinstr. 75, D-64295 Darmstadt, Germany
E-mail: henniger@darmstadt.gmd.de

Abstract
This paper proposes an approach for generating test cases in Concurrent TTCN from a system of asynchronously communicating finite state machines. We give an algorithm for generating a noninterleaving model of prime event structures from a generalized model of asynchronously communicating finite state machines and deal with the generation of test cases from prime event structures.

1 INTRODUCTION

The behaviour of communicating systems can be modelled by means of a set of finite state machines (FSMs) that run concurrently and communicate with each other via First-In-First-Out (FIFO) queues. The formal description techniques Estelle [ISO89] and SDL [ITU92] are based on such a model, extended by additional features.

The behaviour described by an individual state machine is characterized by a set of sequences (i.e. totally ordered sets) of events. The behaviour described by a set of communicating state machines could also be characterized by the set of all possible sequences of events, where events of the individual state machines are intermixed. In fact, this is done in approaches generating a composite state machine and in interleaving semantics definitions.

Testing of Communicating Systems, M. Kim, S. Kang & K. Hong (Eds)
Published by Chapman & Hall © 1997 IFIP

Interleaving models preserve many properties of the original specification and are relatively easy to formalize. However, the combinatorial explosion in the number of possible event sequences (interleavings) forms a major problem in generating interleaving models and renders these approaches infeasible in many practical cases. Knowledge of all possible interleavings is in many cases not necessary: If events occurring at different state machines are independent, then their particular order of occurrences does not change their combined effect (cf. e.g. [Pra86]).

Furthermore, interleaving models hide the independence of events occurring at different state machines. This is a problem in test case generation where often a finite subset of possible event sequences has to be selected: In case of independent events from different state machines, there is no way to control the order of occurrence of these events, and there is no guarantee that a selected interleaving can really be observed.

At present, Concurrent TTCN, an amendment to the test description language TTCN [ISO91] designed to specify test cases in a multi-party testing context, is in the process of being standardized. In conventional TTCN, the behaviour descriptions of all test components have to be interleaved in a single tree. Even if split up into several local trees or test steps attached to each other, the behaviour description of a test case forms a single "evaluation tree" as defined in the operational semantics of conventional TTCN. Concurrent TTCN allows for independently executable test components each of which processes its own evaluation tree [BG94].

All these reasons call for the use of noninterleaving models for test case generation and have initiated recent work on this topic. This paper gives an algorithm for generating a prime event structure model from a generalized model of asynchronously communicating state machines. This algorithm is adaptable to the standardized description techniques Estelle and SDL. Furthermore, the paper deals with test case generation from the generated prime event structure.

The rest of the paper is organized as follows: Section 2 briefly reviews related work. Section 3 contains definitions for the models used. Section 4 gives an algorithm for generating a prime event structure equivalent to a set of asynchronously communicating state machines. Section 5 deals with the generation of test cases in Concurrent TTCN from a prime event structure. Section 6 gives concluding remarks.

2 RELATED WORK

Previous work on test case generation from state machine models mainly focused on the model of a single FSM (see e.g. [Sid90]) or of a single EFSM (extended finite state machine) ([CA90], [UY91], [CZ93], [HUK95], and others). Test case generation from systems of communicating FSMs requires different approaches. Several approaches for generating black-box test cases from systems of communicating FSMs have been proposed.

Methods with explicit test purposes ([GHN93], [WL93], [FJJV96]) ensure the consistency between test cases, specification, and test purposes and offer much flexibility to the test specifier. However, they require considerable manual effort to define appropriate test purposes and do not guarantee a systematic fault coverage.

Methods with implicit test purposes generally ensure a systematic coverage of the specification. However, as they need to explore the possible behaviour, these methods suffer from the state-explosion problem. Different approaches to alleviate the state-explosion problem have been proposed. [LSKP93] pursues an approach similar to program slicing, pruning the given communicating FSMs to contain only a subset of actions; thus yielding a set of smaller, simplified specifications. For synchronously communicating FSMs, [SLU89] and [TK93] generate a composite FSM as an interleaving model by incremental composition and reduction. [AS91], [KCKS96], and [Ulr97] aim at diminishing the state explosion by generating noninterleaving models of the original specification and have inspired the work presented in this paper. [AS91] builds on the reduced reachability analysis approach proposed in [II83] and [KIIY85]. The reduced reachability tree generated by the algorithm presented in [AS91] still contains redundant information. In [KCKS96] a formalized approach is outlined, but no complete algorithm is given. [Ulr97] makes a detour by first transforming the set of communicating FSMs into an 1-safe Petri net, then unfolding the net ([McM95], [ERV96]), and finally constructing a "behaviour machine" as a suitable starting point for test case generation.

We extend the previous work in [AS91], [KCKS96], and [Ulr97] by presenting an algorithm for direct transformation of a set of asynchronously communicating state machines into a noninterleaving model suitable for test generation.

The work on test case generation for concurrent systems benefits from the noninterleaving models and methods developed in the context of verification and concurrency theory, such as [Pra86], [Maz88], [Win88], [PL91], [God96], and others.

3 MODELS

3.1 Models for sequential behaviour

Depending on the level of abstraction, sequential behaviour can be modelled by means of various formalisms such as finite state machines, labelled transition systems, etc. On a high level of abstraction, the behaviour of a sequential discrete event system is characterized by a set of sequences of discrete, observable events, like transmission or reception of messages or time-outs. One distinguishes between events and actions: Actions label events. An event is an occurrence of its action. The same action can be performed various times producing a new, distinguishable event each time.

We use a state machine model similar to the one used in [ZWR+80].

Definition 1 An *input-output state machine (IOSM)* is a quadruple (S, A, T, s_0) where

- S is a finite non-empty set of states,
- A is a finite non-empty set of actions partitioned into a set of input actions A_I and a set of output actions A_O ($A_I \cup A_O = A, A_I \cap A_O = \emptyset$),
- $T \subseteq S \times A \times S$ is the transition relation, and
- $s_0 \in S$ is the initial state. □

An output action is denoted by $!a$; an input action by $?a$. Each $t \in T$ is a transition from the present state to a next state, associated either with an input or an output action. An IOSM is represented graphically by a directed graph where nodes represent states and arcs represent transitions.

3.2 Models for concurrency

Various formalisms have been defined for describing the behaviour of concurrent systems. A recent classification is given in [SNW96]. Models for concurrency are classified into:

- behaviour or system models,
- interleaving or noninterleaving models, and
- linear-time or branching-time models.

Behaviour models focus on describing the behaviour in terms of the order of events, abstracting away from states. In contrast, the so-called system models describe the order of events implicitly, explicitly representing states, which possibly repeat. Interleaving models are those that hide the difference between concurrency between several state machines and nondeterminism inside individual state machines. Noninterleaving models take this difference into account. Branching-time models represent the branching structure of the behaviour, i.e. the points in which choices are taken, while linear-time models do not.

Asynchronously communicating input-output state machines

Communicating state machines can be classified as a system/noninterleaving/branching-time model. The individual IOSMs communicate with each other and with their environment by exchanging messages. In case of synchronous communication, the exchange of a message between two state machines is regarded as a single event. In case of asynchronous communication, the exchange of a message takes a send and a receive event. After sending a message, the message is buffered for some time, generally in a FIFO queue, before it eventually will be received. We assume asynchronous communication via FIFO queues in order to come close to the semantics of Estelle, SDL, and TTCN.

Different queue semantics can be defined. In Estelle, one can choose whether the messages sent to a state machine (module instance, in Estelle terms) from several other state machines shall be interleaved in a shared queue (common queue) or not (individual queue). In SDL, messages sent to a state machine (process instance, in SDL terms) from arbitrary state machines are always interleaved in a shared queue (input port). In SDL, in some cases communication is nearly synchronous: Any message sent via a signal route or via a nondelaying channel to a process instance waiting in a current state without save or priority inputs is received instantaneously, not needing to be buffered.

We assume a generalized model similar to the one used in [ZWR+80] and base the further discussion on this model. A system of asynchronously communicating IOSMs is composed of a set of IOSMs and a set of perfect (i.e. without loss or reordering of messages) FIFO queues that connect IOSMs with each other and with their environment. Each pair of IOSMs may be connected by at most one FIFO queue for each direction. Note that in contrast to [ZWR+80] we do not require that the IOSMs form a closed system. We tolerate open interfaces to the environment. This frees us from manually specifying the behaviour of test components, which we would like to find algorithmically. The input messages from the environment are referred to as trigger messages (cf. [AS91]). Figure 1 shows an example system.

Reachability tree

A reachability tree is a behaviour/interleaving/branching-time model. The reachability tree of a system of asynchronously communicating state machines is a directed tree where

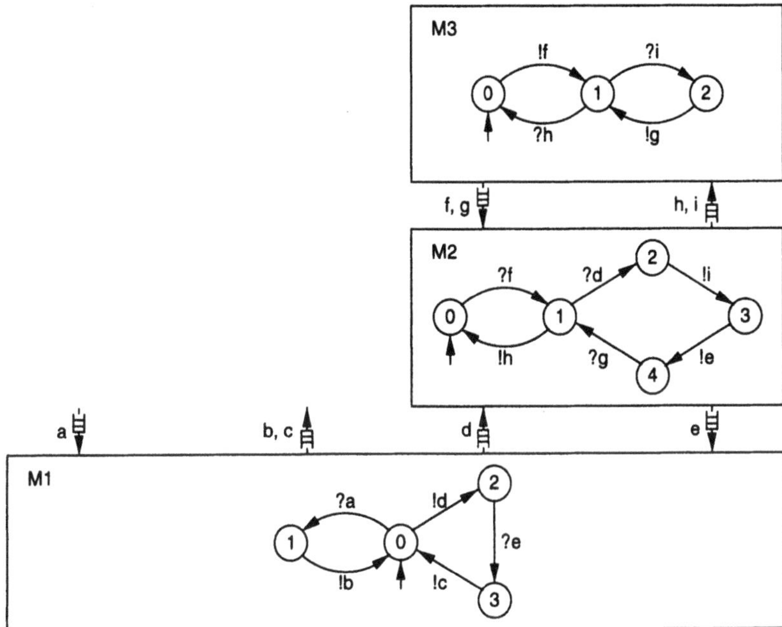

Figure 1 Example 1, a system of asynchronously communicating IOSMs.

the root is the initial global state, the set of nodes is the set of all global states reachable from the initial global state, the set of arcs is a set of events, labelled by actions, and arcs connect subsequent global states.

Figure 2 shows an initial part of the reachability tree for Example 1 computed by the perturbation method [ZWR+80]. For the generalized model with at most one queue for each direction between each pair of IOSMs, a global state can be nicely represented by means of a matrix gs: Each element gs_{ii} on the diagonal represents the current state of IOSM m_i, and each off-diagonal element $gs_{ij}, i \neq j$, represents the contents of the queue from m_i to m_j. "–" represents empty and nonexistent queues. The shaded nodes represent global states reached repeatedly, in which the reachability tree has been cut off. The initial global state consists of all IOSMs in their initial states and all queues empty. We assume that a transition with a trigger message (a message from the environment) is always enabled as soon as the IOSM is in the start state of that transition; the environment (which is the tester, in our case) is assumed to place the trigger message at the head of the queue to the IOSM.

The reachability tree models in an interleaving manner the same behaviour as the system of communicating IOSMs. For test case generation, a behaviour model, describing the order of events more explicitly than the compact system model does, would be very helpful. However, the reachability tree overspecifies the order relation and hides the independence of events in separate subsystems. Furthermore, because of the large number of nodes and arcs in the reachability tree (state explosion), computation of the reachability tree of a system of communicating IOSMs is not feasible in most practical cases. In the

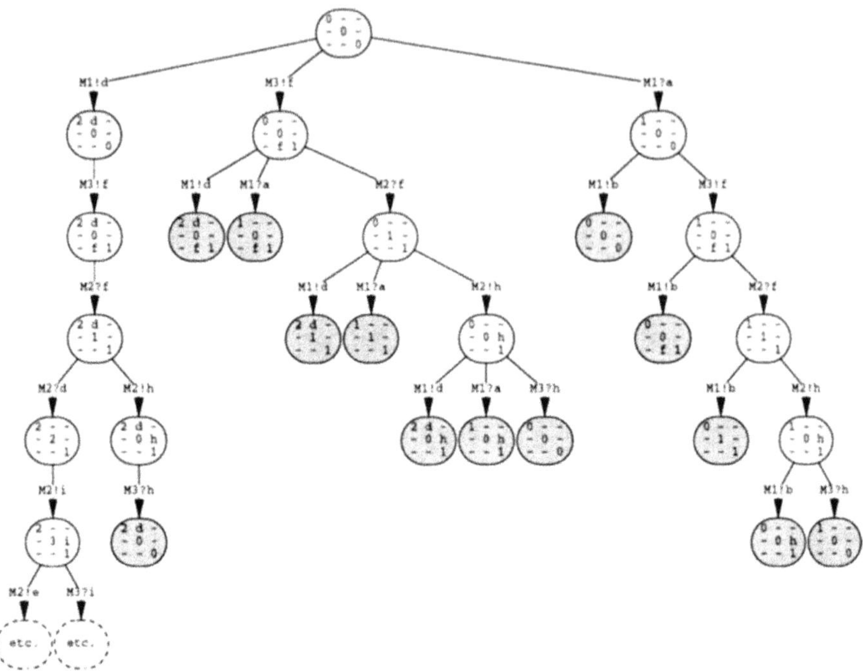

Figure 2 Initial part of the reachability tree for Example 1.

test generation approach described below, the reachability tree will not be computed. Figure 2 has been included to allow a comparison with the noninterleaving behaviour model that we use, the prime event structure (Figure 3).

Prime event structure

Event structures [NPW81] can be classified as behaviour/noninterleaving/branching-time models. Different event structure models have been defined in the literature. An overview can be found in [Kat96].

Definition 2 A *(labelled) prime event structure (PES)* is a quadruple $(E, \preceq, \#, l)$ where

- E is a countable set of events,
- $\preceq \subseteq E \times E$ is a partial order, the causality relation,
- $\# \subseteq E \times E$ is the (irreflexive and symmetric) conflict relation,
- $l : E \rightarrow A$ is the action-labelling function,

such that $\forall e \in E:$

(1) $\{e' \in E \mid e' \preceq e\}$ is finite, and
(2) $\forall e', e'' \in E : (e \# e' \wedge e' \preceq e'') \Rightarrow e \# e''$. □

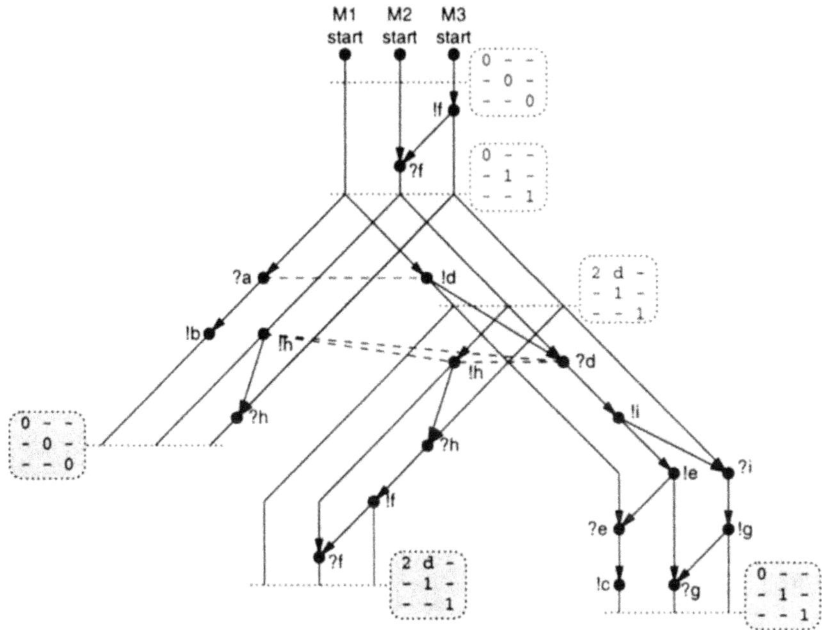

Figure 3 Initial part of the prime event structure for Example 1.

The meaning of $e \preceq e'$ is that if the events e and e' both happen, then e must happen before e'. Condition (1) states that the number of causes of any event is finite. The meaning of $e \# e'$ is that the events e and e' can not happen both. Condition (2) states that if an event e is in conflict with some event e', then it is in conflict with all causal successors of e' (conflict inheritance property). If two events are neither causally related nor in conflict, these events are independent from each other and both can occur in arbitrary order.

Figure 3 shows an initial part of the PES for the example of Figure 1. A PES is represented as a graph where bold-faced points represent events, directed arcs lead to the immediate causal successors of an event, and undirected dashed arcs connect events in immediate conflict. Next to an event e its label $l(e)$ is indicated. The global states indicated at the margin are not an integral part of the PES. They function as labels indicating where behaviour encountered earlier repeats. Events occurring in the same IOSM form a directed tree. The trees for the concurrent IOSMs are drawn with parallel arcs. The PES resembles the "space-time diagram" introduced in [Lam78], however, not with a linear, but a branching "time axis".

4 CONSTRUCTION OF A PRIME EVENT STRUCTURE

4.1 Starting point of the test case generation approach

The starting point of our test case generation approach is a correct specification of the implementation under test (IUT) and of the test context. The test context depends on the chosen test architecture. In the realm of protocol conformance testing, the test context includes the underlying service provider between IUT and lower tester as well as a user above the IUT.

The behaviour of test components needs not to be specified prior to test case generation. It will be derived algorithmically. The test components will be fitted to the open interfaces of the specification of IUT and test context. As for the reachability tree, we assume again that the environment provides for the right trigger messages when they are expected by the specification of IUT and test context. This is an appropriate assumption for conformance testing. It ensures that all expected external behaviour of IUT and test context is covered, while unexpected behaviour (as for robustness tests) is left out.

Correctness of the specification should be checked by validation techniques. In particular, queue overflow is regarded as a potential specification error. Unbounded growth of the number of messages in the queues is a problem in theory and in practice and can be avoided by appropriate design criteria [ZWR+80]. If all queues are bounded, then the specification has a finite (yet probably very large) state space, and analysis algorithms, like the one described below, terminate.

Figure 1 shows a simple example, applying the remote test method [ISO91]. The IOSM M2 models the IUT. M1 models the service provider between IUT and lower tester. M3 models the user above the IUT. The lower tester will be connected to the open interface of M1. This example is interesting as it is still easy to check, yet contains more than two concurrent IOSMs. As Figure 1 provides for only one test component, we have included another simple example that provides for two test components and is more suitable for demonstrating the generation of test cases in Concurrent TTCN. Figure 4 shows the specification of the same IUT, subjected to the distributed test method [ISO91]. Here, the service access point above the IUT is accessible, and the user above the IUT is replaced by an upper tester. Lower tester and upper tester reside in different real systems and have to communicate with each other using test coordination procedures. Abstract test cases for the distributed test method in conventional TTCN often leave the test coordination procedures unspecified. Abstract test cases in Concurrent TTCN have the advantage that the test coordination procedures are included in terms of coordination messages between the individual test components.

4.2 Algorithm for constructing a PES

Below, the algorithm for generating an equivalent PES from a system of asynchronously communicating IOSMs is described in a meta-programming language. To avoid excessive parameter passing, we present the algorithm using global data.

We need some definitions. The *global state gs* of a system of asynchronously communicating IOSMs is a k-tuple $(s_1, \ldots, s_n, q_1, \ldots, q_m)$ where

- $s_1, \ldots, s_n$ are the current states of the IOSMs $m_1, \ldots, m_n$ and
- $q_1, \ldots, q_m$ are the contents of the queues between the IOSMs.

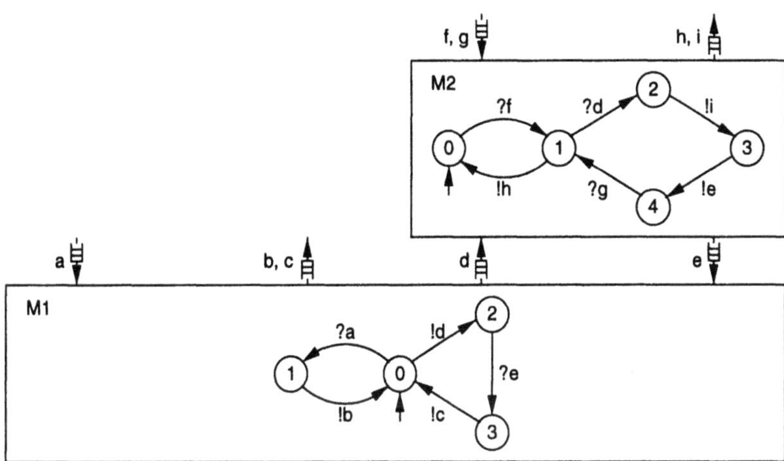

Figure 4 Example 2, a system of asynchronously communicating IOSMs.

A *configuration* C of a PES $(E, \preceq, \#, l)$ is a finite subset of E such that:

- $e \in C \Rightarrow \forall e' \preceq e : e' \in C$ (i.e., C is causally closed), and
- $\forall e, e' \in C : \neg(e \# e')$ (i.e., C is conflict-free).

The *final state* of a configuration C, denoted $fs(C)$, is the global state reached after all events $e \in C$, and no other events, have occurred.

Let gs be the global state of a system of asynchronously communicating IOSMs and let m_i be an IOSM from this system. A transition $t = (s, \mu, s')$ of m_i is *enabled* in gs if the current state of m_i is s and either

- t is an output transition ($\mu = !a$), or
- t is an input transition ($\mu = ?a$) and the message to be received is at the head of the corresponding queue.

$ET_i(gs)$ denotes the *set of enabled transitions* of m_i in gs. A transition $t = (s, \mu, s')$ of m_i is *potentially enabled* in gs if the current state of m_i is s and t is an input transition ($\mu = ?a$) with the corresponding queue empty. $PT_i(gs)$ denotes the *set of potentially enabled transitions* of m_i in gs. $next(gs, t)$ denotes the next global state reached from global state gs on execution of transition t. The different semantics of Estelle and SDL are taken care of in the generic functions computing the next global state and determining the enabled and potentially enabled transitions.

The algorithm updates E, $\preceq$, and l explicitly. The conflict relation $\#$ is implicitly given as any two events of the same IOSM m_i that are not causally related are in conflict to each other. A new event of m_i is denoted e_{ic_i} where c_i is the event counter for m_i.

The data structure *conf* represents a configuration of the PES to which events are appended. *conf* contains the following fields: *fs*, the final state of the configuration; $predecessor_i$ $(1 \leq i \leq n)$, the last event for m_i; $send_\mu$ $(\mu \in A)$, a FIFO queue for send events with label μ; $Wait_i$ $(1 \leq i \leq n)$, the set of potentially enabled transitions that m_i waits for. For breadth-first processing of alternative branches, the algorithm uses a FIFO queue data structure *conf_queue* with the two basic access operations *put* and *get*.

The algorithm can be outlined as follows. First, the data structures are initialized. Construction of the PES begins with the initial configuration containing only "dummy" *start* events, which are needed to make the *append* procedure applicable also for the beginning of the PES. Each enabled transition that is the only transition of its IOSM in the current state is appended to the PES. The global state reached is added to the set of visited global states *Visited*. The order of execution does not matter in case that transitions from several IOSMs can be appended. This is repeated until a global state is reached that has been reached before or until no more transitions without alternatives within the same IOSM are enabled. Then, we have reached a point where the PES branches out. For one IOSM m_i, all enabled alternative transitions are appended to the PES opening up new branches of the PES. If m_i has potentially enabled transitions in the current state, an additional new branch is opened up and the potentially enabled transitions are stored in the *Wait* set. In this new branch, no transitions are appended for m_i unless a potentially enabled transition becomes enabled. If the final state of a configuration in a new branch is new, the information in *conf* about this configuration is put into the FIFO queue *conf_queue*. One after the other, these configurations will be processed as described above for the initial configuration. This continues until there are no more nodes to be investigated in *conf_queue*.

Consider the example in Figure 1 and its PES in Figure 3. In the initial global state, the transitions $(0, ?a, 1)$ and $(0, !d, 2)$ of M1 and $(0, !f, 1)$ of M3 are enabled. $(0, !f, 1)$ is the only potential transition of M3 in this state and is executed first. In the new global state, the transition $(0, ?f, 1)$, the only potential transition of M2 in this state, is executed. Afterwards, no more transitions without alternatives are enabled, and the two conflicting transitions of M1 are executed leading to two new branches, and so on.

pes_construction;
  begin
    for all $i \in \{1, \ldots, n\}$ **do** $c_i := 0$;
    $E := \bigcup_{i:=1}^{n}\{e_{i0}\}$; $\preceq := \emptyset$; $l := \bigcup_{i:=1}^{n}\{(e_{i0}, start)\}$;
    $Visited := \emptyset$;
    $conf.fs := initial$;
    for all $i \in \{1, \ldots, n\}$ **do** $conf.predecessor_i := e_{i0}$;
    $put(conf\_queue, conf)$;
    repeat
      $conf := get(conf\_queue)$;
      while $i \in \{1, \ldots, n\}$ **such that**
        $(|ET_i(conf.fs)| = 1) \wedge (|PT_i(conf.fs)| = 0) \wedge (conf.fs \notin Visited)$ **do begin**
        $Visited := Visited \cup \{conf.fs\}$;
        for $t \in ET_i(conf.fs)$ **do** $append(conf, t)$;
      end;
      for one $i \in \{1, \ldots, n\}$ **such that**
        $(|ET_i(conf.fs)| \geq 1) \wedge (|ET_i(conf.fs) \cup PT_i(conf.fs)| > 1)$ **do begin**
        $branching\_point := conf$;
        if $(conf.Wait_i \neq \emptyset)$ **then** $ET_i := ET_i(conf.fs) \cap conf.Wait_i$

```
        else ETi := ETi(conf.fs);
        for all t ∈ ETi do begin
           append(conf, t);
           if (conf.Waiti ≠ ∅) then conf.Waiti := ∅;
           if conf.fs ∉ Visited then put(conf_queue, conf);
           conf := branching_point;
        end
        if (conf.Waiti = ∅) ∧ (|PTi(conf.fs)| > 0) then begin
           conf.Waiti := PTi(conf.fs);
           put(conf_queue, conf);
        end
     end
   until empty(conf_queue);
end.

append(conf;t);
 begin
   ci := ci + 1;
   E := E ∪ {e_ici};
   ⪯ := ⪯ ∪ {(conf.predecessori, e_ici)};
   with t = (s, μ, s') do begin
      if μ ∈ AO then put(conf.sendμ, e_ici);
      if μ ∈ AI then ⪯ := ⪯ ∪ {(get(conf.sendμ), e_ici)};
   end;
   l := l ∪ {(e_ici, μ)};
   conf.predecessori := e_ici;
   conf.fs := next(conf.fs, t);
 end.
```

4.3 Some properties of the generated model

A discrete event system is in a certain state at any time. The current state of a system depends on which events have happened before, i.e. on the history of the system, and determines which events can happen next, i.e. the possible continuations. In state-oriented models, such as a reachability tree, states of the system are explicitly represented as nodes of the graphical representation. In the PES model, the current global state is not explicitly represented. However, global states of the system are implicitly represented in a distributed manner. Each of the individual IOSMs may be in any "state" along a path of the parallel trees, where all causal predecessors happened before. As an example, consider the left-most path in Figure 3. After performing the event labelled M1?a, IOSM M1 may already perform M1!b while M2 is still waiting for M3 to perform M3!f before M2?f may occur. The matrices at the margin of the PES show the final states of some configurations, reached after all events from above the corresponding dotted lines have occurred. They serve as labels to indicate the possible continuations of the PES.

The algorithm generates only an initial part of the PES. The PES is cut off when behaviour encountered earlier repeats. The generated initial part may be expanded by appending the sub-PES's starting with the corresponding global states to the cut-off

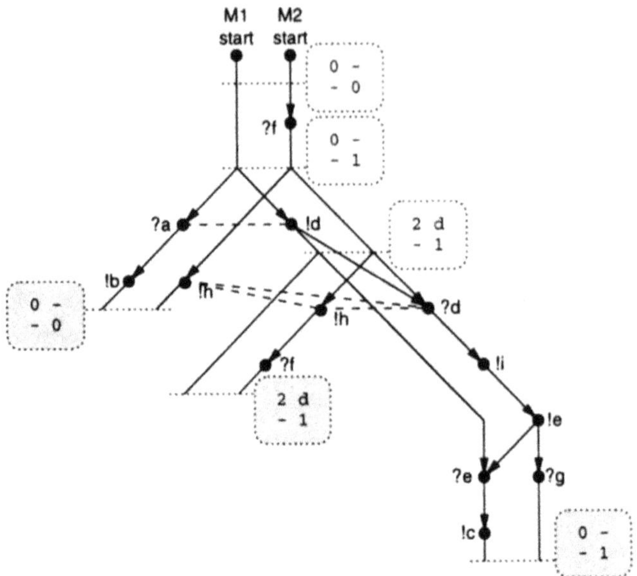

Figure 5 Initial part of the prime event structure for Example 2.

points. A PES $P = (E, \preceq, \#, l)$ of a system of asynchronously communicating IOSMs is *complete* if for every reachable state gs there exists a configuration C such that:

- $fs(C) = gs$ (i.e., gs is represented in P), and
- for every transition (s, μ, s') enabled in gs there exists a configuration $C' = C \cup \{e\}$ such that $e \notin C$ and e is labelled by μ.

The PES obtained by expanding the generated initial part is complete. A proof is omitted here.

A PES does not hide the difference between nondeterminism due to choice of events inside an individual state machine and due to choice of events from different state machines, as interleaving models do. Each branching-point of the PES corresponds to a choice inside an individual state machine. Nondeterminism due to concurrency, i.e. arbitrary order of events occurring at different state machines, does not cause a branching-point in the PES. Paths of the PES represent significantly different behaviour, not only a different order of independent events, as paths of a reachability tree may do.

5 GENERATION OF TEST CASES IN CONCURRENT TTCN

As a test case description contains only events occuring at the PCOs of the test architecture (Figure 6), the PES (Figure 5) needs to be restricted to these events (cf. [BGP89]). This is done by labelling events to be deleted by τ, the nonobservable action. Let A_{PCO}

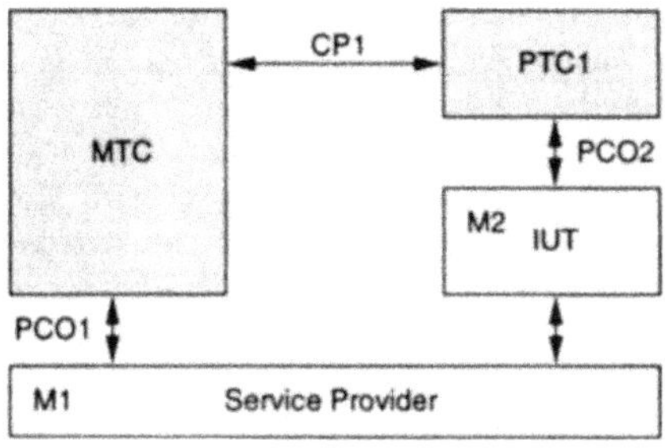

Figure 6 Test architecture for Example 2.

be the set of actions controllable and observable at the PCOs of the test architecture. The projection function is a function $p : A \rightarrow A \setminus A_{PCO} \cup \{\tau\}$ defined by:

$$p(\mu) = \begin{cases} \tau & \text{if } a \notin A_{PCO} \\ \mu & \text{otherwise.} \end{cases}$$

Application of the projection function to the label of each event of a PES results in an order-preserving mapping to a projection of the PES. In our example, the actions M1!d, M2?d, M2!e, and M1?e are not visible at the PCOs and become labelled by τ. The projection can be reduced by skipping events labelled by τ, resulting in a restricted PES (Figure 7). Note that due to the transitivity of the causality relation (as a partial order), there is a directed arc from the event labelled M2!i to the event labelled M1!c in Figure 7. Due to the conflict inheritance property, there are now dashed arcs between M1?a and M2!i and between M2!h and M2!i.

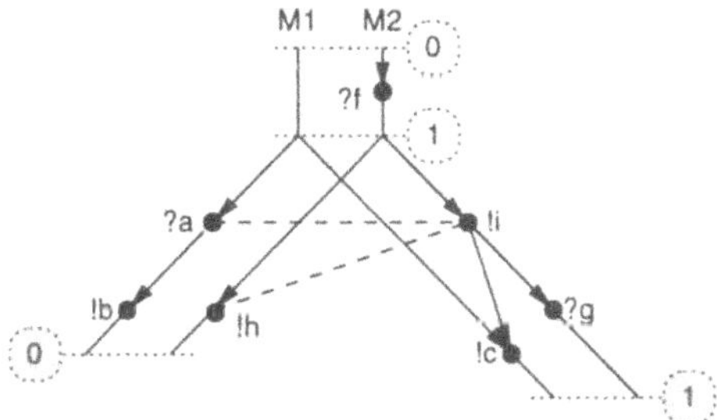

Figure 7 Restricted initial part of the prime event structure for Example 2.

The restricted PES models the behaviour of IUT and test context that is visible at the PCOs. The tester behaviour is the inversion of the restricted PES, i.e. input events are changed to output events and vice-versa. Inversion of inputs and outputs is generally carried out in test generation from asynchronous models.

As each path of the PES represents a significant behaviour, it is desirable that a test suite covers each path of the generated initial part of the PES. We propose to form a test case for each path of the restricted initial part of the PES (hence, a test case generation method with implicit test purposes). If the number of paths is too large, an appropriate subset has to be choosen using extra information from outside the specification. Optimization techniques trying to minimize the number of test events are outside the scope of this paper.

As we assume that the tester is, in general, distributed into a main test component and several parallel test components, the events of the inverted restricted PES have to be separated into behaviour descriptions for the individual test components. This step is carried out together with the selection of test cases by traversing a particular path of the inverted restricted PES following the causality relation and recording events belonging to the different test components in separate behaviour trees. If a single test component comprises events from different concurrent IOSMs, then the interleavings of these events have to be computed now. In the behaviour description of a single test component, concurrency can only be expressed by means of interleavings.

If an event μ of test component tc_i is immediately succeeded by an event μ' of another test component tc_j (e.g., crossing arrow from M2!i to M1!c in Figure 7) and μ and μ' are not transmission and reception of the same message, then a coordination message from tc_i to tc_j is inserted into the test case description. The coordination message informs tc_j that μ has occurred in tc_i. We use a coordination message if and only if it is necessary in order not to loose sight of the global order of events. We assume that the delay of coordination messages is not larger than the delay of messages in PCOs. Otherwise, one could not tell whether μ' has occurred before μ, which would be wrong behaviour, or after μ, which is correct.

If we reach a branching point in the inverted restricted PES, for each test case one of several conflicting events is selected. All conflicting input events (input to the test component, output from IUT or test context) have to be taken into account in the test case description as alternatives leading to an INCONCLUSIVE verdict. As these events are initiated by IUT or test context, the test components cannot prevent their occurrence though they do not fit to the intended test purpose. If a permissible event occurs that conflicts to the one expected according to the test purpose, one has to assign the INCONCLUSIVE verdict and to try to execute the test case later again.

At the beginning of the behaviour descriptions of the main test component for each test case, CREATE constructs, activating the parallel test components, are inserted. At the end of each path of the initial part of inverted restricted PES, a PASS verdict is assigned. Finally, OTHERWISE events, leading to a FAIL verdict, are added to each level of indentation to deal with any unexpected behaviour. Table 5(a) shows the behaviour description of the test case for the path to the left in Figure 7. Table 5(b) shows the behaviour description for the path to the right.

6 CONCLUSIONS

An algorithm for generating a PES equivalent to a system of asynchronously communicating IOSMs has been presented. The algorithm is generic and can be adapted to the semantics of communication over queues used in Estelle and in SDL.

The PES is a suitable starting point for generating test cases in Concurrent TTCN as it specifies the order of events in a noninterleaving manner in a tree structure. How to generate test cases from the PES has been outlined. The approach is applicable for generating multi-party and interoperability test cases.

Table 1 Test cases for Example 2.

(a)

Behaviour	Verdict
CREATE(PTC1:PTC1Tree)	
PCO1?c	INCONC
PCO1?OTHERWISE	FAIL
PCO1!a	
PCO1?b	PASS
PCO1?OTHERWISE	FAIL
PTC1Tree	
PCO2?OTHERWISE	FAIL
PCO2!f	
PCO2?h	PASS
PCO2?i	INCONC
PCO2?OTHERWISE	FAIL

(b)

Behaviour	Verdict
CREATE(PTC1:PTC1Tree)	
PCO1?OTHERWISE	FAIL
CP1?CM1	
PCO1?c	PASS
PCO1?OTHERWISE	FAIL
PTC1Tree	
PCO2?OTHERWISE	FAIL
PCO2!f	
PCO2?i	
CP1!CM1	
PCO2!g	PASS
PCO2?h	INCONC
PCO2?OTHERWISE	FAIL

ACKNOWLEDGEMENTS

The author is grateful to Hasan Ural and Bernd Baumgarten for very helpful discussions.

REFERENCES

[AS91] N. Arakawa and T. Soneoka. A test case generation method for concurrent programs. In J. Kroon, R.J. Heijink, and E. Brinksma, editors, *Protocol Test Systems, IV*, pages 95–106, Leidschendam, The Netherlands, 1991. North-Holland.

[BG94] B. Baumgarten and A. Giessler. *OSI conformance testing methodology and TTCN*. North-Holland, 1994.

[BGP89] B. Baumgarten, A. Giessler, and R. Platten. Test derivation from net models. In J. de Meer, L. Mackert, and W. Effelsberg, editors, *Protocol Test Systems, II*, pages 141–159, Berlin, Germany, 1989. North-Holland.

[CA90] W. Chun and P.D. Amer. Test case generation for protocols specified in Estelle. In J. Quemada, J. Mañas, and E. Vázquez, editors, *Formal Description Techniques, III*, pages 191–206, Madrid, Spain, 1990. North-Holland.

[CZ93] S.T. Chanson and J.-S. Zhu. A unified approach to protocol test sequence generation. In INFOCOM '93 [INF93], pages 106–114.

[dBdR88] J.W. de Bakker and W.-P. de Roever, editors. *Linear Time, Branching Time, and Partial Order in Logics and Models for Concurrency*, Noordwijkerhout, The Netherlands, 1988. Springer-Verlag.

[ERV96] J. Esparza, S. Römer, and W. Vogler. An improvement of McMillan's unfolding algorithm. In T. Margaria and B. Steffen, editors, *Tools and Algorithms for the Construction and Analysis of Systems*, pages 87–106, Passau, Germany, 1996. Springer.

[FJJV96] J.-C. Fernandez, C. Jard, T. Jéron, and C. Viho. Using on-the-fly verification techniques for the generation of test suites. In R. Alur and T.A. Henzinger, editors, *Computer-Aided Verification, 8th International Conference, CAV '96*, pages 348–359, New Brunswick, NJ, USA, 1996. Springer-Verlag.

[GHN93] J. Grabowski, D. Hogrefe, and R. Nahm. Test case generation with test purpose specification by MSCs. In O. Færgemand and A. Sarma, editors, *6th SDL Forum*, pages 253–266, Darmstadt, Germany, 1993. North-Holland.

[God96] P. Godefroid. *Partial-order methods for the verification of concurrent systems: an approach to the state-explosion problem.* Springer-Verlag, 1996.

[HUK95] O. Henniger, A. Ulrich, and H. König. Transformation of Estelle modules aiming at test case generation. In A.R. Cavalli and S. Budkowski, editors, *Protocol Test Systems, VIII*, Evry, France, 1995. Chapman & Hall.

[ICN93] *International Conference on Network Protocols*, San Francisco, CA, USA, 1993.

[II83] M. Itoh and H. Ichikawa. Protocol verification algorithm using reduced reachability analysis. *The Transactions of the IECE of Japan*, E 66(2):88–93, 1983.

[INF93] *Conference on Computer Communications, IEEE INFOCOM '93*, San Francisco, CA, USA, 1993.

[ISO89] Information processing systems – Open Systems Interconnection – Estelle: A formal description technique based on an extended state transition model. International Standard ISO 9074, 1989.

[ISO91] Information technology – Open Systems Interconnection – Conformance testing methodology and framework. International Standard ISO/IEC 9646, 1991.

[ITU92] Specification and description language SDL '92. ITU-T Recommendation Z.100, 1992.

[Kat96] J.-P. Katoen. *Quantitative and qualitative extensions of event structures.* PhD thesis, University of Twente, 1996.

[KCKS96] M.C. Kim, S.T. Chanson, S.W. Kang, and J.H. Shin. An approach for testing asynchronous communicating systems. In B. Baumgarten, H.-J. Burkhardt, and A. Giessler, editors, *Testing of Communicating Systems, IX*, Darmstadt, Germany, 1996. Chapman & Hall.

[KIIY85] M. Kajiwara, H. Ichikawa, M. Itoh, and Y. Yoshida. Specification and verification of switching software. *IEEE Transactions on Communications*, COM-33(3):193–198, March 1985.

[Lam78] L. Lamport. Time, clocks, and the ordering of events in a distributed system. *Communications of the ACM*, 21(7):558–565, 1978.

[LSKP93] D. Lee, K.K. Sabnani, D.M. Kristol, and S. Paul. Conformance testing of protocols specified as communicating FSMs. In INFOCOM '93 [INF93], pages 115–127.

[Maz88] A. Mazurkiewicz. Basic notions of trace theory. In de Bakker and de Roever [dBdR88], pages 284–363.

[McM95] K.L. McMillan. A technique of state space search based on unfolding. *Formal Methods in System Design*, 6(1), 1995.

[NPW81] M. Nielsen, G. Plotkin, and G. Winskel. Petri nets, event structures and domains, part I. *Theoretical Computer Science*, 13:85–108, 1981.

[PL91] D.K. Probst and H.F. Li. Partial-order model checking: a guide for the perplexed. In K. Larsen and A. Skou, editors, *Computer-Aided Verification, 3rd International Conference, CAV '91*, pages 322–331, Aalborg, Denmark, 1991. Springer-Verlag.

[Pra86] V. Pratt. Modeling concurrency with partial orders. *International Journal of Parallel Programming*, 15(1):33–71, 1986.

[Sid90] D.P. Sidhu. Protocol testing: the first ten years, the next ten years. In L. Logrippo, R.L. Probert, and H. Ural, editors, *Protocol Specification, Testing, and Verification, X*, pages 47–68, Ottawa, Ontario, Canada, 1990. North-Holland.

[SLU89] K.K. Sabnani, A.M. Lapone, and M.Ü. Uyar. An algorithmic procedure for checking safety properties of protocols. *IEEE Transactions on Communications*, 37(9), 1989.

[SNW96] V. Sassone, M. Nielsen, and G. Winskel. Models for concurrency: Towards a classification. *Theoretical Computer Science*, pages 297–348, 1996.

[TK93] K.C. Tai and P.V. Koppol. Hierarchy-based incremental analysis of communication protocols. In ICNP '93 [ICN93], pages 318–325.

[Ulr97] A. Ulrich. A description model to support test suite derivation for concurrent systems. In M. Zitterbart, editor, *10. GI/ITG-Fachtagung Kommunikation in Verteilten Systemen*, Braunschweig, Germany, 1997.

[UY91] H. Ural and B. Yang. A test sequence selection method for protocol testing. *IEEE Transactions on Communications*, 39(4):514–523, 1991.

[Win88] G. Winskel. An introduction to event structures. In de Bakker and de Roever [dBdR88], pages 364–397.

[WL93] C.J. Wang and M.T. Liu. Automatic test case generation for Estelle. In ICNP '93 [ICN93], pages 225–232.

[ZWR+80] P. Zafiropulo, C.H. West, H. Rudin, D.D. Cowan, and D. Brand. Towards analyzing and synthesizing protocols. *IEEE Transactions on Communications*, COM-28(4):651–660, April 1980.

Fault detection in embedded components

A. Petrenko and N. Yevtushenko †

CRIM, Centre de Recherche Informatique de Montréal,
1801 Av. McGill College, Montréal, H3A 2N4, Canada,
Phone: (514) 840-1234, Fax: (514) 840-1244, petrenko@crim.ca
† *Tomsk State University,*
36 Lenin str., Tomsk, 634050, Russia,
yevtushenko.rff@elefot.tsu.tomsk.su

Abstract

We address in this paper the problem of detecting faults located in a given component embedded within a composite system. The system is represented as two communicating FSMs, a component FSM inaccessible for testing and a context machine that models the remaining part of the system which is assumed to be correctly implemented. We elaborate a systematic approach for deriving external tests which can detect all predefined types of faults in the embedded component. The approach is based on the construction a proper characterization of the conforming behavior of the component in context, derivation of internal tests and translation into external tests.

Keywords

Communicating FSMs, fault models, conformance testing, embedded testing, test derivation

1 INTRODUCTION

The model of communicating state machines, see e.g., [Boch78], [BrZa83], is widely used for development of complex systems. It serves as an underlying model

Testing of Communicating Systems, M. Kim, S. Kang & K. Hong (Eds)
Published by Chapman & Hall © 1997 IFIP

for description techniques such as Statecharts, ROOM, ESTEREL, SDL. One of the important issue is test derivation from a formal specification in the form of communicating state machines. A straightforward solution is to construct a global composed machine from a reachability graph such that describes the behavior of a system at points accessible for testing and apply existing test derivation methods developed for FSMs. The behavior of a system even consisting of deterministic components may be nondeterministic and a test derivation method which can treat nondeterministic I/O FSMs should be used [LBP94]. This approach suffers from several drawbacks. First, even if each component of the system is given as an I/O FSM, the global I/O machine may not exist due for example, livelocks. A number of verification methods and tools could be used to check properties of the given system, so it is reasonable to assume that tests should be derived from a verified system of communicating state machines such that its composed machine exists. Second, the number of states in the composed machine (assuming that we are able to construct it) may easily trigger tests with a high fault coverage to explode. Two main approaches have been tried to alleviate the test explosion effect.

According to the first approach, systematic test derivation with fault coverage is avoided, transition coverage of individual component machine is attempted instead. This could be achieved a partial exploration of the composed machine either by adopting a random walk [West86], see [LSKP96], by generating a certain part of the entire composed machine comprising transitions chosen for testing [HLS96] or a reduced composed machine [KoTa95]. The advantage of this approach is that the need for global machine construction is obviated. However, the fault detection ability of the approach is unknown.

The second approach is driven by a divide-and-conquer strategy and is closely related to the problem of submodule construction, known also as redesign, plant-controller, or equation solving, where we are required to construct the specification of a submodule X when specifications of the overall system and of all submodules except X are given [MeBo83], [QiLe91], [ABBD95], [LJK95], [HeBr95]. A given system of communicating FSMs is viewed in two parts, one part (an embedded component) is to be tested and the other (context of the component) is assumed to be error-free. The main issue here is how to systematically derive tests tuned for the embedded component (testing in context). The basic idea is to reduce testing in context to testing in isolation so that existing methods could become fully applicable. Once this problem is solved we may similarly proceed deriving tests for the remaining part of the system (the target component and context switch their roles). Since faults usually do not affect all the components of a system the resulting test suite would normally have a high fault coverage, while test explosion effect is alleviated. This approach has been elaborated in [PYLD93], [PYD94], [PYBD96] and [PYB96]. Here we continue that work for providing systematic methods for test derivation from communicating state machines.

The rest of the paper is organized as follows. In Section 2, we briefly summarize the results of [PYBD96] and [PYB96] related to this work. The novel parts are presented in Section 3 and 4. Section 3 gives a method for constructing a so called embedded equivalent of the component in context which explicitly characterizes all implementations conforming to a given specification in context and facilitates test derivation. Section 4 discusses the problem of translating internal tests derived from the embedded equivalent into external tests. Two approaches to solve the problem are proposed. We conclude in Section 5 with a discussion of future work.

2 FRAMEWORK FOR TESTING IN CONTEXT

2.1 Finite state machines

A finite state machine (FSM) is a completely specified initialized (possibly nondeterministic) Mealy machine which can be formally defined as follows. A *finite state machine* A is a 5-tuple (S, X, Y, h, s_0), where S is a set of states with s_0 as the initial state; X - a finite nonempty set of input symbols; Y - a finite nonempty set of output symbols; and h - a behavior function $h: S\times X \rightarrow \wp(S\times Y)\backslash\varnothing$, where $\wp(S\times Y)$ is the powerset of $S\times Y$ [Star72]. The machine A becomes deterministic when $|h(s, x)|=1$ for all $(s, x)\in S\times X$.

We extend the behavior function to a function on the set X^* of all input sequences containing the empty sequence ε, i.e., $h: S\times X^* \rightarrow \wp(S\times Y^*)\backslash\varnothing$. Assume $h(s, \varepsilon) = \{(s, \varepsilon)\}$ for all $s\in S$, and suppose that $h(s, \beta)$ is already specified. Then $h(s, \beta x) = \{ (s', \gamma y) \mid \exists s''\in S\ [(s'', \gamma) \in h(s, \beta) \wedge (s', y) \in h(s'', x)] \}$. Given a sequence α over the alphabet $X\cup Y$, we use α^X to denote the X-projection of α that is obtained by deleting all symbols $y\in Y$ from the sequence α.

The function h^1 is the next state function, while h^2 is the output function of A, where h^1 is the first and h^2 is the second projection of h, i.e., $h^1(s, \alpha) = \{ s' \mid \exists\ \beta \in Y^*\ [(s', \beta) \in h(s, \alpha)] \}$, $h^2(s, \alpha) = \{\beta \mid \exists\ s' \in S\ [(s', \beta) \in h(s, \alpha)] \}$ for all $\alpha\in X^*$. We use $h^1_\beta(s, \alpha)$ to denote the set of states reached by the machine when it executes I/O sequence α/β starting from state s. Given two states s of the FSM A and r of the FSM $B= (T, X, Y, H, t_0)$, and a set $V\subseteq X^*$; state r is said to be a *V-reduction* of s, written $r \leq_V s$, if for all input sequences $\alpha\in V$ the condition $H^2(r, \alpha) \subseteq h^2(s, \alpha)$ holds; r is not a V-reduction of s, $r \not\leq_V s$, if there exists an input sequence $\alpha \in V$ such that $H^2(r, \alpha) \not\subseteq h^2(s, \alpha)$. States s and r are *V-equivalent* states, written $s \cong_V r$, iff $s \leq_V r$ and $r \leq_V s$. On the class of deterministic machines, the above relations coincide. We denote $\leq$ the V-reduction in the case where $V=X^*$, similarly, $\cong$ denotes the equivalence relation. Given two machines, A and B, B is a reduction of A, written $B\leq A$, if the initial state of B is a reduction of the initial state of A. If $B\leq A$ and B is deterministic then it is referred to as a D-reduction of A. Similarly, the equivalence relation between machines is defined. $B\cong A$, iff $B\leq A$ and $A\leq B$. The equivalence and reduction relations serve as conformance relations between implementations and their FSM specifications for deriving test suites with guaranteed fault coverage [SiLe89], [PBD93], [YaLe95], [PYB96a].

A *fault model* is a triple $<A, \sim, \Im>$ [PYB96], where A is a reference specification, a set $\Im$ is the fault domain that is a set of possible implementations defined over the same input alphabet as the specification, and $\sim$ is a conformance relation. In this paper, we consider $\sim \in \{\cong, \leq\}$. A *complete* test suite w.r.t. the fault model is a finite set E of finite input sequences such that for all $B \in \mathfrak{I}$, $B \not\sim A$ implies $B \not\sim_E A$.

If the fault domain is an arbitrary finite set $\mathfrak{I}$ of implementation machines then in order to derive a complete test suite w.r.t. the fault model a traditional method (mutant killing technique) could be used. For each FSM $B \in \mathfrak{I}$, we derive an input sequence that distinguishes B from the reference specification A whenever they are not equivalent (or B is not a reduction of A). The union of input sequences over all machines $B \in \mathfrak{I}$ gives a desired test suite. Because of its complexity, such a solution is feasible for a small number of faults to be detected, for example for single output faults. At the same time, there are certain fault models for which there is no need to explicitly enumerate machines of the fault domain. For these fault models, a complete test suite is derived based on the properties of the specification machine A. As an example, we could mention a classical (black-box) fault model $< A, \cong, \mathfrak{I}_m(X)>$ where A is a completely specified and deterministic FSM, and $\mathfrak{I}_m(X)$ is the set of all FSMs over the input alphabet X of A with at most m states. A number of competing methods exist, see e.g., [SiLe89], [PBD93], [YaLe95]. As is shown in [PYB96], a similar approach can be taken to devise fault models and to derive complete tests for embedded components. In this paper, we propose new methods for testing in context such that allow to obviate an expensive mutant killing technique.

2.2 Model of a system with the embedded component

Many compound systems are typically specified as a collection of communicating FSMs. As noticed in [PYBD96] the system of two communicating FSMs (IUT and context), connected as shown in the upper part of Figure 1, is general enough to discuss problems related to testing an embedded component.

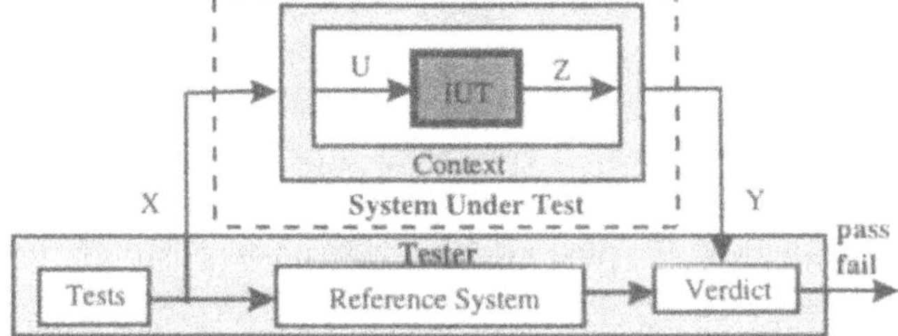

Figure 1 Architecture for testing the embedded component (IUT).

We assume that we are given an FSM *Spec* which represents the behavior of the component (IUT) embedded within the system that should be tested, while a machine C, called the *context* machine, is a composed machine of all components of the system, except the component of interest, that are assumed fault-free. As in [PYB96], we assume that the sets X, U, Z, and Y of actions are pairwise disjoint. Two (deterministic) FSMs are communicating asynchronously via bounded input queues where actions are stored. We assume that the system at hand has a single message in transit, i.e. a next external input x is submitted to the system only after it has produced an external output y to the previous input. Under these assumptions, the collective behavior of two communicating FSMs can be described by means of a *product* machine and a *composed* machine. The product machine $Spec \times C$ is represented by a graph of global states, obtained by performing reachability computation [BrZa83]. It is in fact, a labeled transition system which represents the joint behavior of all components. If the product machine $Spec \times C$ has a cycle labeled only with internal actions from the alphabet $U \cup Z$ then the system falls into livelock when an appropriate input sequence is applied, i.e. the system cannot produce an external output. In this case, the system's behavior cannot be described by an I/O FSM and we conclude that the composed machine does not exist. Otherwise, a composed machine $RS = Spec \circ C$ can be obtained by hiding of all internal actions in the product machine, determinizing the obtained LTS and by pairing inputs with subsequent outputs [PYB96], [PYBD96].

Example. Consider the system [PYB96] of context and component machines, shown in Figure 2. The composed machine $RS = Spec \circ C$ is shown in Figure 2(c).

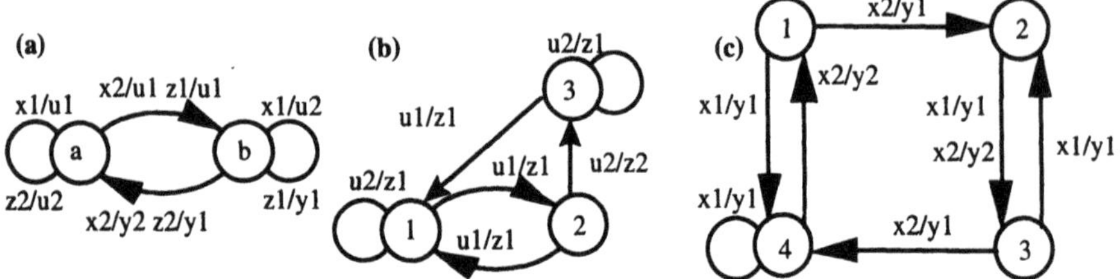

Figure 2 The context C (a), component *Spec* (b), and the composed machine *RS* (c).

2.3 Explicit fault model for testing in context

Testing in context is based on the test architecture shown in Figure 1. We assume that the tester executes test cases simultaneously against the system under test and its specification, called the reference system. The reference system is modeled by the composed machine $RS = Spec \circ C$. The embedded component (IUT) is the target of tests. The context does not need to be tested. Verdicts are produced by a part of the tester called a verdict machine. The verdict machine produces the verdict **fail** and enters a state FAIL when output actions of a system under test and reference system do not coincide or the system under test falls into livelock. No communication between the component and context can be observed or controlled.

Based on the test architecture (Figure 1), we define a fault model for deriving complete test suites as in [PYB96]. Let $\mathfrak{I}_m(U, Z)$ denote the set of all implementation FSMs *Imp* over alphabets U and Z with at most m states such that $Imp \circ C$ exists. Then the triple $<RS, \cong, \mathfrak{I}_m(U, Z) \circ C>$ where $\mathfrak{I}_m(U, Z) \circ C = \{Imp \mid Imp \in \mathfrak{I}_m(U, Z)\}$, is called the *explicit fault model* for testing in context. In this paper, we attempt to elaborate a systematic method for deriving a test suite complete w.r.t. the explicit fault model. In [PYB96], we have considered a number of fault models relevant for testing in context, however this problem was left open.

2.4 Approximation of the component's behavior

To systematically derive tests for the embedded component we need a complete and concise characterization of detectable and undetectable faults. This is what we call the approximation of component's behavior in the given context C [PYBD96], [PYB96] that completely describes the permissible behavior of the embedded component w.r.t. any external input sequence. Below we briefly summarize its construction.

A trace of the embedded component is permissible if it is a valid trace of its specification *Spec*. If it is not in *Spec* then, depending on a particular external input sequence, it may be permissible or forbidden. The verdict machine producing the *fail*-verdict in response to the external input sequence indicates that the behavior of the component is forbidden. We formalize the notions of permissible and forbidden traces of the embedded component as follows.

A trace $\beta/\gamma \in U^*/Z^*$ is *forbidden* w.r.t. an external input sequence $\alpha \in X^*$ if there exists a prefix $\beta_1...\beta_k/\gamma_1...\gamma_k$ of β/γ such that for an appropriate prefix $\alpha_1...\alpha_k$ of the sequence α it holds that the U-projection of the output sequence of the context C to $\alpha_1\gamma_1...\alpha_k\gamma_k$ is equal to $\beta_1...\beta_k$ while its Y-projection is not equal to the output sequence of the reference system *RS* to $\alpha_1...\alpha_k$. Trace β/γ is said to be *permissible* w.r.t. the external input sequence α otherwise. Trace β/γ is permissible if it is permissible w.r.t. all external input sequences. In other words, the trace β/γ is forbidden w.r.t. an external input sequence α if every system composed of a component that contains trace β/γ is not equivalent to the reference system *RS* w.r.t. α, i.e. α can be considered as an external test detecting any nonconforming implementation of *Spec* with trace β/γ.

The idea of constructing the approximation is based on the test architecture presented in Figure 1. To capture all possible behavior of the embedded component we replace it with a chaos machine *Ch* over the alphabets U and Z that has just one state [PYB96]. The chaos machine is nondeterministic, it produces all possible outputs z in response to each input u. We construct the product machine $Ch \times C \times RS \times Ver$ as an LTS, hide all actions Y and verdicts in the obtained LTS and determinize it. The resulting LTS is transformed back to an FSM, denoted $[[Spec]]_C$, in alphabets $X \cup U$ and $Z \cup \{null, fail\}$. Any global state where the verdict machine is in a fail-state is a designated state FAIL of $[[Spec]]_C$. An external input x is coupled with the output *fail* and labels a transition to the state FAIL if all subsequent internal actions lead to the state FAIL; otherwise it is coupled with the output *null*. The remaining internal inputs u are paired with the internal outputs z. "Don't care" transitions of the obtained FSM are specified as transitions to another designated state TRAP. Specifically, if an external input x causes a "don't care" transition from a particular state then the machine has a transition to the state TRAP labeled *x/null*, for an input u a corresponding transition to the TRAP state is labeled with input u and each internal output $z \in Z$. Intuitively, the TRAP state indicates that any behavior of the component machine when the FSM $[[Spec]]_C$ trapped to this state, is permissible since it cannot be executed. Any behavior leading to the FAIL state is forbidden, since it results in a wrong external output. For more details on the construction of the approximation of the component in context the reader is referred to [PYBD96]. Figure 3 shows the approximation $[[Spec]]_C$ for our example. The

FSM $[[Spec]]_C$ captures the most essential for testing aspect of the behavior of the whole system shown in Figure 1. In particular, the verdict machine in response to a particular external input sequence produces the *fail*-verdict in a current global state of the system if and only if the FSM $[[Spec]]_C$ reaches the state FAIL. This property of the approximation is formally stated as follows.

Proposition 2.1. Given the approximation $[[Spec]]_C = (S, X \cup U, Z \cup \{fail, null\}, h, s_0)$ and trace $\beta/\gamma \in (U/Z)^*$, the trace β/γ is forbidden iff there exists an I/O sequence α/δ of $[[Spec]]_C$ with the $(U \cup Z)$-projection β/γ that takes $[[Spec]]_C$ from the initial state to the state FAIL.

Given a forbidden trace β/γ, we denote $\alpha(\beta/\gamma)$ the input part of an I/O sequence of $[[Spec]]_C$ that has the $(U \cup Z)$-projection β/γ and takes $[[Spec]]_C$ from the initial state to the state FAIL. The trace β/γ is forbidden w.r.t. the X-projection of α. The approximation of the component in context characterizes the relationship between deviations in the behavior of the embedded component and external input sequences capable of revealing a fault through the context. However, its shortcoming is that existing test derivation methods cannot be directly applied to derive external tests. At the same time, as we are going to demonstrate in the subsequent section, it can be further transformed into another machine allowing for a direct use of these methods.

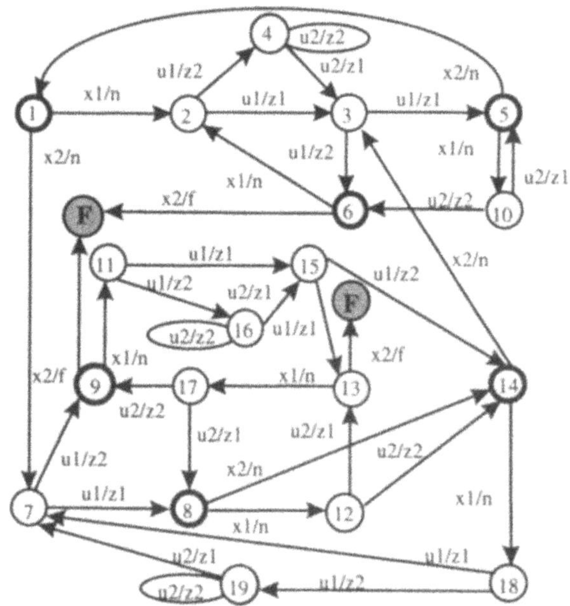

Figure 3 The approximation of *Spec* in context. State TRAP as well as its incoming transitions are not shown, state **F** is the FAIL state.

3 EMBEDDED EQUIVALENT OF A COMPONENT MACHINE

In order to use regular methods for test derivation we now would like to transform the approximation $[[Spec]]_C$ into an FSM such that all its I/O sequences in alphabets U and Z of *Spec* are permissible w.r.t. every possible external input sequence.

Equivalently, we define a machine by excluding from the set $(U/Z)^*$ all traces β/γ such that are forbidden w.r.t. some external input sequence in X^*. Let Tr be the set of traces of a machine and $[[Spec]]_C = (S, X \cup U, Z \cup \{fail, null\}, h, s_0)$.

An FSM is said to be the *embedded equivalent* of the component *Spec* in context *C*, denoted $EE = (P, U, Z \cup \{fail\}, H, p_0)$ if its traces in $Tr(EE)$ over the inputs U and outputs Z satisfy the conditions of Proposition 2.1, namely:

$$\forall \beta/\gamma \in (U/Z)^* \ (\beta/\gamma \text{ is forbidden}) \Leftrightarrow \beta/\gamma \notin Tr(EE) \vee H^1_\gamma(p_0, \beta) = \{\text{FAIL}\}.$$

The idea of transforming the approximation $[[Spec]]_C$ into the embedded equivalent is to hide all external inputs X and to group its states into subsets such that all external inputs cause transitions in the FSM $[[Spec]]_C$ within the same subset, making sure that all forbidden traces are removed. The situation is somewhat similar to a classical problem of determinizing a nondeterministic finite automaton (the subset construction) [HoUl79], where all non-observable actions have to be removed while preserving all the traces of a given automaton. In fact, as in our case, all states reached from a given state through internal transitions (corresponding to external inputs) could be merged to form a single state of resulting machine. The essential difference is that in our case, we should retain only traces that are permissible w.r.t. all external input sequences, i.e. that are common for all states reached from the same state after non-observable actions. In other words, we should determine the intersection of such traces for each state instead of collapsing traces. As the intersection may sometimes become empty we use a designated output *fail* and state FAIL in the embedded equivalent to indicate that a certain common trace can no longer be extended, since there exists an external input sequence that "forbids" any extension. To formalize the procedure we need the following definition.

Given the FSM $[[Spec]]_C = (S, X \cup U, Z \cup \{fail, null\}, h, s_0)$, a set B of states of $[[Spec]]_C$ is said to be *closed* (w.r.t. external inputs) if $h^1(s, x) \subseteq B$ holds for every $s \in B$ and $x \in X$. For a subset $B \subseteq S$, a minimal by inclusion closed set including B is called the *closure* of B.

We present the procedure using our example (Figure 3). The closure of the initial state of $[[Spec]]_C$ is the set {1, 2, 7} which is the initial state of an FSM *EE*. In $[[Spec]]_C$, inputs u_1 and u_2 cause transitions to the TRAP state from state 1. State 2 has the following transitions: 2-u_1/z_1->3, 2-u_1/z_2->4. State 7: 7- u_1/z_1->8, 7- u_1/z_2->9. Both states have transitions caused by input u_2 to state TRAP.

Consider input u_1. We have $\bigcap_{s \in \{1,2,7\}} h^2(s, u_1) = \{z_1, z_2\}$.

For the output z_1, we find the union of states $\bigcup_{s \in \{1,2,7\}} h^1_{z_1}(s, u_1) = \{3, 8, \text{TRAP}\}$. The closure of {3, 8, TRAP} is the set {3, 8, 12, 14, 18, TRAP}, since from state 8 there are transitions on external inputs leading to 12 and 14; as well as from 14 to 3 and 18. This a new state in the FSM *EE*. As a result, the FSM *EE* has a transition

{1, 2, 7}-u_1/z_1->{3, 8, 12, 14, 18, TRAP}.

Consider now output z_2. $\bigcup_{s \in \{1,2,7\}} h^1_{z_2}(s, u_1) = \{4, 9, \text{TRAP}\}$. The closure of the set {4,

9} is the set {4, 9, 11, FAIL, TRAP}. The presence of state FAIL in the obtained set means that the I/O sequence u_1/z_2 is forbidden. As a result, the FSM *EE* has no transition from the state {1, 2, 7} labeled with u_1/z_2.

Take next input u_2. $\bigcap_{s\in\{1,2,7\}} h^2(s, u_2) = \{z_1, z_2\}$. We have $\bigcup_{s\in\{1,2,7\}} h^1_{z_1}(s, u_2) = \bigcup_{s\in\{1,2,7\}} h^1_{z_2}(s, u_2) = \{\text{TRAP}\}$. Thus, there is a "don't care" transition in the FSM *EE*, namely, {1, 2, 7}-$u_2/z_1,z_2$->{TRAP}. In a similar way we proceed with a newly obtained state {3, 8, 12, 14, 18, TRAP}. The final result, i.e. the FSM *EE*, is shown in Figure 4. As the example shows, the procedure of constructing the embedded equivalent is quite straightforward and we do not elaborate it further to save space for other results.

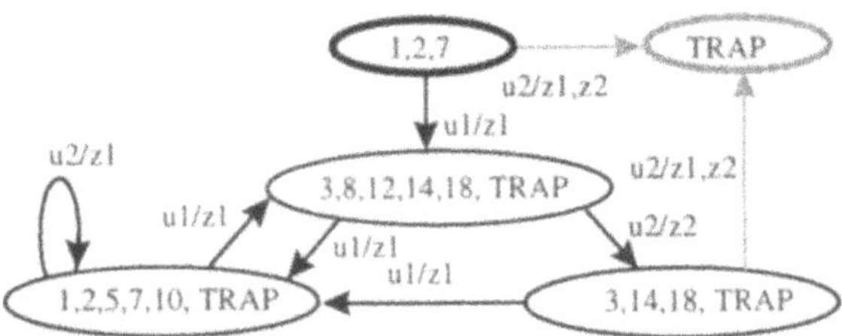

Figure 4 The embedded equivalent *EE*.

The embedded equivalent of the component in context explicitly characterizes all implementations conforming to a given specification *Spec* in context *C*.

Theorem 3.1. Given the specification *Spec* of the component, the context *C*, and an implementation FSM *Imp* over the same alphabets *U* and *Z*, as *Spec*, let $Imp\circ C$ be the composed machine. Then $Imp\circ C$ is equivalent to $RS = Spec\circ C$ iff $Imp \leq EE$.

Proof. Let *Imp* be a reduction of *EE*. Suppose that the FSM $Imp\circ C$ is not equivalent to the machine *RS*. Then there exists an external input sequence $\delta\in X^*$ such that $Imp\circ C$ is not equivalent to *RS* w.r.t. this sequence, i.e. the pair β/γ of sequences β and γ that are induced by δ at the inputs of *Imp* and the context *C* is forbidden w.r.t. δ. Thus, the trace β/γ of *Imp* is not an I/O sequence of *EE*; therefore *Imp* is not a reduction of *EE*. A contradiction.

Suppose now that the FSM $Imp\circ C$ is equivalent to *RS* but *Imp* is not a reduction of *EE* w.r.t. an appropriate input sequence β, i.e. the output sequence γ of *Imp* to β is not in the set of output sequences of *EE* to β. Then, by definition of *EE*, there exists a sequence $\delta\in X^*$ such that the trace β/γ is forbidden w.r.t. δ, i.e. the FSMs $Imp\circ C$ and *RS* are not equivalent w.r.t. δ. A contradiction. ❒

We know that a similar characterization of conforming implementations can be obtained based on the most general solution to the equation $G\circ C \equiv Spec\circ C$ with *G* being a free variable [PYB96]. As discussed in [PYB96], based on the solution *G* "local" tests to test the component in isolation can be derived, however, these tests are not easy to translate into external tests. Unlike the general solution *G*, the embedded equivalent gives an effective answer to the problem of test translation, as we are about to demonstrate.

Let $EE = (P, U, Z \cup \{fail\}, H, p_0)$. By the definition of the embedded equivalent, each trace in $Tr(EE) = \bigcup_{\beta \in U^*} H^2(p_0, \beta)$ is permissible w.r.t. any external input sequence while for any other trace $\beta/\gamma \in (U/Z)^* \setminus Tr(EE)$ there exists a sequence $\alpha(\beta/\gamma)$ such that β/γ is forbidden w.r.t. $\alpha(\beta/\gamma)$. Consider an arbitrary sequence $\beta \in U^*$. If the set $H^2(p_0,\beta)$ contains all possible output sequences of the same length as β, no sequence $\alpha(\beta/\gamma)$ exists. We use $Z^{|\beta|}$ to denote the set of all sequences in Z^* which have the length of β. Then for each sequence $\gamma \in Z^{|\beta|} \setminus H^2(p_0,\beta)$, a sequence $\alpha(\beta/\gamma)$ exists. Its X-projection is the external test $\alpha(\beta/\gamma)^x$ that can detect an erroneous behavior of the embedded component with the trace β/γ. If now we find at least one sequence $\alpha(\beta/\gamma)$ for each $\gamma \in Z^{|\beta|} \setminus H^2(p_0,\beta)$ and derive the X-projection we have an external test suite which detects all faults internally revealed by the sequence β (in the following section we elaborate a proper method for finding sequences $\alpha(\beta/\gamma)^x$).

At the first sight, the price of this solution seems high since the number of sequences in the set $Z^{|\beta|} \setminus H^2(p_0,\beta)$ is exponential. The following observation helps us to drastically reduce it. Any extension of a forbidden trace (β/γ) is forbidden as well, therefore if we have already found a sequence $\alpha'(\beta'/\gamma')$ for a prefix β' of the sequence β there is no need to consider any extension of β'/γ'. The question comes now how we could choose input sequences $\beta \in U^*$ based on the given embedded equivalent.

Consider the fault model $F = \langle EE, \leq, \Im_m(U, Z)\rangle$, where $\Im_m(U, Z)$ is the set of all possible implementations with up to m states over the alphabets U and Z, where $m \geq n$, the number of states in the given specification of the embedded component *Spec*. There exists a method for deriving a test suite complete w.r.t. this fault model [PYB96a]. We have the following result.

Theorem 3.2. Given an FSM $EE = (P, U, Z \cup \{fail\}, H, p_0)$, let T be a complete test suite w.r.t. the fault model $\langle EE, \leq, \Im_m(U, Z)\rangle$. Then the set

$$E = \{\alpha(\beta/\gamma)^x \mid \beta \in T \;\&\; \gamma \in Z^{|\beta|} \setminus H^2(p_0, \beta)\}$$

is a complete test suite w.r.t. the explicit fault model $\langle RS, \cong, \Im_m(U, Z) \circ C\rangle$.

Consider our working example. The specification of the embedded component *Spec* has three states (Figure 2). We assume that no fault in the component increases the number of states, i.e. $m = 3$. The method of [PYB96a] applied to the FSM *EE* (Figure 4) produces the following test suite:

$$T = \{u_1u_1u_1u_1u_2;\ u_1u_1u_1u_2u_1u_2;\ u_1u_1u_2u_1u_2;\ u_1u_1u_2u_2u_2;\ u_1u_2u_1u_1u_2;\ u_1u_2u_1u_2u_2\}$$

It is complete w.r.t. the fault model $\langle EE, \leq, \Im_3(U, Z)\rangle$ and once translated into external sequences (see next section) it is complete w.r.t. the explicit fault model $\langle RS, \cong, \Im_3(U, Z) \circ C\rangle$.

In practical situations, we are often ready to sacrifice complete coverage of all output and transition faults for shorter tests. We may consider, for example, the fault model $\langle EE, \leq, \Im_{Spec}\rangle$, where $\Im_{Spec}$ denotes the set of all FSMs that are mutants of the FSM *Spec* with output faults. In our example, we construct a transition tour of the FSM *EE*: $u_1u_1u_1u_2u_1u_2$ (there is no need to cover any transition to the TRAP state since they are not executable in context). It is just one of the six sequences in the test suite T. Note that this sequence does not cover all the transitions of the original specification of the component *Spec* (Figure 2b). At the same time, not each transition tour of the latter covers all the transitions of the former.

4 TRANSLATION OF INTERNAL TESTS INTO EXTERNAL TESTS

Once the embedded equivalent of a given component in context is constructed, an internal test suite could be produced based on a chosen fault model (e.g. output or transition faults). Tests are internal and should be translated into external tests which could be applied to the context. Theorem 3.2 suggests how this could be done. Let $\beta \in U^*$ be an internal input sequence, i.e. internal test. Applied to the FSM $EE = (P, U, Z \cup \{fail\}, H, p_0)$ the sequence β produces the set of output sequences $H^2(p_0, \beta)$. Each trace β/δ such that $\delta \in H^2(p_0, \beta)$ is permissible w.r.t. any external input sequence. At the same time for any trace β/γ such that $\gamma \in Z^{|\beta|} \setminus H^2(p_0, \beta)$, there exists a sequence $\alpha(\beta/\gamma)$ such that the trace β/γ is forbidden w.r.t. $\alpha(\beta/\gamma)^x$. Once found, the sequence $\alpha(\beta/\gamma)^x$ is an external test which forces the context to execute the internal test β provided that an IUT executes the trace β/γ. If we find a sequence $\alpha(\beta/\gamma)^x$ for all $\gamma \in Z^{|\beta|} \setminus H^2(p_0, \beta)$ then we have a set of external tests corresponding to a single internal test β. To execute the internal test β against a particular implementation of the component one external test suffices, but since we do not know much about an IUT we should use all of them.

The key issue is then to find for a given test $\beta \in U^*$ all (or at least one) sequences $\alpha(\beta/\gamma)$ for every trace β/γ, where $\gamma \in Z^{|\beta|} \setminus H^2(p_0, \beta)$. This could actually be solved by constructing a synchronous product of the approximation and an FSM representing all the forbidden traces β/γ. The approach is very similar to that of finding from a specification a test covering a given test purpose for testing an isolated implementation, see for example, [FJJV96]. In that approach, test derivation is based on a depth-first traversal of the product. In our case, a forbidden trace serves as a test purpose and the approximation $[[Spec]]_C$ with two distinct types of alphabets X and U plays the role of a specification. The procedure is thus slightly more involved:

1. We construct an FSM, called a *test machine* $A(\beta)$, such that has a distinct state for each trace α/δ, where α is a proper prefix of β and $\delta \in H^2(p_0, \alpha)$, as well as two designated states FAIL and TRAP. Transitions are defined in the following way. For each trace $\alpha u/\delta z$ such that αu is a prefix of β and $\delta z \notin H^2(p_0, \alpha u)$, we define a transition from a corresponding state to the FAIL state labeled with u/z. The FAIL state has looping transitions labeled with all pairs u/z, $u \in U$ and $z \in Z$. All traces β/δ, where $\delta \in H^2(p_0, \beta)$, take the test machine to the TRAP state. Since each state of the test machine accepts at most one input u of the sequence β, we define transitions to the TRAP state for all the remaining internal inputs in U and all internal outputs in Z. Once the TRAP state is reached, no forbidden trace can be found for the internal

test β. To synchronize the test machine with the approximation we should also equalize their alphabets. In particular, we augment the test machine with the external inputs in X, they cause looping transitions at each state with the *null* output.

2. Given the two FSMs, $A(\beta) = (Q, X \cup U, Z \cup \{null\}, g, q_0)$ and $[[Spec]]_C = (S, X \cup U, Z \cup \{fail, null\}, h, s_0)$ we are interested in input sequences that simultaneously lead them to the FAIL states. In the former machine such a sequence causes a forbidden trace, while in the latter, its x-projection is the external test for this forbidden trace. We construct the synchronous product of $A(\beta)$ and $[[Spec]]_C$ as an FSM $A(\beta) \times [[Spec]]_C = (Q \times S, X \cup U, Z \cup \{fail, null\}, g \times h, q_0 s_0)$, where $g \times h(qs, a) = \{ [(q_b^1(q, a), h_b^1(s, a)), b] \mid b \in g^2(q, a) \cap h^2(s, a) \}$ if $g^2(q, a) \cap h^2(s, a) \neq \varnothing$ and $g \times h(qs, a) = \{(\text{FAIL}, \text{FAIL}), fail\}$ if $g^2(q, a) \cap h^2(s, a) = \varnothing$. By definition of $A(\beta)$, $g^2(q, a) \cap h^2(s, a) = \varnothing$ implies $a \in X$, $g^2(q, a) = \{null\}$, and $h^2(s, a) = \{fail\}$.

3. We find all traces of the synchronous product $A(\beta) \times [[Spec]]_C$ from the initial global state to the global state (FAIL, FAIL). To shorten the length of a trace we could skip looping transitions finding shortest paths from the initial state.

4. For a particular forbidden trace β/γ different sequences $\alpha(\beta/\gamma)$ may be found, it is sufficient to choose one of them for each forbidden trace. In other words, for each forbidden trace β/γ, we find one trace α/β with the $(U \cup Z)$-projection β/γ that takes the FSM from its initial state to the state (FAIL, FAIL). Alternatively, we could optimize the number of external tests by solving a set cover problem [John74]. Finally, we find the x-projection of the obtained sequences.

We illustrate the construction using our example. Consider, as an example, the internal test $u_1 u_1$. Figure 5 shows the test machine constructed for this test. There are two forbidden traces, u_1/z_2 and $u_1/z_1 u_1/z_2$, the test machine enters the FAIL state after these traces.

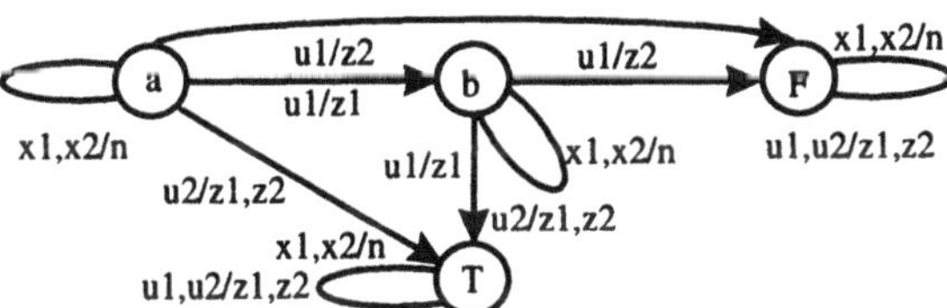

Figure 5 The test machine derived from the input sequence $u_1 u_1$.

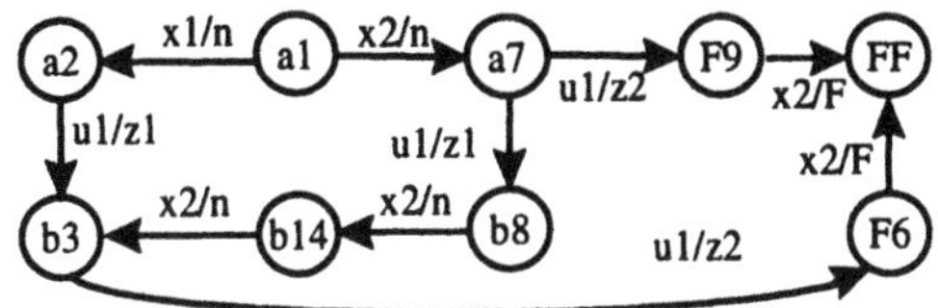

Figure 6 A fragment of the machine $A(\beta) \times [[Spec]]_C$ for $\beta = u_1 u_1$.

A fragment of the product of the test machine and approximation (Figure 3) which contains the necessary traces is shown in Figure 6. For the forbidden trace u_1/z_2 we have a single sequence $x_2 u_1 x_2$, and for $u_1/z_1 u_1/z_2$ we have two sequences $x_1 u_1 u_1 x_2$ and $x_2 u_1 x_2 x_2 u_1 x_2$ reaching the state (FAIL, FAIL). Accordingly, there are two possible solutions $\{x_2 x_2, x_1 x_2\}$ and $\{x_2 x_2, x_2 x_2 x_2 x_2\} = \{x_2 x_2 x_2 x_2\}$. To execute the internal test u_1u_1 we may use a single external test $x_2x_2x_2x_2$ or two tests x_2x_2 and x_1x_2.

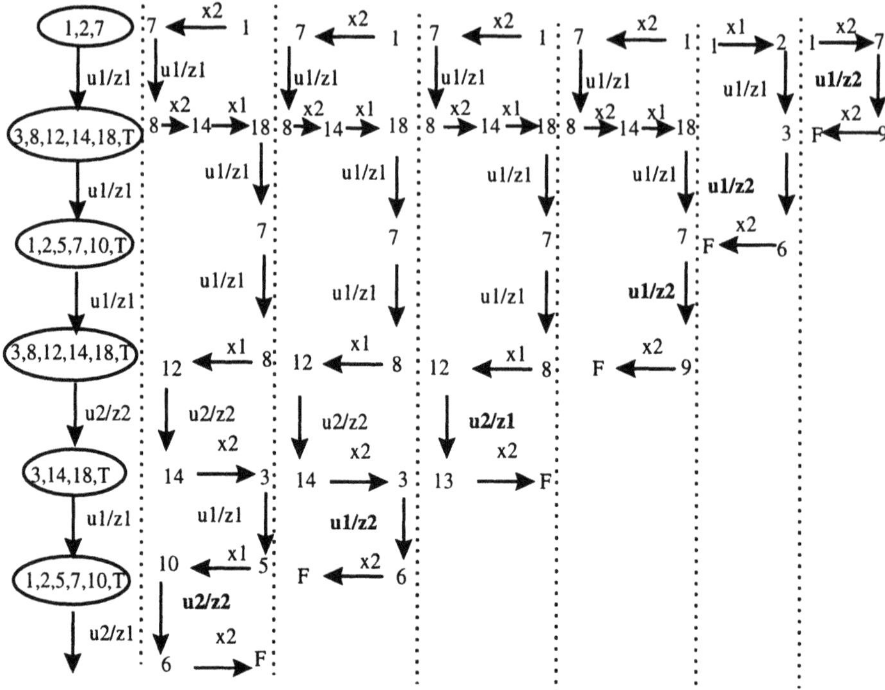

Figure 7 Translating the internal test $u_1u_1u_1u_2u_1u_2$ into external tests.

The above procedure allows us to find not only a sequence $\alpha(\beta/\gamma)^x$ for each forbidden trace β/γ but also to translate a given internal test β into a number of external tests by deriving them for each forbidden trace which an IUT can execute when the test β is internally applied. It facilitates the optimization of the number of external tests since it can deliver the set of all minimal external tests for each internal test. However, the approach relies on the product of the approximation and test machine which may have too many states to be even constructed. To reduce the complexity of test translation we need a simpler method for finding a single sequence $\alpha(\beta/\gamma)$ for a given trace β/γ and not all of them.

The idea of such a method is based on the fact that states of the embedded equivalent constructed as subsets of states of the approximation allow us to backtrack a sequence $\alpha(\beta/\gamma)$ for a given trace β/γ starting from the final state FAIL in the approximation. The backtracking procedure is illustrated in Figure 7 for the transition tour $u_1u_1u_1u_2u_1u_2$ of the embedded equivalent (Figure 4). The transition graph presents transitions in the embedded equivalent caused by this internal test. The columns correspond to all forbidden traces the test can cause in an IUT. A forbidden part u/z of each trace leading eventually the FAIL state (F) is depicted in bold. Consider the longest forbidden trace $u_1/z_1u_1/z_1u_1/z_1u_2/z_2u_1/z_1u_2/z_2$. The suffix u_2/z_2 is executed in the approximation from one of the states {1,2,5,7,10, TRAP}. By direct inspection of Figure 3 we find that it is state 10. FAIL state is reached from state 10 with u_2/z_2x_2 through state 6. Backtracking continues until the initial state 1 is reached. The x-projection gives the test $x_2x_2x_1x_1x_2x_1x_2$ for the considered forbidden trace.

As a result, to execute a single internal test, a transition tour of the FSM *EE* $u_1u_1u_1u_2u_1u_2$, the following external input sequences are required:

$$\{x_1x_2;\ x_2x_2x_1x_2,\ x_2x_2x_1x_1x_2x_2;\ x_2x_2x_1x_1x_2x_1x_2\}.$$

Four sequences of the total length of 19 external test events are needed to detect all output faults in the embedded component.

Next we apply the backtracking procedure to the internal test suite T complete w.r.t. the fault model $<EE, \leq, \mathfrak{I}_3(U, Z)>$ (see Section 4) and obtain the following external test suite complete w.r.t. the explicit fault model $<RS, \cong, \mathfrak{I}_3(U, Z)\circ C)>$:

$$\{x_1x_1x_1x_1x_1x_2;\ x_1x_1x_1x_2;\ x_1x_1x_2x_2x_1x_2;\ x_1x_1x_2x_2x_2;\ x_1x_2x_1x_1x_2;\ x_1x_2x_1x_2;\ x_2x_2x_1x_1x_2x_1x_2;$$
$$x_2x_2x_1x_1x_2x_2;\ x_2x_1x_1x_1x_2;\ x_2x_1x_1x_2;\ x_2x_1x_2x_1x_1x_2;\ x_2x_1x_2x_2x_2;\ x_2x_2x_1x_2\}.$$

The total length is 67, as is indicated in [PYB96], where such a test suite was obtained by an ad hoc procedure. Note that in this particular example, translation of a complete internal test suite into an external one almost doubles the length of tests. The increase depends, of course, on how "transparent" is the context to signals from/to the embedded component.

5 CONCLUSION

In this paper, we have considered the problem of test derivation aimed at detecting faults in a component embedded within a given system modeled by communicating state machines assuming that the rest of the system has no faults. The presented results are based on a general framework for testing in context elaborated in the previous work [PYBD96], [PYB96], [PYD94], [PYLD93]. We have demonstrated that tests which detect all predefined (transition or output) faults can be systematically derived through the following steps. First, we construct a so called approximation of the component in context, which characterizes the behavior of any implementation of the component. This step was elaborated in our previous papers. New procedures proposed in this paper are as follows. The approximation is transformed into an embedded equivalent of the component. The latter contains the behavior of any conforming implementation and is used to derive internal tests complete with respect to a chosen fault model. An existing method for deriving tests from a nondeterministic FSM and reduction relation between an implementation and its specification can be applied at this point. Since we assume that no access is possible to the embedded component internal tests have to be translated into external tests applied at available test access points. Two approaches have been elaborated to solve the last problem. Compared to the published results, we have elaborated a systematic approach which leads to better results, i.e. shorter tests with the same fault coverage guarantee.

Possible future work is related to generalization of this approach to nondeterministic communicating state machines and extended finite state machines. It would also be interesting to see whether the constructions used in our approach could be further simplified to treat real-size specifications. More research is required to merge the two approaches, the one elaborated in this work and the other based on a partial exploration of a composed machine while preserving their advantages.

Acknowledgments. This work was partially supported by the NSERC grants OGP0194381 and STRGP200.

6 REFERENCES

[ABBD95] Aziz, A., Balarin, F., Brayton, R. K., DiBenedetto M. D., and Saldanha, A. (1995) Supervisory control of finite state machines, *Proceedings of the 7th International Conference CAV'95,* pp. 279-292.

[Boch78] Bochmann, G. v. (1978) Finite state descriptions of communication protocols, *Computer Networks,* 2.

[BrZa83] Brand, D., and Zafiropulo, P. (1983) On communicating finite state machines. *Journal of ACM*, **30**, **2**, 323-42.

[FJJV96] Fernandez, J. C., Jard, C., Jeron, T., and Viho, G. (1996) Using on-the-fly verification techniques for the generation of test suites, *Proceedings of the 8th International Conference CAV'96.*

[HeBr95] Heerink, L. and Brinksma, E. (1995) Validation in context, *Proceedings of the 15th IFIP International Symposium on Protocol Specification, Testing, and Verification*, Chapman & Hall.

[HLS96] Huang, S., Lee, D., and Staskauskas, M. (1996) Validation-based test sequence generation for networks of extended finite state machines, *the Proceedings of the IFIP 1st Joint International Conference FORTE/PSTV,* Chapman & Hall, pp. 403-418.

[HoUl79] Hopcroft, J. E., and Ullman J. D. (1979) *Introduction to automata theory, languages, and computation*, Addison-Wesley, New York.

[John74] Johnson, D.S. (1974) Approximation algorithms for combinatorial problems. *Journal Comput. Syst. Sci.*, 9, pp. 256-78.

[KoTa95] Koppol, P. V., Tai, K. C. (1995) Conformance testing of protocols specified as labeled transition systems, *Proceedings of the 8th International Workshop on Protocol Test Systems (IWPTS'95)*, pp. 143-158.

[LBP94] Luo, G., Bochmann, G. v., and Petrenko, A. (1994) Test selection based on communicating nondeterministic finite state machines using a generalized Wp-method. *IEEE Trans. on Soft. Eng.*, SE-20, 2, 149-62.

[LJK95] Lin, B., de Jong G., and Kolks, T. (1995) Hierarchical optimization of asynchronous circuits, *Proceedings of the 32nd DAC,* pp. 712-717.

[LSKP96] Lee, D., Sabnani, K. K., Kristol, D. M., and Paul S. (1996) Conformance testing of protocols specified as communicating finite state machines - a guided random walk based approach, *IEEE Trans. on Communication,* vol. 44, 5.

[MeBo83] Merlin, P., and Bochmann, G. v. (1983) On the construction of submodule specifications and communication protocols, *ACM Trans. on Programming Languages and Systems,* Vol. 5, No. 1, pp. 1-25.

[PBD93] Petrenko, A., Bochmann, G. v., and Dssouli, R. (1993) Conformance relations and test derivation, Invited Paper, *Proceedings of the 6th International Workshop on Protocol Test Systems (IWPTS'93)*, pp.157-178.

[PYBD96] Petrenko, A., Yevtushenko, N., Bochmann, G. v., and Dssouli, R. (1996) Testing in context: framework and test derivation, *Computer Communications Journal*, Special issue on Protocol Engineering, 19, pp.1236-1249.

[PYB96] Petrenko, A., Yevtushenko, N., and Bochmann, G. v. (1996) Fault models for testing in context, *Proceedings of the IFIP 1st Joint International Conference FORTE/PSTV*, Chapman & Hall, pp. 163-178.

[PYB96a] Petrenko, A., Yevtushenko, N., and Bochmann, G. v. (1996) Testing deterministic implementations from nondeterministic fsm specifications, *Proceedings of the 9th IWTCS'96*, Chapman & Hall, pp.125-140.

[PYD94] Petrenko, A., Yevtushenko, N., and Dssouli, R. (1994) Testing strategies for communicating fsms, *Proceedings of the 7th IWTCS'94*, pp. 193-208.

[PYLD93] Petrenko, A., Yevtushenko, N., Lebedev, A., and Das, A. (1993) Nondeterministic state machines in protocol conformance testing, *Proceedings of the 6th IWPTS*, pp. 363-378.

[QiLe91] Qin, H., and Lewis, P.(1991) Factorization of finite state machines under strong and observational equivalencies, *Journal of Formal Aspects of Computing*, Vol. 3, pp. 284-307.

[Star72] Starke, P.H. (1972) *Abstract automata*. North-Holland/American Elsevier.

[SiLe89] Sidhu, D. P., and Leung, T. K. (1989) Formal methods for protocol testing: a detailed study, *IEEE Trans. on Soft. Eng.*, SE-15, 4, pp.413-426.

[West86] West, C. (1986) Protocol validation by random state exploration, *Proceedings of the 6th ISPSTV*.

[YaLe95] Yannakakis, M., and Lee, D. (1995) Testing finite state machines: fault detection, *Journal of Computer and System Sciences*, 50, pp. 209-227.

7 BIOGRAPHY

Alexandre Petrenko received the Diploma degree in electrical and computer engineering from Riga Polytechnic Institute and the Ph.D. in computer science from the Institute of Electronics and Computer Science, Riga, USSR. In 1996, he has joined CRIM, Centre de Recherche Informatique de Montréal, Canada. He is also an adjunct professor of the Université de Montréal, where he was a visiting professor/researcher from 1992 to 1996. From 1982 to 1992, he was the head of a research department of the Institute of Electronics and Computer Science in Riga. From 1979 to 1982, he was with the Networking Task Force of the International Institute for Applied Systems Analysis (IIASA), Vienna, Austria. His current research interests include high-speed networks, communication software engineering, formal methods, conformance testing, and testability.

Nina Yevtushenko received the Diploma degree in radio-physics in 1971 and Ph.D. in computer science in 1983, both from the Tomsk State University, Russia. She is currently a Professor at that University. Her research interests include the automata and FSM theory and testing problems.

A Pragmatic Approach to Generating Test Sequences for Embedded Systems

Luiz Paula Lima Jr.[1] *and Ana R. Cavalli*
Institut National des Télécommunications
9, rue Charles Fourier - 91011 Evry Cedex - France
Tel: (+33 1) 60 76 44 74 Fax: (+ 33 1) 60 76 47 11
{lima|Ana.Cavalli}@hugo.int-evry.fr

Abstract

Application architectures have evolved to distributed architectures where applications are no longer seen as software blocks, but rather as cooperating software components, possibly distributed over the network. Some of the application's components may have already been thoroughly tested while others have not. This paper presents a pragmatic solution to component testing by means of controlling the composition process in order to identify global transitions that reflect the component's behaviour. The application of the proposed method is illustrated by an example based on the handling of a telephone call.

Keywords

Test generation, component testing, embedded systems, distributed architectures, automata composition.

1. This work was funded by CNPq.

Testing of Communicating Systems, M. Kim, S. Kang & K. Hong (Eds)
Published by Chapman & Hall © 1997 IFIP

1 INTRODUCTION - COMPLEX SYSTEMS

Application platforms have evolved from monolithic architectures to distributed ones where a system is seen as an open set of interworking components. These systems are complex for their components are usually hierarchically organized and may have a certain degree of autonomy. In these systems, all external events can affect any part of its internal state. This is the primary motivation for vigorous testing, but for all except the most trivial systems, exhaustive testing is impossible [1]. In other words, this increasing complexity of computer systems and their communication protocols can no longer be handled by traditionally informal or *ad hoc* methods for conformance and interoperability testing [2]. Since we have neither the mathematical tools nor the intellectual capacity to model and test the complete behaviour of large discrete systems, we must either be content with acceptable levels of confidence regarding their correctness or try to find out other ways to tackle their complexity.

Abstraction is one of the most prevalent techniques to deal with complexity. In the domain of conformance testing, for instance, a common abstraction is not to consider system's internal signals when generating test sequences. Of course, the choice of the level of abstraction (or what are the primitive components in a system) is relatively arbitrary and is largely up to the discretion of the observer of the system [1].

The application of embedded testing techniques is also an important aspect to consider when simplifying the validation of these systems. But current experience has shown that the embedded nature of components make the current type of automatic test generation useless. This has been the case for the GSM-MAP protocol [3], and for the SSCOP protocol environment for AAL5 (ATM Adaptation Layer) [4]. For instance, the application of these new techniques to the interaction between SSCOP and Q2130 on top, and their relation with the Q2931 signalling protocol (to which they provide the SSCF service), would be of particular interest.

In this paper, we present a pragmatic solution to component testing of complex systems by means of abstraction (defining a composition algorithm that removes internal actions) and by means of controlling this composition process in order to identify global transitions that reflect the behaviour of the component under test. The paper is organized as follows. Section 2 introduces the idea of embedded testing, underlining its relevance in the context of complex systems. Section 3 suggests a test architecture for embedded systems together with basic definitions and assumptions. Our embedded testing techniques are based on an algorithm for automaton composition that is detailed in Section 4. Section 5 presents a method for deriving test sequences for complex systems using, basically, goal-oriented techniques and our tools. Conclusions are drawn in Section 6.

2 EMBEDDED TESTING BASICS

It is widely accepted that testing is a crucial phase in the development of complex systems such as communication protocols [2]. Nevertheless, there is a strong need for systematic methods for testing these systems since the existing methods for test derivation from *Labelled Transition Systems* (LTS) and *Input/Output Finite State Machines* (I/OFSM) (based on the "black-box" representation of the *implementation under test* - IUT) are not adequate in this context. In fact, some of the system components may have already been thoroughly tested or a certain level of confidence may have been assigned to them so that they no longer need to be subject of test. Test derivation methods that generate test sequences for only a subset of the system components are called "*embedded testing methods*" [5] or "*gray-box testing methods*" [2] or even "*methods for testing in context.*" [6]

Example 1. Consider the system depicted in Figure 1. Assume that module *C* is known to be faultless and that module *I* must be tested[1]. Let us also assume that we do not have access to *I*'s internal interfaces (since the implementation of the system is given as a black box). Therefore, internal signals sent to/from *I* may reach the environment after passing through *C*. Module *C* acts as a kind of "filter" and system responses to environment stimuli must be correctly interpreted in order to verify that module *I* works as specified. ❑

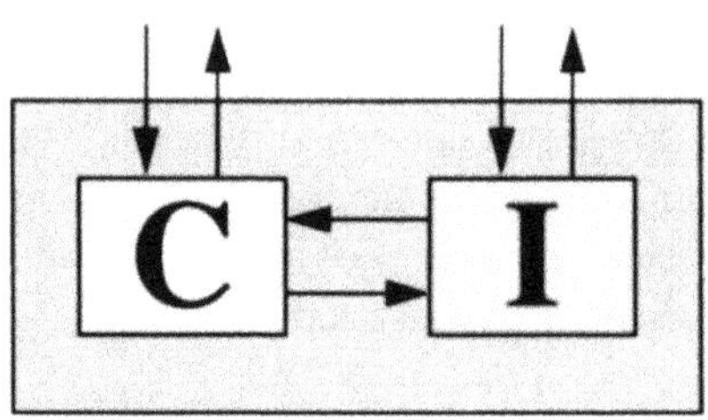

FIGURE 1. A generic complex system.

Traditional methods for testing in isolation turn out to be inadequate for, basically, two reasons:

- Module *I* can neither be "removed" from the system (in order to be tested in isolation) nor can it give access to its internal interfaces. Traditional methods are then obliged to test the system as a whole.
- Obviously, testing the whole system would test module *I* as well, but then we would have (unnecessarily) tested a part of *C*'s behaviour that is independent of *I*. This happens, because the system's global behaviour is likely to contain behaviour that only concerns module *C*.

Embedded testing represents situations that occur very frequently in protocol conformance testing, functional testing of digital circuits (specially, multiprocessor networks) as well as in testing of object-oriented packages[2]. Although the details of each component implementation may remain hidden, to be able to test such sys-

1. Modules C and I may be viewed as the composition of all machines of the system that are not under test and that are subject of testing, respectively.

tems, we must have information about the component configuration (or structure) within the system. Embedded testing methods take advantage of the information about the configuration of the complex system components.

To sum up, embedded testing is concerned with testing a system component when the tester does not have direct access to its interfaces. The access is then made through another process which acts as a sort of "filter." According to [5], if "control and observation are applied through one or more OSI implementations which are above the protocol(s) to be tested, the [testing] methods are called embedded."

Pragmatic embedded testing techniques identify parts of the system's global behaviour that reflect the behaviour of the component under test, and then performs tests on only those parts. Intuitively, the set of test sequences to test the whole system contains redundant or unnecessary elements that we would like to avoid when testing.

3 TEST ARCHITECTURE FOR EMBEDDED SYSTEMS

3.1 Preliminary Definitions

An *input-output finite state machine (I/OFSM)* is a tuple: $\langle S, I, O, \delta, \lambda, s_0\rangle$ where $\boldsymbol{S}, \boldsymbol{I}, \boldsymbol{O}$ are finite, non-empty sets of states, inputs and outputs respectively; s_0 is the initial state; $\delta : S \times I \rightarrow 2^S$ is the state transition function and $\lambda : S \times I \rightarrow 2^O$ is the output function.

One test is referred to as a *test case*. A *test suite* is a set of test cases that tests all conformance requirements.

The action of the conceptual tester involves interactions with the *System Under Test (SUT)*. These can, in theory, be observed and controlled from several different points that are called *Points of Control and Observation* (*PCOs*). PCOs can be modelled as two queues: an output queue for control of test events to be sent to the Implementation Under Test (IUT); and an input queue for the observation of test events received from the IUT.

For testing purposes, a complex system may be divided in two subsystems: a non-empty set of *components under test* (or simply *component*, or *IUT*), and a (possibly empty) set of components that are not concerned by testing (the *context*). The problem of testing a system with an empty context reduces to the traditional problem of testing in isolation.

2. In this work, we are particularly interested in ODP-like systems where different objects communicate within an arbitrary configuration and where we do not intend to test the entire system, but only some of its components.

3.2 Architecture

The *test architecture* is the description of the environment in which the component is tested. It describes the relevant aspects of how the component is embedded in other systems during the testing process, and how it communicates via these embedding systems with the tester (see Figure 2).

A test architecture consists of [5]:

- a tester;
- an implementation under test (IUT);
- a test context;
- points of control and observation (PCOs);
- implementation access points (IAPs, also "interfaces").

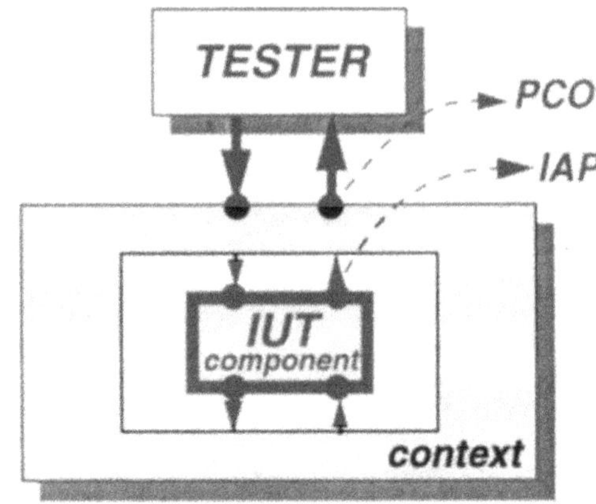

FIGURE 2. Generic test architecture.

In the ideal test architecture (for testing in isolation), the IAPs and the PCOs coincide, the test context is empty, and the tester interacts directly with the IUT. This is rarely the case in real systems, though. The *System Under Test* (SUT) is composed of the IUT and the test context.

The tester is equipped with a timer that is started when a signal is sent to the SUT. On receipt of a response from the SUT, this signal is checked with respect to the test case. After a time out period, if no signal is received, then a fail verdict is issued. Input data for the tester consists of the test suite which guides all testing activities expressing what signals should be sent to the SUT, and what the expected responses are. The test suite represents the reference system in the tester.

3.3 Hypothesis

In order to be able to employ the embedded method for test derivation, described in Section 5, we make the assumption that the context is correctly implemented and that a faulty component implementation does not increase the number of states of the global machine. The latter is a variation on a common hypothesis for testing in isolation [7] that makes it possible to evaluate the test coverage in embedded testing.

The IUT interacts with its context through synchronous communication with input queues of finite size. This implies that a next input x is only submitted to the system after it has produced an external output y in response to the previous input (*I/O ordering constraint* [6]).

The SUT is "reactive" in the sense that one input signal can trigger one or more outputs which are simultaneously sent back to the environment. That is, an output (or set of outputs) must be identified as a response of the system to a particular given stimulus or input.

4 TRACE-KEEPING ALGORITHM FOR AUTOMATON COMPOSITION

The generation of test sequences from formal specifications of systems has been traditionally based on the exhaustive simulation of the specification in order to obtain an automaton that represents the global behaviour of the system. Since it is impossible, in most of the cases, to deal with the size of the automaton that represents the complete behaviour of these systems, a reasonable approach is to simulate the execution of the specification by controlling the range of values assigned to each internal variable and each parameter of input messages. The closer this range is to the real one, the more realistic and the larger the test will be. Obviously, there is always a compromise between accuracy (completeness) of the automaton and its size. But, even with an automaton of a "computable" size, the process of test sequence derivation may not be able to cope with that automaton in a reasonable period of time.

To date, to generate test sequences, what we have done is to take the "big" automaton (that is, the one which is as close to the specification as possible) and then, through the definition of view points (PCOs), abstract the signals which are irrelevant in the current consideration or view point. Then, we may proceed by minimizing the automaton using an algorithm (described in [3]) which removes all internal signals (if the choice of the PCOs is well done). We thereby obtain an automaton that corresponds to the "big one," but abstracting details we do not yet want to consider (see Figure 3a). In general, this automaton has a reasonable size, and therefore it can be used as input for the process of deriving test sequences.

However, even with a "big" automaton generated by simulation, the "reduced" one is often simpler than we would like. Producing an even "bigger" automaton would, in principle, result in a bigger "reduced" automaton, but in many cases a "bigger" automaton just cannot be generated due to storage, memory or computational limitations.

To solve these problems we will consider the use of *composition algorithms* (see Figure 3b). The idea is to avoid the initial automaton size explosion by dividing (which is often already done, if we are dealing with modular systems) the specification into smaller, interrelated modules which are then simulated to produce more complete or smaller automata. The simple composition of these automata, without taking into account any kind of abstraction (PCOs), would lead to the "big" automaton of the traditional case that corresponds to the Cartesian product of the two automata. However, if we use information about our abstraction level we are able to compose them and at the same time avoid the explosion of the model. In other

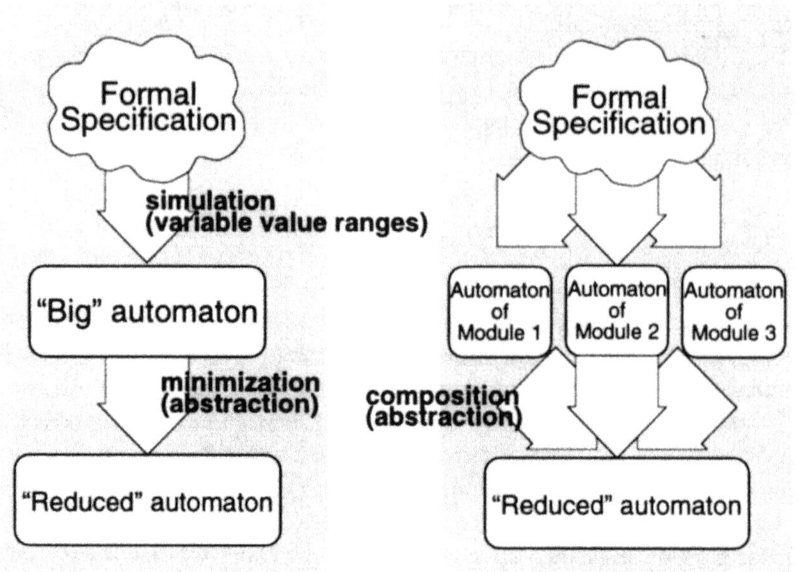

FIGURE 3. Approaches to produce the reduced automaton that will be used as input to the test derivation process.

words, we compose the automata removing internal signals which are not part of what we want to consider for the moment. Composition is done through simulation in order to avoid the generation of unreachable states.

4.1 Definitions

Before proceeding, let us define some useful terms. Let A_1 and A_2 be two I/OFSMs and Π be the cartesian product of A_1 and A_2.

Definition 1: A *global state* is a state of Π. The set of all global states is denoted by $\Gamma = \{\sigma | \sigma \text{ is a state of } \Pi\}$.

Definition 2: A *reachable global state* is a global state that is attained during the joint execution of A_1 and A_2. The set of all reachable global states is denoted by $\mathfrak{R} = \{\rho | (\rho \in \Gamma, \rho \text{ is attained during the joint execution of A1 and A2})\}$.

Definition 3: An *unreachable global state* is a global state that never happens in the joint execution of two machines.

Definition 4: A *stable global state* is the global state that Π reaches after sending a response to environment and before receiving another signal from it. Let us denote the set of all stable global states by

$\Sigma = \{\tau | (\tau \in \Re, \tau \text{ is attained just after sending a signal to the environment})\}$

Definition 5: A *transient global state* is a reachable state which is not stable. Many internal message exchanges and state changes can take place after receiving a signal from the environment and before sending back a response to it. These intermediary states are called transient global states.

The relation between Γ, $\Re$ and Σ is given in Figure 4. The set of transient global states is given by $\Re - \Sigma$.

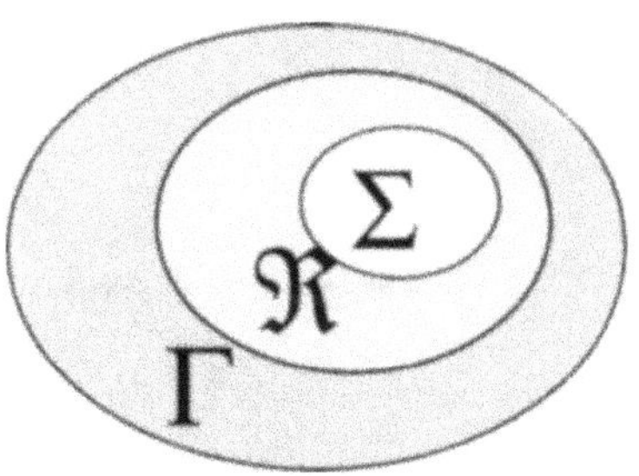

FIGURE 4. Relation between sets of states.

Example 2. Let us consider the automata depicted in Figure 5. Using composition without consideration of internal signals, we would obtain an automaton that is the cartesian product of the two first automata (Figure 6). Internal signals in the cartesian product machine can be hidden using algorithms like the one describe in [3].

Considering *ia* and *ib* internal signals, the automaton that corresponds to the global behaviour, after hiding these signals is depicted in Figure 7.

However, if we compose both automata already taking into account information about the internal signals, we will obtain the same result with the advantage of not producing the intermediary large automaton which corresponds to the cartesian product.

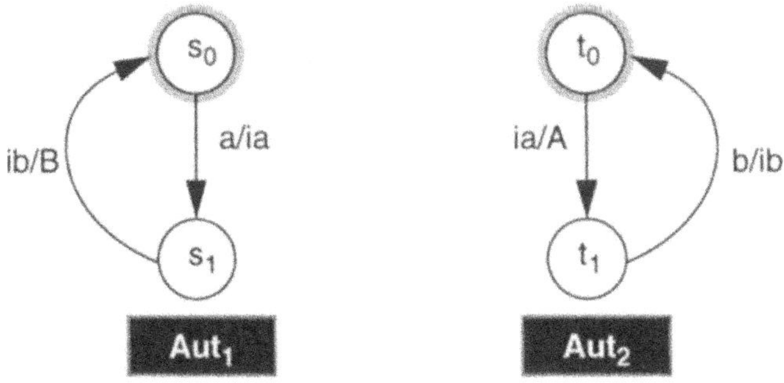

FIGURE 5. Simple example for automata composition.

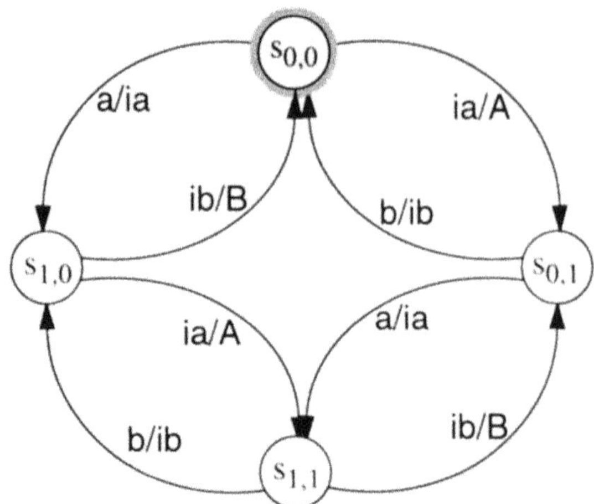

FIGURE 6. Cartesian product of Aut_1 and Aut_2.

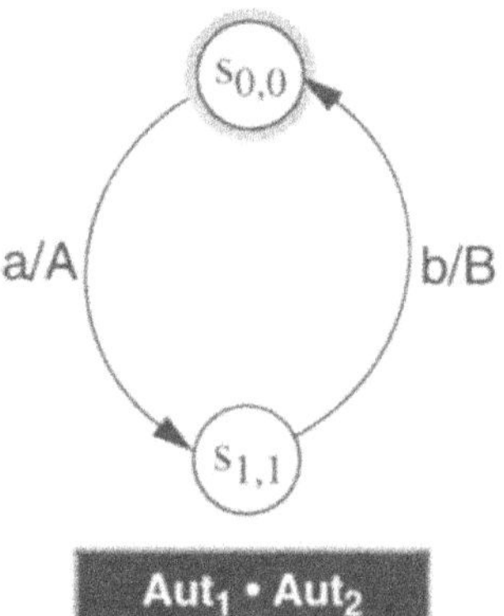

FIGURE 7. Automaton representing the joint external behaviour of Aut_1 and Aut_2.

In this example, $\Gamma = \{s_{0,0}, s_{1,0}, s_{1,1}, s_{0,1}\}$, $\Re = \{s_{0,0}, s_{1,0}, s_{1,1}\}$, and $\Sigma = \{s_{0,0}, s_{1,1}\}$. There is only one transient state ($s_{1,0}$) and one unreachable state ($s_{0,1}$). ❑

4.2 Composition Algorithm

In this section we describe an algorithm to compute the global composition of two automata while removing internal actions. The algorithm is made as modular as possible, so it may be implemented in a distributed fashion. (Our current implementation is centralized, however it serves our purpose.)

4.2.1 Input data and object configuration

Let A_1 and A_2 be the two automata we intend to compose, let **E** be the set of the signals exchanged with the environment[1] and $k \in \{1, 2\}$.

The diagram of Figure 8 shows the object configuration used in the composition

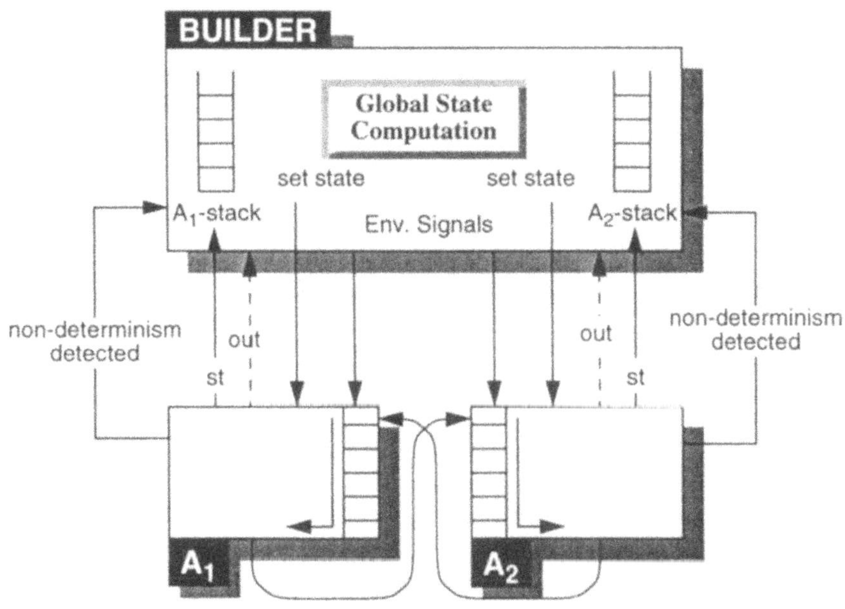

FIGURE 8. Object configuration in the automata composition algorithm.

process. There are basically three objects that communicate by means of message passing: *objects* A_k, and the *builder.*

An A_k *objects* implements automata behaviour. Incoming signals are placed in an input queue and consumed as soon as possible. They cause an outgoing transition

1. **E** corresponds to the PCO definition.

from the current state to be traversed producing an output either to the *builder* object or a peer object (A_k). Also, each state change is reported back to the builder that records them in individual stacks, so it will be able to keep track of all reachable global states during subsequent steps.

The *builder* is the object that controls the composition process and gathers results from objects A_k. These results are used to build up a composite transition, say *trc*, which is instantiated at the end of each step. Getting an output signal from either A_k means that it has obtained all the necessary information to instantiate *trc* and that it can advance to the next step.

4.2.2 The Composition Process

Initially, A_1 and A_2 are set to their initial states which correspond to the global initial state. The builder is aware of which signals could be processed by each machine in each state. It then sends a signal to, say A1, and waits for a response from either A_1 or A_2. Meanwhile, many massage exchanges may take place between A_1 and A_2 until they reach a stable state (when their input queues are both empty and an external signal is sent back to the builder). In order to compute subsequent global states, all reachable states must be saved by the builder in its stacks[1]. A global transition is then instantiated from:

- A_1 and A_2's initial states (before sending the signal);
- the signal sent to the system;
- the system's response; and
- the composite stable state that is composed of the states reached by each machine.

This procedure is repeated until there are no more unvisited outgoing transitions from the current global state whose inputs belong to **E**.

Upon receipt of a signal, each A_k object changes its internal state and sends a signal to another object (a peer object or the builder).

This approach differs from the synchronous product described in [8] and [9]. In fact, while in the synchronous product a transition belongs to the product machine if it can be traversed in the two components or if it can be traversed only in the specification [9], in our composition algorithm, each environment signal is sent to and received from either the context or the component machine, and what is modelled is their joint execution with internal signals being exchanged between them. However, the algorithm of Section 4.2 can be used to obtain the same result as the synchronous product, composing an artificial context that makes visible only some parts of

1. All reachable states are potentially stable states (reachable states may be transient or stable, according to Section 4.1). That is why they must be saved in the builder, so that the builder will be able to get back to them later on.

the component behaviour. An additional advantage over the synchronous product is better control over the observation (i.e. only input - or output - sequences may be observed, if desired).

4.2.3 Extensions - Transition Marking and Behaviour Exploration

The algorithm also includes a complex scheme of transition marking, is also needed to tackle the following issues:

- Multiple (simultaneous) outputs (i.e. when one signal from the environment stimulates several simultaneous outputs);
- Live-lock detection (if components exchange messages indefinitely);
- Simultaneous triggering of multiple transitions (with simultaneous state changes).

If the machines are non-deterministic, then a mechanism of behaviour exploration guarantees that all possible branches are examined. A_1 or A_2 warn the builder when there is a non-deterministic choice for the last input, so that the builder will send the same signal a second time and a different transition will then be traversed. As a result, non-deterministic machines are usually produced.

4.2.4 Errors and Warning Messages

There are basically two undesired situations that may happen during the composition process and that are reported back as errors or warnings:

1. **Incompatibility errors:** A_k was not expecting a given internal signal from its peer machine at its current state. In this case, the internal signal is simply "forwarded" to the environment (builder) that instantiates a global transition with an error message (for it contains an internal output signal).
2. **Unreachability warnings:** During the joint execution of both machines some transitions of either machine may not be traversed and some states even may not be visited. This means that a part of the machine behaviour was not exercised in the joint execution. This kind of information can be useful, for instance, for feature interaction detection [10].

In the first case, either the machines were not designed to work together or they are badly specified. In the second case, however, there may be represented situations where the component presents (additional) functionalities that are not used by its context (or vice-versa).

4.3 Example: *Subscriber Connection Unit (SCU)* and *Subscriber*

Let us use the described algorithm to compose the I/OFSMs presented in Figure 9 (internal signals are underlined in both automata). These machines represent the behaviour of a telecommunication system that is composed of two processes: the *Subscriber* and the *Subscriber Connection Unit.* They specify the handling of the arrival of a telephone call and are composed of states whose names are given in Table 1.

TABLE 1. State names for SCU and Subscriber.

SCU		Subscriber	
State number	**State name**	**State number**	**State name**
S_0	idle	T_0	idle
S_1	wait_for_answer	T_1	ringing
S_2	conversation	T_2	wait_for_stop_ringing
S_3	control_by_called	T_3	conversation
S_4	fault	T_4	control_by_called
		T_5	fault

The composition algorithm proceeds as follows: the builder sets both machines to their initial states (assume that the initial global state is S_0T_0). From these states, there is only one external signal that can be treated by the SCU, namely, *call_arriving* (*call* is an internal signal). Because there is a non-deterministic choice for signal *call_arriving*, assume it traverses transition [S_0,"call_arriving/call",S_1]. Since the output *call* is an internal signal, it is sent to the Subscriber causing a state change (from T_0 to T_1) and a *NULL* signal to be sent back to the builder. Upon receipt of a signal from the Subscriber, the builder understands that the system has reached a stable state and that a new global transition can be instantiated (in this case, [S_0T_0,"call_arriving/NULL",S_1T_1]). A new reachable global state S_1T_1 is saved in the builder for later analysis.

Since there is another non-traversed transition from state S_0T_0 with an external input (transition [S_0,"call_arriving/NULL",S_0]), both machines are reset to their respective states (S_0 and T_0) and *call_arriving* is sent again to the SCU which now traverses transition [S_0,"call_arriving/NULL",S_0], and another global transition ([S_0T_0,"call_arriving/NULL",S_0T_0]) is instantiated.

Since there is no other non-traversed transition from state S_0T_0 with an external input, a new global state is computed from the set of reachable global states (S_1T_1)

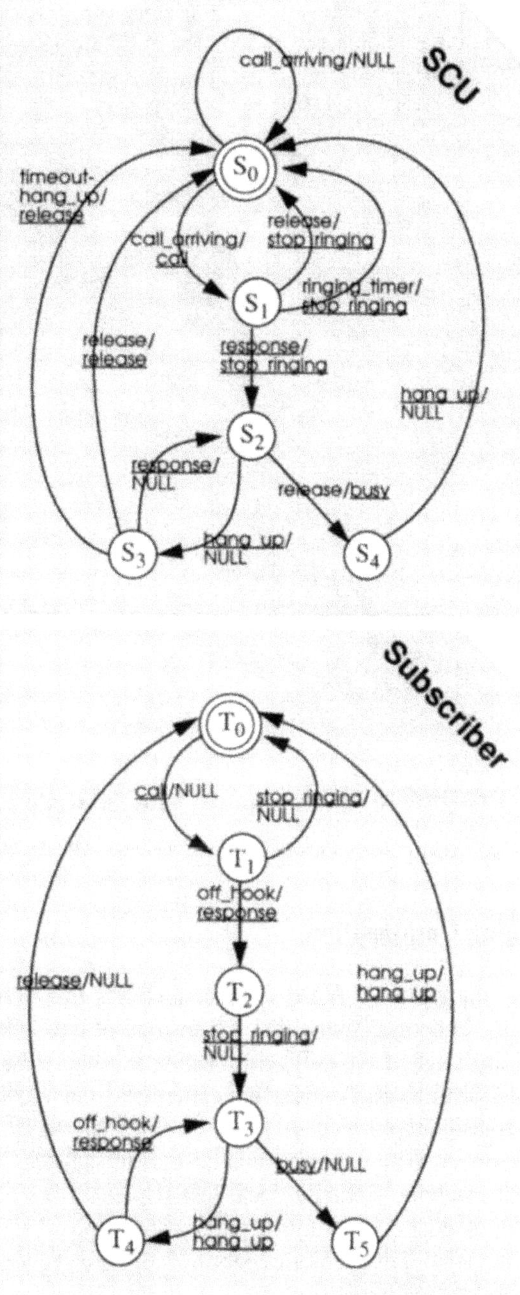

FIGURE 9. Input automata used as input examples for the composition algorithm.

and the process continues until no other global state can be obtained (the set is empty).

The composite automaton obtained is depicted in Figure 10.

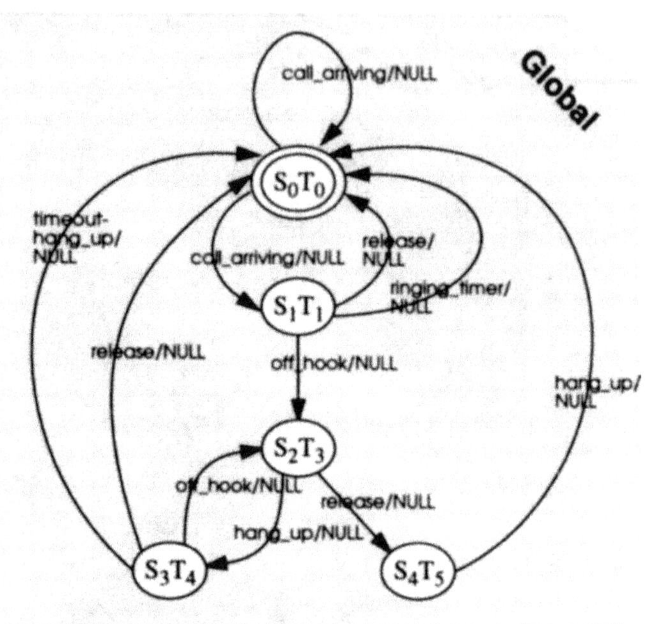

FIGURE 10. UCS composed with Subscriber.

4.4 Trace-keeping Composition

Each transition in the global I/OFSM (i.e. the I/OFSM that describes the global behaviour of the SUT) comes from either a transition of only one component or a combination of transitions of the two components. It is therefore possible to keep track of global I/OFSM transitions that were generated from a transition of the IUT, and to use these transitions for testing only the component under test. Doing so, it becomes easy to distinguish relevant transitions from unnecessary or redundant ones in the global machine. Actually, local test sequences are not "translated" in terms of global test sequences, but rather parts of the global behaviour that reflect the behaviour of the local transitions are identified and test sequences are generated for only those global transitions.

In order to better understand trace-keeping composition, we introduce the concept of equivalence in context as defined in [6].

Definition 6: Let "•" represent the composition operation as described in

Section 4.2. Two machines M_1 and M_2 are *equivalent* in a context C if and only if the joint execution of M_1 and C does not contain live-locks (i.e. the composite machine $M_1 \bullet C$ exists); and $M_1 \bullet C$ is equivalent to $M_2 \bullet C$.

An important question at this point is whether testing a global transition is equivalent to testing a corresponding transition of the component machine. The answer is not straight forward. Let C be the context machine, $Spec$ be the component specification and Imp the component implementation. Assume that global transition t ($t \in C \bullet Spec$) was generated by the composition of t_C belonging to the context C and t_S belonging to the component machine $Spec$ (other cases where many implementation/context transitions originate a single global transition are analogous). The absence of transition t in the global machine $C \bullet Imp$ means that the implementation is faulty (since the context is correctly implemented - see Section 3.3). However, if transition $t \in C \bullet Imp$, then either $t_S \in Imp$ or the implementation did something which is equivalent in the context[1].

Example 3. Consider the I/OFSMs depicted in Figure 11 (internal signals are underlined). Although *Imp* is generally considered to be a faulty implementation of *Spec*, it is not, actually, in the context of *C*, because the composition $C \bullet Spec$ is equivalent to $C \bullet Imp$. Therefore, if the global transition labelled *a/d* exists in the composite machine, we cannot affirm that the transition labelled *ia/id* belongs to the implementation. Nevertheless, we are still able to state whether the implementation has a set of equivalent transitions in that context. ❑

This observation leads us to the following conclusion: if there are at least two different paths composed of transitions labelled *internal/internal* (an internal input and an internal output) that lead to the same state in the context machine, then, intuitively, the implementation is free to take the path it wants without changing the aspect of the global behaviour (since all message exchanges are internal). Otherwise, the implementation would be obliged to take the unique existing path in order to preserve the global behaviour.

---

1. We do not consider here the problem of *latent faults* pointed out in [6], since our testing methods apply a signature to the arrival state of the transition in order to check its correctness.

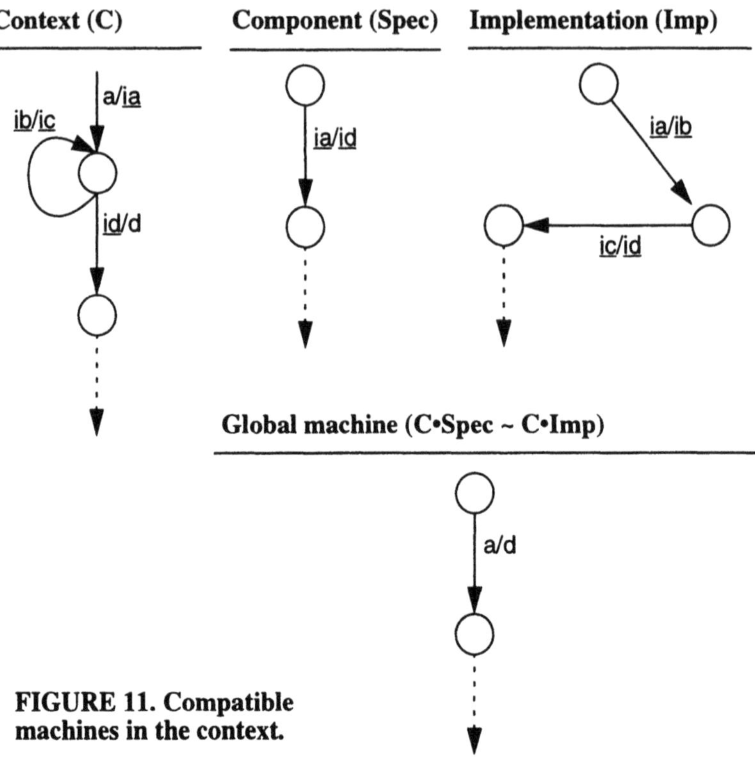

FIGURE 11. Compatible machines in the context.

Example 4. If we consider the Subscriber from the example of Section 4.3 to be our component under test, we observe that,

... in order to test component transition...	**... we should test global transition...**
(T_0,"call/NULL",T_1)	(S_0T_0,"call_arriving/NULL",S_1T_1)
(T_1,"stop_ringing/NULL",T_0)	(S_1T_1,"release/NULL",S_0T_0) *and* (S_1T_1,"ringing_timer/NULL",S_0T_0)
(T_1,"off_hook/response",T_2) *and* (T_2,"stop_ringing/NULL",T_3)	(S_1T_1,"off_hook/NULL",S_2T_3)
(T_3,"busy/NULL",T_5)	(S_2T_3,"release/NULL",S_4T_5)
(T_3,"hang_up/hang_up",T_4)	(S_2T_3,"hang_up/NULL",S_3T_4)
(T_5,"hang_up/hang_up",T_0)	(S_4T_5,"hang_up/NULL",S_0T_0)

... in order to test component transition...	**... we should test global transition...**
(T_4,"release/NULL",T_0)	(S_3T_4,"timeout-hang_up/NULL",S_0T_0) *and* (S_3T_4,"release/NULL",S_0T_0)
(T_4,"off_hook/response",T_3)	(S_3T_4,"off_hook/NULL",S_2T_3)

This is true since there are no alternative paths in SCU whose transitions are all labelled *internal/internal* and provided that the context (SCU) is correctly implemented, which is one of our assumptions (Section 3.3). ❑

5 TEST SEQUENCE GENERATION FOR EMBEDDED SYSTEMS

Goal-oriented testing techniques consist of selecting a subset of the global system's behaviour that is likely to be faulty or that is critical within the system and generating test sequences for only those parts. In general, this selection is made in an *ad hoc* manner by human experts that identify the portions of the system's behaviour that might be subject of testing. Obviously, the system is only partially tested and this technique guarantees a behaviour coverage with regard to the subsystem investigated [8].

In this section, the idea is basically to couple together goal-oriented techniques and trace-keeping composition in order to generate test sequences that concern only the component under test. Using the trace-keeping algorithm of Section 4.2 we can automatically identify the parts of the system that reflect the component's behaviour following which we can use goal-oriented techniques to test this subsystem.
In a non-optimized test generation method each transition is tested in the following manner:

1. Use the shortest path to set the system to the initial state of the transition at hand;
2. Send an input signal and check system's output;
3. Check if the system moved to the correct state.

Many techniques to improve this method have been suggested in the literature and they basically consist of finding a path including all system transitions (in the traditional approach). Since we do not need to test all system transitions, we would be glad to find a path traversing only global transitions that affect the component under test. However, the set of transitions that reflect the component's behaviour in the global machine may not form a (strongly) connected I/OFSM. Therefore, some global transitions that do not concern the component itself may have to be kept when generating the test sequences (this is a problem for goal-oriented testing techniques in general).

We are currently working on a tool called TESTGEN developed at INT in order to incorporate test generation for embedded components. It uses the I/OFSM of the global system (without the internal actions) and a list of transitions to be tested as input data, and it generates test sequences for only the transitions belonging to that list. The tests are performed in two different ways: 1) by defining a tour that starts and ends at the initial state and includes the transitions that define the test purposes (transitions that concern the component under test) or 2) by defining a tour that includes these transitions but also the signatures of the arriving states. In the first case, test sequences are shorter but only detect output faults. In the second case, we are able to detect output and transfer faults. Both are optimized.

6 CONCLUSION

In this paper we have presented a pragmatic approach to generating test sequences for embedded components of complex systems. The approach proposed is based on: 1) the definition of a composition procedure that allows the abstraction of the internal signals exchanged between the processes that compose the system, whilst preserving the exchanges between the system and its environment. The trace-keeping composition algorithm that was defined allows the identification of parts of the global system specification that reflect the component's behaviour; 2) goal-oriented testing. The transitions that reflect the component's behaviour specification can be used to build up test objectives that only test the component's implementation.

This approach presents the following advantages: it is not necessary to test the system as a whole (as is the case for traditional methods); it is possible to test the component's behaviour in context and to detect if the component's implementation conforms to its specification. It is also possible to detect if the system implementation includes an embedded component that is equivalent in context to the component specification.

7 REFERENCES

[1] G. Booch, *Object Oriented Analysis and Design with Applications,* 2nd Edition The Benjamin/Cummings Publishing Company, 1994.

[2] PT, *Component Testing for Mobile and Broadband Telecommunications - COIMBRA,* COPERNICUS Project Proposal, Feb. 1996.

[3] R. Anido, A. Cavalli, T. Macavei, L. P. Lima, M. Clatin, M. Phalippou, *Testing a real protocol with the aid of verification techniques,* XXII SEMISH - Brazil, Aug. 1996, pp. 237-248.

[4] A. Cavalli, B. Lee and T. Macavei, *Test generation for the SSCOP-ATM networks protocol,* SDL Forum'97, September 1997, France.

[5] ISO, *Information Technology, Open Systems Interconnection, Conformance Testing Methodology and Framework,* International Standard IS-9646. ISO, 1991.

[6] A. Petrenko, N. Yevtushenko, G. v. Bochman, *Fault models for testing in context,* Proceedings of FORTE - Kaiserslatern, Germany - 8-11 Oct. 96.

[7] O. Charles, R. Groz, *Formalisation d'Hypothèses pour l'Évaluation de la Couverture de Test,* Proceedings of CFIP'96 - Rabat, Morocco, 14-17 Oct. 1996.

[8] Algayres, B; Lejeune, Y. and Hugonnet, F. *GOAL: Observing SDL Behaviours with GEODE,* Proceedings of the 7th SDL Forum, Oslo, Norway, 26-29 Sept. 1995 - pp. 223-230.

[9] J-C Fernandez, C. Jard, T. Jéron, G. Viho, *Using on-the-fly verification techniques for the generation of test suites,* Summer School MOVEP'96, Nantes, France, 18-21 Jun. 1996.

[10] L. P. Lima, A. Cavalli, *Service Validation - An Embedded Testing Approach,* Proceedings of EUNICE Summer School - Lausanne, Switzerland, 23-27 Jun. 1996.

8 BIOGRAPHY

Luiz Paula Lima Jr. is currently a PhD student at INT (*Institut National des Télécommunications*), Evry, France and he received his MSC degree in 1994 at UNICAMP (State University of Campinas), Brazil. His current research interests include object-oriented distributed systems and platforms (ODP/CORBA) and testing methods for these architectures.

Ana Rosa Cavalli received the Doctorat d'État es Mathematics and Computer Science in 1984 from the University of Paris VII, Paris, France. From 1985 to 1990, she was a staff research member at the CNET (*Centre National d'Etudes des Télécommunications*), where she worked on software engineering and formal description techniques. Since 1990, she joined the INT (*Institut National des Télécommunications*) as professor. Her research interests include formal description techniques, validation of protocols and services, computing methodology and testing methods for distributed architectures.

PART EIGHT

Tools and Environments

The European initiative for the development of Infrastructural Tools: the INTOOL programme.

P.COUSIN
European Project Officer for the European Commission and EOTC; Technical Director - CETECOM France
2, rue Jacques Monod - F-91893 Orsay Cedex France
Tel: +33169351313 Fax: +33169351314; cousinph@aol.com

Abstract

This paper presents the INTOOL programme of the European Commission which gave financial support to the development of Infrastructural Tools. Its aim was to improve the quality of test system components, to speed up and to reduce the costs of global test services development process in Information Technology and Telecommunications within Europe. This initiative, of a horizontal nature, was the political and technical follow up of the multiple sectoral support given by the CTS (Conformance Testing Services) programme during 12 years and involving more than 100 Million Ecus. Details are given on the political and technical background which gave justification to the launching of the INTOOL programme in 1994. The achievements of the programme are presented.

Keywords

European Commission, generic, Intool, CATG, compiler, open architecture

Testing of Communicating Systems, M. Kim, S. Kang & K. Hong (Eds)
Published by Chapman & Hall © 1997 IFIP

1 INTRODUCTION

Increasingly there is a drive to make communications protocol testing more cost-effective. One facet of this is to increase the use of automated tools in an efficient manner. Early tools tended to be created as bespoke software to solve a specific problem. However, the market for specific test tools is rather limited and insufficient to justify the high cost of these early developments. Thus, there is now a strong movement in the market towards using more generic tools which can be tailored to solve a whole range of problems.

What is tending to happen at present is that the dominant tool suppliers are creating their own suites of generic tools which can be combined in many different ways but only with tools from the one supplier. Different suites of tools from different suppliers remain incompatible. Thus, test tool users are getting locked into using tools from a single supplier and this is hindering the development of a truly competitive test tool market in Europe. Users want to be able to select the best tools for the job with the confidence that different tools from different suppliers can be used in combination.

In order firstly to have a genuinely open and competitive test tools market and secondly to have maximum flexibility in using automated tools in communications testing (whether for conformance testing or interoperability testing or both), it is desirable to be able to select different tools from different suppliers and use them in combination. In order to ensure this, it is necessary to have agreed specifications for key interfaces between different tools.

To address these issues the European Commission launched the Infrastructural Tool programme (INTOOL) and have been supporting three INTOOL projects during 1995 and 1996 in order to support the development of generic tools to facilitate the use of automation in the testing infrastructure in Europe.

2 BACKGROUND

The achievement of a single European market will increasingly rely on the removal of technical barriers to trade.

Since 1983, considerable efforts have been made by the European Community to develop one of the key ingredients required to promote the objectives of economic integration in this field : a standardisation policy aimed at opening the markets to the free circulation of goods and the implementation of trans-European services.

Much progress has been accomplished in the area of European standardisation for several industrial sectors by the European Standardisation Organisations (ESO: CEN-CENELEC-ETSI). At the same time, it was apparent from early on that testing and certification had a vital role to play as a necessary complement to standardisation actions if standards were to be implemented in practice.

Testing and certification represent an essential component of Community standardisation policy. In many fields the complexity of standards is magnified to a degree where it becomes difficult, if not impossible, to implement the standards without creating technical divergence which will ultimately result in a lack of interworking or a mismatch to the initial specifications. Therefore, the need for an adequate guarantee that products conform to standards- to unique interpretation of standards- emerges as a decisive condition for building up confidence in the standardised products.

Consequently, in 1985, the Commission of European Communities launched the conformance test services (CTS) programme, covering only the IT&T field to provide tools and facilities to meet the growing market for truly interoperable IT&T systems.

The basic idea of the CTS programme was to establish real testing services for the market, capable of verifying the conformity of products to the reference standards, based on the principle of a standard testing methodology thus leading to comparability of results and, eventually, mutual recognition of test reports and certificates. The organisations providing the services are called testing centres or testing laboratories.

Since 1985, six calls for proposals have been launched to invite interested and qualified organisations to set up testing service at reduced risk (50% contribution), each call resulting in a set of new projects launched for an average duration of 30 months.

The current map of the CTS programme includes 50 testing centres offering testing services across Europe for about 60 technical areas. Each service is offered by at least two centres in Europe: although only two are funded, often the number of centres offering the same service is larger. The CTS experience, however, involving, as it did, multiple contacts with relevant technical committees of the ESOs, as well as the members of the testing community, has highlighted the following problems which were not fully addressed within the CTS programme:

- INTERPRETATION AND IMPLEMENTATION OF STANDARDS: the development process of standards does not always make it possible to guarantee a rapid and effective implementation of the conformity assessment procedures against these standards under optimum economic conditions.
- COST AND TIME REQUIRED TO DEVELOP NEW TOOLS: the development cost of new tools is too high and all the more so where the product is integrated and has hidden functions. The deadlines for their development are often too long.
- NECESSARY NEW THINKING ON METHODS: the product validation methods in certain complex fields still remain confined to the field of theoretical studies.

Bearing these aspects in mind, in 1992, the EC launched a study where the objective was to analyse the different steps between the availability of standards and the availability of maintainable conformance testing services for that standard. The main result was the definition of a model identifying processes, key entities and **areas where the application of productivity tools may contribute to improve efficiency and economy**.

Following such a line, in June 1994, the EC gave support to the development and delivery of infrastructural tool(s) generally accessible within Europe with the aims of improving the quality of test system components and/or speeding up the global and reducing the cost of test services development process and/or facilitating the establishment of conformance testing services in Information Technology and Telecommunications.

The call for proposals was limited to a well defined list of domains.

- **CATG : Computer Aided Test Generation;**
- **GCI: Generic Compiler or Interpreter;**
- **OTE: Open Test Environment.**

Three projects involving several European companies were established. They terminated in March 1997 and the public results are now available. A free CD-ROM gathering all the information and giving the public domain specifications is available on demand from the author of this paper.

3 THE CATG PROJECT

It is apparent from many different sources that there is a great deal of work to do to produce a complete set of Abstract Test Suites from a base standard or specification notation starting point. Once these are produced, there is a considerable ongoing maintenance requirement. As in many other areas, computer technology may be applied to increase efficiency and effectiveness. Years of experience in the domain of test suite generation in TTCN(Tree and Tabular Combined Notation) and in SDL (Specification Description Language) have been concretised in the CATG project which put a commercial tool on the market.

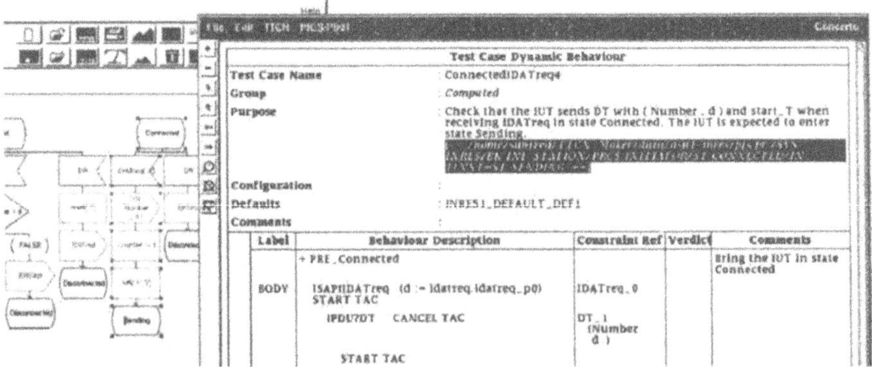

Figure 1 Exemple of TTCN Test Case generated from SDL description.

4 THE GCI PROJECT

This project worked on a model which would help to develop TTCN compilers, the final objective being to directly implement standards coming from standardisation bodies already drafted in TTCN into test systems.

Existing TTCN test systems can normally be thought of as containing two parts; one that handles the interpretation and execution of the TTCN, and one that adapts this TTCN execution for use with the particular system under test. The intention of the GCI interface is to allow the separation of these parts, so that the same TTCN compiler/interpreter can be reused in a number of different test systems

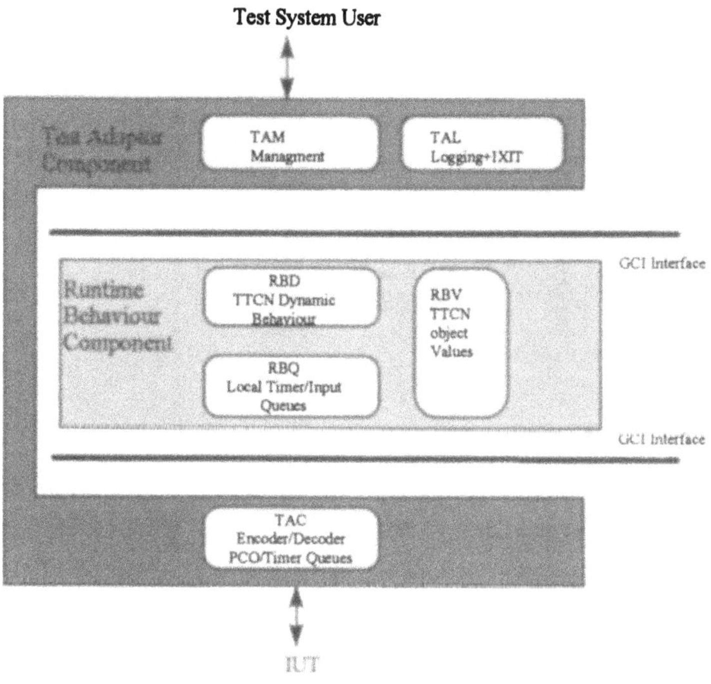

Figure 2 The GCI Interface model.

5 THE OTE PROJECT

The basic idea was to define an open architecture where generic reusable components and existing dedicated products can co-operate. The project has provided a set of specifications (OTE architecture, objects, communications protocols) as well as commercially available tools. The **OTE Architecture** is an open environment dedicated to the distributed test process. It includes the interface to access and manage the OTE objects and the interface for tool communication. The architecture can be implemented with several technologies (Corba, Proprietary solutions etc). **Objects Specifications** offer a common format for the objects involved in testing. Interchange of information between the actors of the test process is easy through OTE. **OTE Piloting Protocol** (PMI) is the powerful solution to control several test equipment involved simultaneously in distributed testing. PMI covers all the phases of the test process and the aspects of remote control of the test equipment.

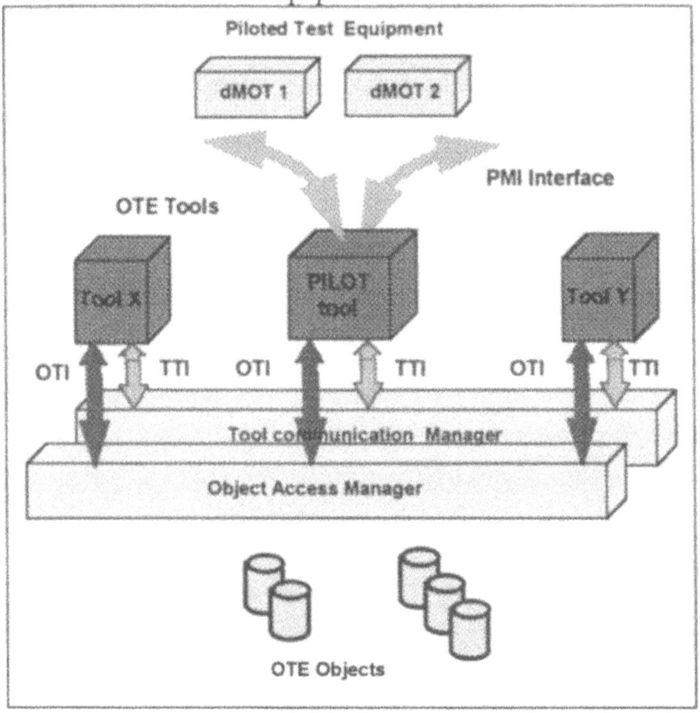

Figure 3 the OTE architecture.

6 BIOGRAPHY

Philippe COUSIN - Technical Director - CETECOM France
15 years experience in Telecommunications, Research and Testing including in the past 5 years a broad insight into ICT and other industrial matters at the European policy-making level at the European Commission and EOTC (European Organisation for Testing and Certification). EC and EOTC Project Officer of CTS and INTOOL projects. Several years of operational activities including within a

testing laboratory and research activities in the development of testing facilities and methodologies (TTCN, SDL, ISO9646). Representative of France Telecom and European Commission in Standardisation bodies and pre-normative various working groups. Project Leader of Standardisation Project teams.

HARPO: TESTING TOOLS DEVELOPMENT

E. Algaba, M. Monedero, E. Pérez, O. Valcárcel
Telefónica I+D
Emilio Vargas, 6
28043 - Madrid (Spain)
algaba@tid.es, mmr@tid.es, emilla@tid.es, ovr@tid.es

Abstract
This paper briefly introduces the HARPO testing tool generation toolkit. The features of the different tools included within HARPO are shown: automatic test generator, TTCN compiler, PICS editor, as well as their role in the testing tools derivation process. Finally, the main features derived from the operation of a testing tool obtained with HARPO are defined.

Keywords
ISO IS-9646, TTCN, ATS, ETS, SDL, MSC, PICS, PIXIT

1 INTRODUCTION

The ISO conformance testing methodology (ISO IS-9646, OSI Conformance Testing Methodology and Framework) is widely accepted as the main framework in telecommunication systems testing. This methodology includes general concepts on conformance testing and test methods, the test specification language TTCN (ISO IS-9646 Part 3), and the process of specifying, implementing and executing a test

Testing of Communicating Systems, M. Kim, S. Kang & K. Hong (Eds)
Published by Chapman & Hall © 1997 IFIP

campaign.

The HARPO toolkit was developed according to this methodology, although its application scope can be extended to other currently more useful kind of tests: interoperability, load, traffic, end to end testing (interworking), etc. HARPO is a set of tools based on both formal system and test specifications of the protocols under test. It has been designed to automatize the process of obtaining the final executable testing tool as much as possible.

This article deals with the methodology and functionality of the HARPO development toolkit.

2 HARPO: TEST DERIVATION AND OPERATION

The HARPO toolkit allows the generation and operation of test suites, with the purpose of automatizing as much as possible the process of specification and implementation of testing tools for protocols, services and communication systems in general. HARPO is composed of a set of tools which automatize not only the process of specifying a test suite, but also that of obtaining an executable version to be run on a final execution platform. The whole process is shown in figure 1.

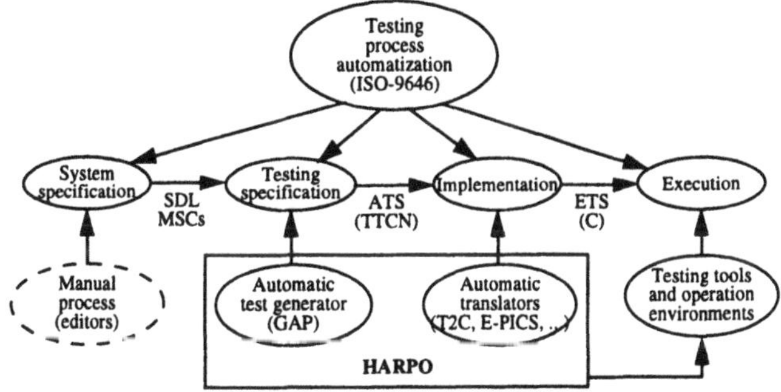

Figure 1 HARPO methodology and testing process.

From the user's point of view, HARPO offers two different set of functionalities: a set of tools to develop the testing tool, including a test generator (GAP), and TTCN translators (TTCN compiler, PICS editor, etc.), and an operation environment for the generated testing tool, using the distributed execution architecture defined in HARPO.

2.1 ATS generation: GAP

The automatic test generator, GAP (figure 2-a), enables its user to derive TTCN test suites automatically. The inputs to this tool are the formal specification of the system written in SDL (ITU-T Z.100 and Z.105) for which test cases are to be automatically derived, and the definition of the test purposes that guide the derivation process, written in MSC notation (ITU-T Z.120) extended to allow the definition of open behaviour patterns for the test purposes.

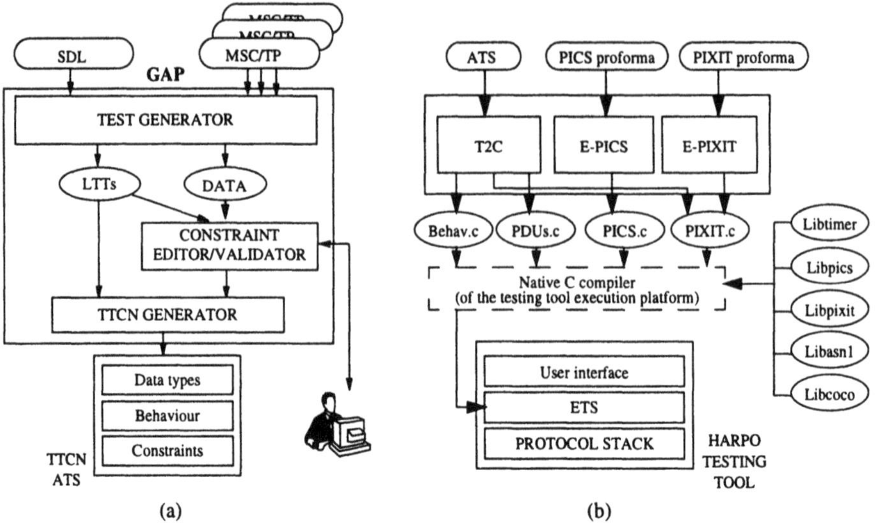

Figure 2 Test generation and ETS derivation subsystem architecture.

Each test purpose is simulated against the SDL specification, allowing GAP to generate several test cases for it, comprising all possible behaviours that fit the test purpose. Data types and constraints are also generated. Constraints with associated predicates (gathered while simulating) may need the intervention of the user to fill in the appropriate values to verify these conditions.

The output of this HARPO subsystem is a complete compilable test suite in TTCN, (behaviour, data type definitions, constraints, etc.). The system does also provide coverage measures achieved by the tests with respect to the formal specification of the system under test. State, signal and transition coverage measures are computed, as well as incremental measures, which allow for the comparison of different test suites in terms of quality.

The GAP tool highly automatizes the process of generating test specifications. Usage of this tool provides enormous advantages with respect to a manual specification of test cases: costs are significantly reduced and the quantity and quality of the generated tests are increased.

2.2 ETS derivation and operation

The ETS generation subsystem (figure 2-b) takes the TTCN ATS (concurrent or not), PICS and PIXIT definitions as its input, in order to derive an executable version of the test suite. It comprises three tools: a TTCN compiler (T2C), a PICS-proforma editor (E-PICS) and a PIXIT-proforma editor (E-PIXIT).

E-PICS provides editing and error checking capabilities, and translates the proforma into C code. E-PIXIT provides editing capabilities and translates the proforma into C code. T2C is also able to generate the C code for the PIXIT-proforma automatically, from the information included in the ATS. The proformas (in C code) are included in the executable testing tool. Thus, they can be dynamically managed in run-time (of the testing tool), i.e., building and initializing the proforma, reading and/or writing individual values of PICS and PIXIT, performing static conformance review, etc., allowing the parameterization of the executable test suite (selection expressions, external parameters, etc.).

T2C translates a TTCN-MP ATS to the C code that implements it. T2C is ATS independent and generates C code for the dynamic part, coding, decoding, building, identifying and matching of constraints (including CMs, if concurrent) for tabular and ASN.1 definitions.

Several libraries provide auxiliary functions to the generated code: timer and pics and pixit proformas management, tabular and ASN.1 auxiliary coding/decoding functions, etc. They depend on the platform(s) on which the testing tool is to be executed.

The architecture of a HARPO generated testing tool is depicted in figure 3.

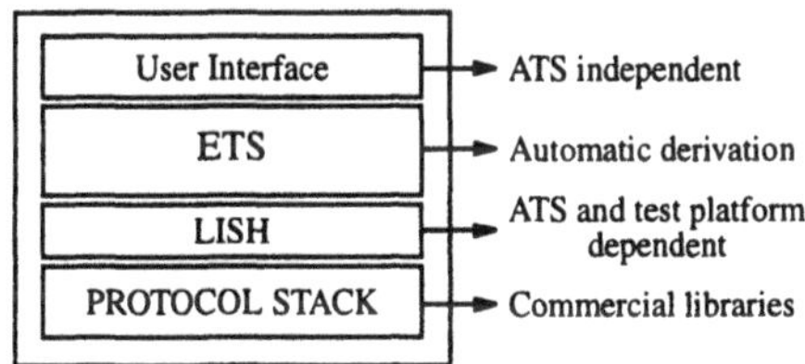

Figure 3 Architecture of a HARPO testing tool.

The LISH support library in figure 3 adapts the interfaces between the automatically generated code and the commercial protocol stack available in the execution platform.

HARPO provides the auxiliary libraries and local, remote and distributed (concurrent) user interfaces for several platforms, including WS with SunOS and Solaris, Chameleon-32 and Chameleon-Open from Tekelec, PC with Windows95 and PT-500 from HP.

Non concurrent testing tools, can be operated locally in the test execution platform or remotely from another machine. HARPO uses a client/server architecture to allow ETS remote operation. In this case, the user interface of the testing tool (figure 3) is no more a graphical user interface (local operation) but the server process of the client/server model.

The same architecture does perfectly suit the distributed operation environment imposed by concurrent TTCN, where the ETS is split in separated processes running in different test platforms. An example of concurrent testing tool is depicted in figure 4. This example shows a HARPO testing tool distributed in three different test platforms.

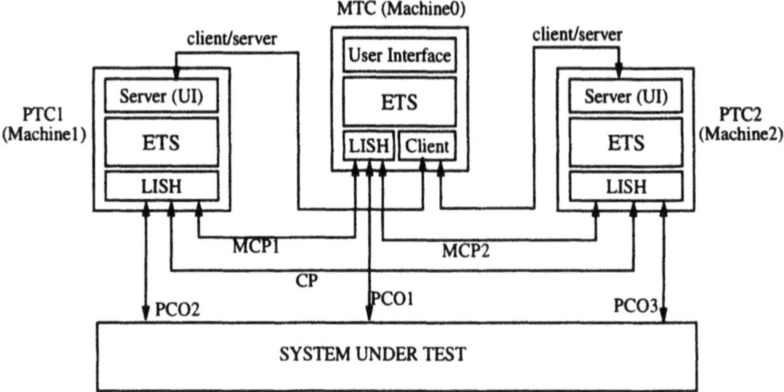

Figure 4 Example of concurrent operation environment.

Since code generated by HARPO allows distributed operation for concurrent test cases, the elements included in such HARPO testing tool must correspond to those defined in the TTCN specification. The ETS is splitted in different components (MTC and PTCs), in such a way that it is possible to build small HARPO testing tools to implement each ETS part, which are coordinated through the defined coordination points (MCPs and CPs) using the corresponding coordination messages (CMs). Concurrent operation and coordination is carried out within LISH, which is responsible of two different matters: PCOs and CPs management. The user interfaces of the PTCs become server processes in communication with the client running in the MTC.

To sum up, developing an executable testing tool is greatly automatized using the HARPO ETS derivation subsystem, thus reducing the development time and the complexity of the process.

3 CONCLUSIONS

The HARPO toolkit provides a high level of automatization in the testing tools development and operation process due to the automatic test generation subsystem (complete TTCN test suite, few user inputs required), the ETS derivation subsystem (data types and constraints handling for tabular and ASN.1, ATS and test platform independent, PICS and PIXIT management embedded in the ETS) and its final testing tool operation environment (local, remote and distributed --concurrent-- testing architecture based on client/server technologies, generic dynamic user interface).

The tools provided by HARPO, following ISO-9646 standard, impact directly on the productivity of the testing tools development and operation:

- Reducing the testing specification and implementation time.
- Generating many more test cases than in a manual process. Quality measures of the generated test suites are available (coverage).

- Easy maintenance and updating of generated testing tools.
- High automatization degree in the complete testing tool development and operation process.
- The C-code produced by HARPO can be easily run on a wide variety of hardware platforms (general purpose computers, protocol analysers, etc.).

HARPO has been successfully used to develop a high number of testing tools: Multilink X.25, ISDN-FAX G4 (transport, session and application protocols), X.32, SS7 (TUP, ISUP, ISDN Access), ISDN interworking (EURESCOM P-412), Core-INAP/SSP (Intelligent Network), ATM UNI (ATM layer), ISDN Supplementary Services, TBR3&4.

These tools are running in many different hardware platforms: WS with SunOS and Solaris, Chameleon-32 and Chameleon-Open from Tekelec, PC with Windows95, PT-500 from HP, etc.

4 BIOGRAPHY

Enrique Algaba is the project manager of testing engineering group in Telefónica I+D (R&D Labs.). Since joining Telefónica I+D in 1988, he has been engaged in the research and development of protocol testing tools and automatizing the testing process.

Miguel Monedero joined Telefónica I+D (R&D Labs.) testing engineering group in 1995. He has been working since then in the fields of test generation and testing automatization.

Esteban Pérez joined Telefónica I+D (R&D Labs.) testing engineering group in 1993 and has been engaged with test generation techniques and in automatizing the development and operation of protocol testing tools.

Oscar Valcárcel began working in Telefónica I+D (R&D Labs.) in 1988, joining the public telephone system development group. In 1992 he was assigned to the testing engineering group, where he has been devoted to automatizing the development and operation of protocol and service testing tools.

Application of a TTCN based conformance test environment on the Internet email protocol

Jun Bi and Jianping Wu
Dept. of Computer Science, Tsinghua Univ.,
Dahong Xiong, Caixin Wu, and Jinwen Cao
Institute of China Electronic System Engineering,
Beijing, P.R.China

Abstract

This paper presents the conformance testing of the Internet email protocol SMTP using an integrated test system PITS. With TTCN based test execution and flexible reference implementation, PITS could test both the OSI and Internet protocols. In this paper, we discuss two methods for testing SMTP and the design of TTCN based SMTP test suite.

Keywords

Conformance testing, protocol test system, SMTP, TTCN

1 INTRODUCTION

The protocol engineering makes it possible to apply formal methods and certain automated tools during the protocol development life cycle. Although it specifically intended for the development of OSI protocols and services, it is

Testing of Communicating Systems, M. Kim, S. Kang & K. Hong (Eds)
Published by Chapman & Hall © 1997 IFIP

possible to have a much broader scope of application for TCP/IP protocols. Today, Internet had been widely accepted as the embryo of global information infrastructure. Therefore, the reliable communication between TCP/IP products is important in the future information highway. Without conformance testing, how could we find the errors in the routers, e-mail systems, and other devices we used? That is why we should do the conformance testing as while as the interpretability testing for TCP/IP products. However, over the last decade, there is little research effort in the formal specification, validation and testing for TCP/IP protocols. We developed an integrated testing environment PITS and the TTCN based test suite of SMTP (Simple Mail Transfer Protocol). PITS is implemented using the Sun Sparc workstation and Solaris 2.4 operating system. With this system, we could test different TCP/IP protocols on the basis of their TTCN test suite and corresponding reference implementation (RI).

This paper is structured as follows, section 2 gives a brief overview of PITS. Section 3 introduces the test organization and TTCN test suite of SMTP.

2 A TTCN BASED TEST ENVIRONMENT

Many earlier test systems are designed for single protocol or single test method. Therefore, their capability is limited. By our experience, the key of a test system is the test suite (TS) and test execution (TE) mechanism. In recent years, ISO has gradually developed the test suites for their standard protocols, and these test suites are described in TTCN. The Protocol Integrated Test System (PITS), developed by Tsinghua University, aims to provide a basic platform for developing protocol testing, and at the same time provides real test system for testing network protocols. Figure 1 shows the main processing flow in PITS.

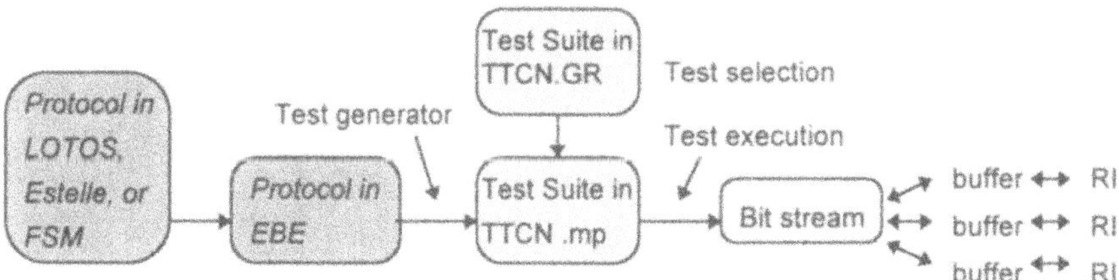

Figure 1 The processing flow in PITS.

As above, this system organizes its testing process on the basis of TTCN test cases, and uses parallel interpreting to raise test efficiency. To generate test cases, we designed a test generator. It could derive the TTCN.MP test suite from the protocol specification of EBE (External Behavior Expression), which could be obtained from the other format of protocol specifications, such as FSM, LOTOS, and Estelle [1]. For the standard TTCN.GR test suites (e.g. ISO/IEC 8882), the

TTCN editor could translate it into TTCN.MP. After test selecting on the basis of PICS and PIXIT, the test cases are interpreted and then executed by TE step by step. TE an engine which interacts with other components of PITS, and controls test process according to the content of TS and simultaneously generates all information required to produce test report. The bit stream generated by TE will be sent to the corresponding buffer and RI, at last. RI is the lower support communicating with IUT. So with the suitable test suites and RIs, PITS could test different protocols by different methods.

3 CONFORMANCE TESTING OF SMTP

3.1 The difference between OSI and TCP/IP testing

SMTP is defined in RFC 821 and RFC 822. The objective of SMTP is to send and receive email reliably and efficiently using the client/server mode. Another important feature of SMTP is its capability to relay mail across transport service environments. Comparing with the peer-to-peer OSI protocols, the client/server and relay are the important modes in the TCP/IP suites. So, when we design the test architecture, we must consider these modes. For the client/server protocols (e.g. FTP, telnet), the IUT of client or server has different protocol functions. Thus we must design different test suites for this asymmetric architecture. Although some products implement both the functions of client and server, we have to classify the two test objectives in the test suite and test them separately in practice. For the relay function in some protocols, we have to control and observe the test events from the two sides of the IUT.

3.2 Test the function of sending and receiving email

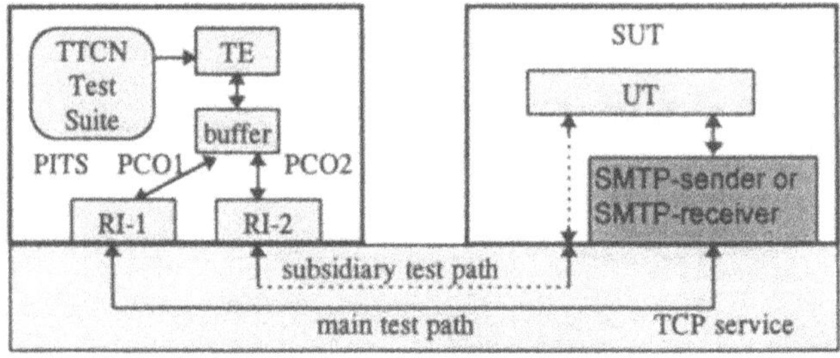

Figure 2 Test architecture of sending and receiving email.

For testing the function of sending and receiving email, we adopt the distributed test method defined in ISO 9646. Figure 2 shows this testing architecture. We implement an upper tester (UT) above IUT. For testing SMTP-receiver, UT only reports the status of IUT. When we test SMTP-sender, UT will make the IUT send email actively and act as the SMTP client. We use two TCP connections (by UNIX socket) as test paths. TE communicates with IUT by main test path (MTP), and with UT by subsidiary test path (STP). The MTP is for "regular" test events, and STP is for "out-of-baud" information (for TE sending test control message to UT or getting response from UT). We implement the TCP-RI as a C++ class. The test events will be sent to RI from buffer according to the PCO identifiers in this class. RI and buffer could communicate by means of the following message: STARTTEST (Start a test case execution); STOPTEST (Stop a test case execution); FRAME_SEND_OUT (TE sends a ASP/PDU); FRAME_RECEIVE (TE receives a ASP/PDU); QUIT (Quit the execution of a test case).

3.3 Test the function of relaying email

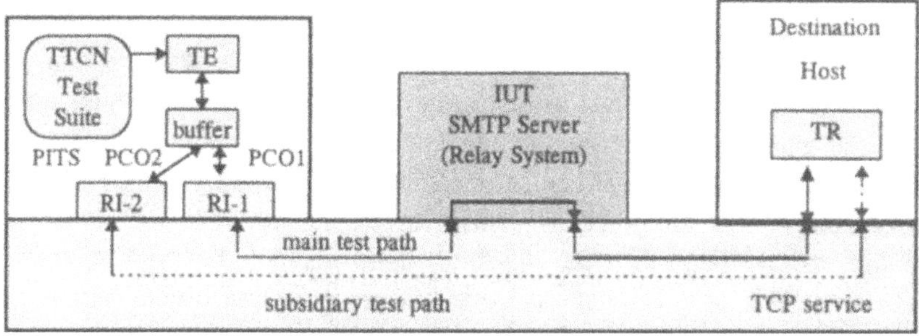

Figure 3 Test architecture of relaying email.

In ISO 9646, a number of standard abstract test methods for end system have been defined. Although there are two relay system test methods in ISO 9646, the capability of YL method is too simple to be put into use in practice and YT has two test systems so the test coordination of these two testers would be a big problem. Referring to the distributed method for end system and YL for relay system in ISO 9646, we present a new test method named as "distributed loop-back". It had been used in the SMTP relay system testing. We implement a test responder (TR) in the mail destination host send/receive the email to/from PITS via IUT. In test system, there are two PCOs for the control and observation from both sides of IUT. When TE sends a email from PCO1 to destination host, the TR will get the relayed events from IUT and return it to the test system through STP. Then this returned message could be obtained by TE from PCO2. Because the events of PCO1 and PCO2 are both from the lower interface of IUT, TR is a conceptual lower tester (LT). However, TR has not the function of test execution and it works under the control of TE. In this architecture, we use only one TE for executing the events of two PCOs, so the problem of the coordination of two LTs is solved. When test executing, the test events for different PCO could be distinguished by the buffer and be sent to the corresponding RI. It makes the test process continuously and high-efficiently. It is just the advantage of this method.

3.4 Design of TTCN based SMTP test suite

There are lots of test generation research results ([2][3][4][5][6] etc.). When we design the SMTP test suite, we use EBE as the protocol model. EBE specifies only the external behavior of a protocol in terms of the input/output sequence and their logical (function and predicate) relations. First, we refine a SMTP external state machine. For 3 protocol functions, we define 3 EBEs: SMTP-SEND-EBE, SMTP-RECV-EBE, and SMTP-RELAY-EBE. Each EBE is a four-tuple <S, s0, T, R>. Here S is the external state set, s0 is initial external state, T is the transition set of S and R is the logic relation set of T. The transition represents most of the SMTP commands: DATA, HELO, MAIL, RCPT, RSET, SEND, SOML, SAML, VRFY, EXPN, HELP, NOOP, QUIT, and TURN. Notice that the "DATA" here is a series of lines sent from the sender to the receiver with no response expected until the last line is sent. For each command there are three possible outcomes: Success (S), Failure (F), and Error (E). The test sequence derivation method is used to identify associations between inputs and outputs through the interaction paths and their I/O subpaths. Then the generic test cases specified in TTCN.GR format can be generated from these I/O subpaths [1]. In this test suite, there are three sub-suites for testing the function of sending, receiving, and relaying. In each sub-suite, there are 8 test groups to test the procedure of each protocol state and one test group to test system parameters. There are 89 test cases in total. For example, when we test the relay function in data transformation phase, we use the following TTCN.GR test case (see table 1). Here IUT is the mail relay system and the test method is shown in figure 3. In this test case, there are two PCOs. TE will control TR to send PDUs at PCO2 then receive the IUT response PDUs from PCO2 and the relayed PDUs from PCO1 by TR. Test verdict will be gotten according to the events from both PCOs.

Table 1 A TTCN based test case of SMTP

Test Case Dynamic Behavior					
Test Case Name: IT_01					
Group : **SMTP/RELAY/IT**					
Purpose : To verity SMTP mail server (server@Relay>) could relay a mail from <sender@Remote> to <receiver@Local> in information transfer phase.					
Default :					
Comments :					
Nr.	**L.**	**Behavior Description**	**Constraint Ref.**	**Verdict**	**Comments**
1		+ Pre_IT			Preamble
2		PCO2 ! Body	Body_01		(1)
3		PCO2 ? 250	250_01		250 ok
4		PCO1 ? MailFrom	MailFrom_10		(2)
5		PCO1 ! 250	250_01		250 ok
6		PCO1 ? RcptTo	RcptTo_01		(3)
7		PCO1 ! 250	250_01		250 ok
8		PCO1 ? Data	Data_01		DATA
9		PCO1 ! 354	354_01		354

10		PCO1 ? Body	Body_02		(4)
11		PCO1 ! 250	250_01	(PASS)	250 ok
12		+ CHK_IT			Check
13		PCO1 ? Otherwise		FAIL	
14		PCO1 ? Timeout T01		FAIL	
15		PCO1 ? Otherwise		FAIL	
16		PCO1 ? Timeout T01		FAIL	
17		PCO1 ? Otherwise		FAIL	
18		PCO1 ? Timeout T01		FAIL	
19		PCO1 ? Otherwise		FAIL	
20		PCO1 ? Timeout T01		FAIL	
21		PCO1 ? Otherwise		FAIL	
22		PCO1 ? Timeout T01		FAIL	
23		PCO2 ? Otherwise		FAIL	
24		PCO2 ? Timeout T01		FAIL	

Detail Comments:
(1) <sender@Remote> (2) MAIL FROM: <server@Relay>
(3) RCPT to: <receiver@Local> (4) Mail Body = stamps + Body_01

4. CONCLUSION

In this paper, we introduce some experience with SMTP testing. The method presented in this paper is also available for another Internet mail protocol like MIME, since the mail mechanisms of MIME and SMTP are same. However, in MIME test suite, the PDU declaration part will be much different from the text mail test suite. PITS had been implemented using Sun Sparc workstation and Solaris 2.4. It had been used in practical testing activity for many OSI and TCP/IP protocols. Our further work focuses on the development of conformance testing for other Internet protocols, such as the routing protocol OSPF v2.

REFERENCES

[1] J. Wu and S.T.Chanson, Test sequence derivation based on external behavior expression, *in 2nd IFIP IWPTS*, 1989.

[2] O.Henniger, B.Sarikaya, et al., Test Suite Generation for Application Layer Protocol from Formal Specifications in Estelle, in O.Rafiq, editors, *6th IFIP IWPTS*, 1993.

[3] G.v.Bochmann, et al., Fault coverage of test based on finite state models, in T.Mizuno, et. al., editors, *7th IFIP IWPTS,* 1994.

[4] S.Kang and M.Kim, Test sequence generation for adaptive interoperability testing, in A.Cavalli, et. al., editors, *8th IFIP IWPTS,* 1995.

[5] A.Petrenko, et. al., Testing deterministic implementations from non-deterministic FSM specifications, in B.Baumgarten, et. al., editors, *9th IFIP IWTCS,* 1996.

[6] M.C.Kim, et.al., An approach for testing asynchronous communicating systems, in B.Baumgarten, et. al., editors, *9th IFIP IWTCS,* 1996.

The INTOOL/CATG European project: development of an industrial tool in the field of Computer Aided Test Generation

Etienne Desécures
Sema Group
56, rue Roger Salengro - 94126 Fontenay-sous-Bois Cedex, France
Tel: +33 1 43 94 58 37 E-mail: Etienne.Desecures@sema-taa.fr

Laurent Boullier
France Télécom - CNET
Technopole Anticipa - 22307 Lannion Cedex, France
Tel: +33 2 96 05 25 17 E-mail: laurent.boullier@lannion.cnet.fr

Bernard Péquignot
Clemessy
18, rue de Thann - B.P. 2499 - 68057 Mulhouse Cedex, France
Tel: +33 3 89 32 32 32 E-mail: b.pequignot@clemessy.fr

Abstract

This paper presents the work and the results of the INTOOL/CATG project, initiated by the EC to provide tools in the field of Computer Aided Test Generation of TTCN test suites. After a survey of potential users' expectation, a tool has been developed and integrated in a complete test design factory covering several aspects of the test life-cycle from SDL editing to ATS completion and maintenance. The technical approach is based on the generation principles of TVEDA, a prototype developed by France Télécom, and on the use of efficient software components. Validation experiments have been carried out and have addressed ISDN and TETRA protocols.

Keywords

Conformance testing, generation of test suites, traceability, SDL, ASN.1, TTCN

1 INTRODUCTION

The INTOOL/CATG (Computer Aided Test Generation) project has been launched in the framework of the "INFRASTRUCTURAL TOOLS" programme of the European Commission. This programme aimed at developing tools to meet the

Testing of Communicating Systems, M. Kim, S. Kang & K. Hong (Eds)
Published by Chapman & Hall © 1997 IFIP

following objectives:
- improve the quality of test system components,
- speed up and reduce the cost of global test services development,
- facilitate the setting of Conformance Testing Services in Information Technology and Telecommunications.

Apart from CATG, the INTOOL programme includes two other projects: **Generic Compilers and Interpreters** (definition of an architecture to ensure independence of TTCN compilers against test tools) and **Open Test Environment** (definition of a standardised architecture and process for conformance test tools).

The goal of the INTOOL/CATG project was to provide a tool achieving automatic test suite generation from protocol specifications available in SDL. The technical approach which has been selected after a call for participation from the European Commission, is based on the industrialisation of the TVEDA prototype (see [6]) which has been developed by France Télécom - CNET. The project aimed also to investigate to which domains the tool is applicable.

The selected consortium includes various complementary expertise and contributions from the following companies: Sema Group, France Télécom CNET, Clemessy, Alcatel Titn Answare, Alcatel Switzerland, Sema Group Telecom and Siemens. Supervision of the INTOOL programme has been assigned to EOTC, the European Organisation for Testing and Certification.

2 FUNCTIONALITIES DEFINITION

Basic tool features

The test generation technique to be implemented in TTCN Maker corresponds to the technique developed for TVEDA. Support of the last versions of the languages, that is to say SDL92 and TTCN edition 2, is an evident requirement for an industrial tool. Operational properties, such as independence from any other TTCN tool, and ability to process real-life cases, which implies high execution speed and efficient memory management, have also been defined at this step. In addition, within the limits allowed by the resources of the project, some freedom was left to provide additional features according to the needs of potential users.

Identification of possible improvements according to users' needs.

In order to evaluate which precise features could be included in TTCN Maker, a detail review of the features which were available in TVEDA has been made, together with a review of the features which could be added. Sixteen potential improvements listed in table 1 have been identified. In order to make requirement choices in accordance with the users' needs, the project has prepared a questionnaire and carried on a survey among the potential users of TTCN Maker, i.e. a fairly large number of companies and institutions involved in telecommunication testing.

The questionnaire was divided into three parts: 1) who are the users, 2) what is their general background in the field of testing, and what are their general needs in terms of software tools, 3) what kind of improvements to the TVEDA prototype they would suggest for TTCN Maker.

The consortium has received 17 answers, the majority of which comes from Specification, Development and Qualification people. According to the responses, it appears that this tool should be able to have complete SDL as input and should be able to generate a complete TTCN ATS with ASN.1 descriptions. The tool has to be provided with a methodological guide and additional functions to select the SDL parts that have to be tested, to structure the test suite and to show the testing architecture. The improvements that have been selected to be implemented in TTCN Maker appears in normal style. The selection took into account the priorities expressed by the responders, the technical feasibility and the budget constraints.

Table 1 Improvements against TVEDA definition.

Considered as important by users	Considered as less important
Produce ASN.1 constraint tables	Coverage evaluation tool
Produce TTCN declaration parts	Traceability links from tests to spec.
TTCN compliant to ETSI style guide	*Accept GDMO input*
Take into account ASN.1 data types	*Take into account other SDL constructs*
Provide a methodological guide	*Remove dynamic process restriction*
Develop a user friendly interface	*Take into account abstract data types*
Remove SDL single process restriction	
Produce complete TTCN test steps	
Produce concurrent TTCN	
More complete TTCN constraints	

3 TOOL FEATURES

As shown by figure 1 presenting TTCN Maker architecture at high level, two versions of the tool have been developed: a stand-alone version, completely independent from any other tool, and a version integrated in a complete test design factory.

Implemented functionalities for the standalone version
The standalone version, invoked through a Unix command line, processes a specification complying with the Z.105 recommendation (SDL combined with ASN.1). From a syntactic point of view the whole Z.105 definition is supported. Nevertheless a good result depends on the semantic correctness of the specification and the conformity to constraints detailed in the user manual of the tool. Moreover, ASN.1 data types and their corresponding values are interpreted. The generation is piloted by a parameter file containing a list of parameters and their values enabling to define the IUT control on the tester, EFSM transformations, and the form of the ATS.

Each ATS produced by TTCN Maker is composed of:
- a Test Suite Overview consistent with the dynamic part,
- a complete declarations part,
- a constraints part to be reviewed,
- a dynamic part containing test cases, test steps and defaults. Test steps are generated empty.

Declarations and constraints are expressed in ASN.1. The format of the generated ATS is TTCN.MP compliant to the delivery 8.4 of TTCN edition 2.

In addition to the ATS, TTCN Maker produces a report file. This report obeys a precise syntax which enables its post-processing by external tools, e.g. test coverage and traceability tools, and contains information related to the generation (generation parameters, errors and warnings, textual cross-reference from the generated ATS to the source SDL).

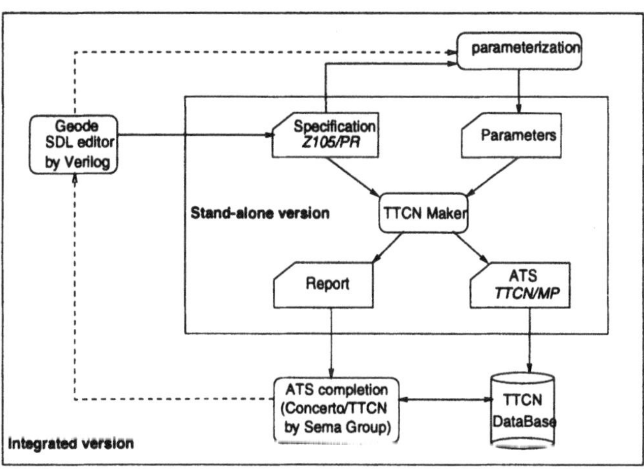

Figure 1 TTCN Maker architecture.

Implemented functionalities for the integrated version

The version of TTCN Maker integrated to Concerto differs from the stand-alone version on the following points:
- selection of the SDL input file and selection of the parameters through a simple graphical user interface,
- TTCN ATS stored in the Concerto/TTCN database,
- on-line help available.

In addition, when coupled with the Geode SDL editor, TTCN Maker enables to select the SDL process to be tested directly in Geode, and sets hypertext links from the generated test cases to the corresponding transitions. Figure 2 gives a simple example of an SDL branch displayed with the GEODE editor and the test case automatically derived from this transition by TTCN Maker, and displayed with the Concerto/TTCN editor. Notice the automatically produced test purpose and comments parts, and the hypertext link attached to the test purpose.

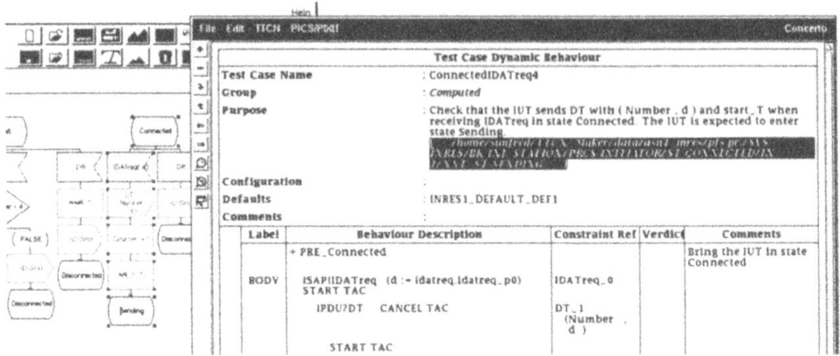

Figure 2 Highlighted transition in Geode and test case in Concerto/TTCN.

4 GENERATION METHOD

Inspired from TVEDA, TTCN Maker is based on a syntactic technique which produces test cases by a simple syntactic transformation of the specification. In this technique, emphasis is put on heuristics which allow an automatic computation of the test purposes.

Constraints on the SDL

The syntactic technique puts some constraints on the SDL constructs which can be handled by the tool. These constraints are of two kinds:

- restrictions on SDL constructs. The tool ignores several constructs.
- restrictions on SDL style. The syntactic technique generates test cases for a single SDL process. Therefore the tool cannot compute the combined behaviour of several processes.

Test purposes

One of the main features of TTCN Maker, is the automatic computation of the test purposes. For doing this, the tool incorporates some heuristics, based on the syntactic structure of the SDL specification. Basically, one test case is produced in order to check each branch of the SDL. This basic test selection mechanism can be modified by the user of TTCN Maker in the following ways:

- restrict the part of the specification that is tested: choice of states, input signals, etc.
- enumerate data values which appear in the SDL branches: test several values of parameters on input signals, or on the contrary merge test cases corresponding to several branches if the observable behaviour is the same in these branches, etc.

These heuristics have been defined after observation of the strategy used by human experts in order to define the test cases.

Test generation process

Once the test purposes have been selected, TTCN Maker computes the behaviour of the test cases. This is done in several steps.

Step 1: Analysis of the SDL specification, and computation of an abstract extended finite state machine (EFSM). During this step, the test architecture is taken into account. The level of abstraction at which the specification is tested can be influenced, through the decision to "control" or not variables or signals.

Step 2: Production of test cases in an internal abstract format. Test cases are produces, which have the following structure:

• call to a preamble which brings the IUT into the start state of the tested transition.

• send and receive events corresponding to the tested transitions. At this level, non determinism is taken into account. Non determinism can result from uncontrolled variables or from uncontrolled events. Constraints are also produced in an incomplete way , using heuristics where possible.

• call to a check sequence which is supposed to check the final state of the tested transition and bring back the IUT in its initial state.

Step 3: Formatting of TTCN suite. The test suite is produced in TTCN according to the *ad hoc* generation parameters. This includes computation of the test suite structure and production of the different TTCN parts.

5 EXPERIMENTS

Numerous experiments performed by CNET with the TVEDA prototype and mentioned in [5] have shown that the TTCN Maker method is particularly suited for protocols up to level 4 of the OSI model. In addition two main experiments were conducted within the INTOOL/CATG project to give an assessment on TTCN Maker itself: an experiment on a part of the TETRA specification developed by ETSI, and an experiment on Q2931 (Broadband ISDN).

The purpose of the Q2931 experiment, performed by Clemessy, was to give an evaluation of the TTCN Maker tool by using it to generate a complete executable test suite. The experiment focused particularly on the work necessary to complete the test suite generated with TTCN Maker to be able to run it. The tested protocol was the L3AS protocol Q2931 Network Side, developed in the RACE 2 / TRIBUNE project and compliant with ATM Forum V3.1. This protocol has been specified in SDL by KPN Research Netherlands and implemented by Clemessy with the Geode tool by Verilog. The ATS generated by TTCN Maker has been compared to an already existing ATS for the same protocol, supplied by KPN. After manual completion the ATS has been successfully compiled. The tester is the C-BIT, a tool developed by Clemessy running on Sun workstations.

This experiment has shown that by generating a complete skeleton of an ATS, with detailed test purposes, TTCN Maker simplifies the work of the test developer:

• The tedious tasks like defining the PDUs, the ASPs, the PCOs or the Timers are achieved automatically by the tool.

• The generation of the dynamic part of the ATS and of a basic constraint for each ASP and PDU allows the developer to concentrate on his most important work : the completion of the constraints and test steps (based on the generated test purposes and on his test objectives) which determines the efficiency of his test suite.

6 CONCLUSION

The experiments have shown that TTCN Maker reached the following results:

- fast generation of test suite skeletons, including message and data type declaration, test purpose creation, test and default behaviours, etc.,
- reducing the time necessary to create a test suite even if an SDL model is not previously available,
- high level of performance due to the syntactic method, whereas simulation techniques often face memory problems because of the so-called "state-explosion" problem,
- high level of parameterisation in order to capture user experience and add it to the benefit of using automated test suite production,
- available either as a completely independent software, or integrated within a test design factory,
- being one of the very first tools accepting Z105 specification including ASN.1, allowing to use SDL specification where the data type declaration is the same as the actual protocol specification, thus producing correct type definition in the ATS.

Nevertheless further useful improvements addressing more complete generation and full processing of the SDL can be performed. In parallel to the INTOOL/CATG project, conclusive research studies based on the connection of TVEDA and SDL verification techniques have been done by France Télécom - CNET (see [7]). This gives the direction for the next versions of the tool.

7 REFERENCES

[1] CCITT, Recommendation Z100: Specification and Description Language.

[2] ITU-T Standardization Sector, Z105 Recommendation.

[3] ISO/IEC 9646-3 (1992): "Information technology - Open systems interconnection - Conformance testing methodology and framework - Part 3: The Tree and Tabular Combined Notation".

[4] ETR 141: Methods for Testing and Specification (MTS); Protocol and profile conformance testing specifications; The TTCN style guide.

[5] E. André, B. Baumgarten, A. Cavalli, M. Gill, O. Henniger, R. Kuhn and R. Lai, Application of Testing Methodologies in Industry - Minutes of the IWTCS'96 Panel Discussion, GMD, December 1996.

[6] L. Doldi, V. Encontre, J.-C. Fernandez, T. Jéron, S. Le Bricquir, N . Texier, and M. Phalippou (1996) Assessment of automatic generation methods of conformance test suites in an industrial context, in *Testing of Communicating Systems* (proceedings of IWTCS'96).

[7] M. Clatin, R. Groz, M. Phalippou and R. Thummel, Two approaches linking a test generation tool with verification techniques, in *Conference proceedings of IWPTS'95.*

PART NINE

Applications of Protocol Testing

Modeling and Testing of Protocol Systems

David Lee
Bell Laboratories, Lucent Technologies
600 Mountain Avenue, Murray Hill, NJ 07974
Tel: (908) 582-5872; *Fax:* (908) 582-5857
E-mail: lee@research.bell-labs.com

David Su
National Institute of Standards and Technology
Bldg 820, W. Diamond, Gaithersburg, MD 20899
Tel: (301) 975 6194; *Fax:* (301) 926-9675
E-mail: dsu@nist.gov

Abstract

Presented are conformance testing of the control portions of protocol systems, which can be modeled by finite state machines and checking sequences that can be used to verify the structural isomorphism of an implementation of the protocol system with its finite state machine-based specification. However, finite state machines are not powerful enough to model data portions associated with many real systems such as Personal HandyPhone Systems and 5*ESS*

Testing of Communicating Systems, M. Kim, S. Kang & K. Hong (Eds)
Published by Chapman & Hall © 1997 IFIP

Intelligent Network Application Protocols. Extended finite state machines with variables and means to test them are also presented. Practical systems like ATM Traffic Management Protocols often contain parameters in the input/output cells; they increase the observability of the system but complicate their testing. Our model is further extended to communicating parameterized extended finite state machines for the test generation.

Keywords
Network protocol, finite state machine, extended finite state machine, conformance testing, checking sequence, complete test set, directed graph, covering path

1 INTRODUCTION

A finite state machine contains a finite number of states and produces outputs on state transitions after receiving inputs. Finite state machines have been widely used to model systems in diverse areas such as sequential circuits, some types of programs, and more recently, network protocols. This motivated early on research into the problem of testing finite state machines to discover aspects of their behavior and to ensure their correct functioning. In a testing problem we have a specification machine, which is a design of a system, and an implementation machine, which is a "black box" for which we can only observe its input/output (I/O) behavior, we want to test whether the implementation conforms to the specification. This is called *conformance testing* or *fault detection*. A test sequence that solves this problem is called a *checking sequence*.

There is an extensive literature on testing finite state machines, the fault detection problem in particular, dating back to the 50's. Moore's seminal 1956 paper on "gedanken-experiments" [Moore, 1956]. introduced the framework for testing problems. Among other fundamental problems, he posed the conformance testing problem, proposed an approach, and asked for a better solution. A partial answer was offered by Hennie in an influential paper [Hennie, 1964] in 1964: he showed that if the machine has a distinguishing sequence of length L then one can construct a checking sequence of length polynomial in L and the size of the machine. Unfortunately, not every machine has a distinguishing sequence. Hennie also gave another nontrivial construction of checking sequences in case a machine does not have a distinguishing sequence; in general however, his checking sequences are exponentially long. Several papers were published in the 60's on testing problems, motivated mainly by automata theory and testing switching circuits. Kohavi's book gives a good exposition of the major results [Kohavi, 1978]. During the late 60's and early 70's there were a lot of activities in the Soviet literature, which are apparently not well known in the West. An important paper on fault detection was by Vasilevskii [Vasilevskii, 1973] who proved polynomial upper and lower bounds on the length of checking sequences. However, the upper bound was obtained by an existence proof, and he did not present an algorithm for constructing efficiently checking sequences. For machines with a reliable reset, i.e., at any moment the machine can be taken to an initial state, Chow developed a method that constructs a checking sequence in polynomial time [Chow, 1978]. For machines without reset, a randomized polynomial time algorithm was reported in [Yannakakis and Lee, 1995]. Yet deterministic polynomial time algorithms remain open.

After introducing some basic concepts of finite state machine, we discuss various techniques for constructing checking sequences, using status messages, reliable reset, distinguishing sequences, identifying sequences, characterization sets, transition tours and UIO sequences, and finally a randomized polynomial time algorithm.

Finite state machines model well control portions of protocols. However, practice systems often contain variables and their operations depend on variable values; finite state machines are not powerful enough to model in a succinct way such physical systems. In the second part of the paper, we use extended finite state machines, which are finite state machines extended with variables, to model systems, including Personal HandyPhone System (PHS) and Intelligent Network Application Protocols (INAP), and to generate tests.

Finally, we further extend the model to communicating parameterized extended finite state machines to discuss testing of ATM Traffic Management Protocols.

2 FINITE STATE MACHINES

Finite state systems can usually be modeled by *Mealy* machines that produce outputs on their state transitions after receiving inputs.

Definition 1 *A finite state machine (FSM) M is a quintuple* $M = (I, O, S, \delta, \lambda)$ *where I, O, and S are finite and nonempty sets of input symbols, output symbols, and states, respectively.* $\delta : S \times I \rightarrow S$ *is the state* *transition function; and $\lambda : S \times I \to O$ is the output function. When the machine is in a current state s in S and receives an input a from I it moves to the next state specified by $\delta(s, a)$ and produces an output given by $\lambda(s, a)$.*

We denote the number of states, inputs, and outputs by $n = |S|$, $p = |I|$, and $q = |O|$, respectively. An FSM can be represented by a *state transition diagram*, a directed graph whose vertices correspond to the states of the machine and whose edges correspond to the state transitions; each edge is labeled with the input and output associated with the transition. For the FSM in Figure 1, suppose that the machine is currently in state s_1. Upon input b, the machine moves to state s_2 and outputs 1. We extend the transition function δ and output function λ from input symbols to strings as follows: for an initial state s_1, an input sequence $x = a_1, \cdots, a_k$ takes the machine successively to states $s_{i+1} = \delta(s_i, a_i)$, $i = 1, \cdots, k$, with the final state $\delta(s_1, x) = s_{k+1}$, and produces an output sequence $\lambda(s_1, x) = b_1, \cdots, b_k$, where $b_i = \lambda(s_i, a_i)$, $i = 1, \cdots, k$. Suppose that the machine in Figure 1 is in state s_1. Input sequence *abb* takes the machine through states s_1, s_2, and s_3, and outputs 011.

Two states s_i and s_j are *equivalent* if and only if for every input sequence the machine will produce the same output sequence regardless of whether s_i or s_j is the initial state; i.e., for an arbitrary input sequence x, $\lambda(s_i, x) = \lambda(s_j, x)$. Otherwise, the two states are *inequivalent*, and there exists an input sequence x such that $\lambda(s_i, x) \neq \lambda(s_j, x)$; in this case, such an input sequence is called a *separating* sequence of the two inequivalent states. For two states in different machines with the same input and output sets, equivalence is defined similarly. Two machines M and M' are *equivalent* if and only for every state in M there is a corresponding equivalent state in M', and vice versa. Given a machine, we can "merge" equivalent states and construct a *minimized* (reduced) machine which is equivalent to the given machine and no two states are equivalent. We can construct in polynomial time a minimized machine and also obtain separating sequences for each pair of states [Kohavi, 1978]. A *separating family* of sequences for a machine of n states is a collection of n sets Z_i, $i = 1, \cdots, n$, of sequences (one set for each state) such that for every pair of states s_i, s_j there is an input string α that: (1) separates them, i.e., $\lambda(s_i, \alpha) \neq \lambda(s_j, \alpha)$; and (2) α is a prefix of some sequence in Z_i and some sequence in Z_j. We call Z_i the *separating* set of state s_i, and the elements of Z_i its separating sequences. Each Z_i has no more than $n - 1$ sequences and of length no more than $n - 1$ [Lee and Yannakakis, 1996a].

Given an FSM A of n states and separating families of sequences Z_i for each state s_i and an FSM B of the same input and output symbols, we say that a state q_i of B is *similar* to a state s_i of A if it agrees (gives the same output) on all sequences in the separating set Z_i of s_i. A key property is that q_i can be similar to at most one state of A. Let us say that an FSM B of no more than n states is *similar* to A, if for each state s_i of A, the machine B has a corresponding state q_i similar to it. Note that then all the q_i's must be distinct, and since B has at most n states, there is a one-to-one correspondence between similar states of A and B. Furthermore, two machines with the same input and output sets are *isomorphic* if they are identical except for a renaming of states. The ultimate goal of testing of systems modeled by finite state machines is to check if an implementation machine B is isomorphic to a specification machine A. Often we first check their similarity and then isomorphism.

Given a complete description of a *specification* machine A, We want to determine whether an *implementation* machine B, which is a "black-box", is isomorphic to A. Obviously, without any assumptions the problem is impossible to solve; for any test sequence we can easily construct a machine B, which is not equivalent to A but produces the same outputs as A for the given test sequence. There is a number of natural assumptions that are usually made in the literature in order for the test to be at all possible. (1) Specification machine A is strongly connected. There is a path between every pair of states; otherwise, during a test some states may not be reachable. (2) Machine A is reduced. Otherwise, we can always minimize it first. (3) Implementation machine B does not change during the experiment and has the same input alphabet as A. (4) Machine B has no more states than A. Assumption (4) deserves a comment. An upper bound must be placed on the number of states of B; otherwise, no matter how long the test sequence is, it is possible that the test does not reach the faulty states or transitions in B, and this condition will not be detected. The usual assumption made in the literature, and which we will also adopt is that the faults do not increase the number of states of the machine. In other words, under this assumption, the faults are of two types: "output faults"; i.e., one or more transitions may produce wrong outputs, and "transfer faults"; i.e., transitions may go to wrong next states. Under these assumptions, we want to design an experiment that tests whether B is isomorphic to A. With the above four assumptions, it is well known [Moore, 1956] that we only have to check if B is equivalent to A.

Suppose that the implementation machine B starts from an unknown state and that we want to check whether it is isomorphic to A. We first apply a sequence that is supposed to bring B (if it is correct) to a known state s_1 that is the initial state for the main part of the test, and such a sequence is called a *homing sequence* [Kohavi, 1978]. Then we verify that B is isomorphic to A using a *checking sequence*, which is to be defined in the sequel. However, if B is not isomorphic to A, then the homing sequence may or may not bring B to s_1; in either case, a checking sequence will detect faults: a discrepancy between the outputs from B and the expected outputs from A will be observed. From now on we assume that a homing sequence has taken the implementation machine B to a supposedly initial state s_1 before we conduct a conformance test.

Definition 2 *Let A be a specification FSM with n states and initial state s_1. A* checking sequence *for A is an input sequence x that distinguishes A from all other machines with n states; i.e., every (implementation) machine B with at most n states that is not isomorphic to A produces on input x a different output than that produced by A starting from s_1.*

All the proposed methods for checking experiments have the same basic structure. We want to make sure that every transition of the specification FSM A is correctly implemented in FSM B; so for every transition of A, say from state s_i to state s_j on input a, we want to apply an input sequence that transfers the machine to s_i, apply input a, and then verify that the end state is s_j by applying appropriate inputs. The methods differ by the types of subsequences they use to verify that the machine is in a right state. This can be accomplished by status messages, separating family of sequences, characterizing sequences, distinguishing sequences, UIO sequences, and identifying sequences. Furthermore, these sequences can be generated deterministically or randomly. The following subsections illustrate various test generation techniques.

2.1 Status messages and reset

A *status message* tells us the current state of a machine. Conceptually, we can imagine that there is a special input *status*, and upon receiving this input, the machine outputs its current state and stays there. Such status messages do exist in practice. In protocol testing, one might be able to dump and observe variable values which represent the states of a protocol machine.

With a status message, the machine is highly observable at any moment. We say that the status message is *reliable* if it is guaranteed to work reliably in the implementation machine B; i.e., it outputs the current state without changing it. Suppose the status message is reliable. Then a checking sequence can be easily obtained by simply constructing a covering path of the transition diagram of the specification machine A, and applying the status message at each state visited [Naito and Tsunoyama, 1981; Uyar and Dahbura, 1986]. Since each state is checked with its status message, we verify whether B is similar to A. Furthermore, every transition is tested because its output is observed explicitly, and its start and end state are verified by their status messages; thus such a covering path provides a checking sequence. If the status message is not reliable, then we can still obtain a checking sequence by applying the status message twice in a row for each state s_i at some point during the experiment when the covering path visits s_i; we only need to have this double application of the status message once for each state and have a single application in the rest of the visits. The double application of the status message ensures that it works properly for every state.

For example, consider the specification machine A in Figure 1, starting at state s_1. We have a covering path from input sequence $x = ababab$. Let s denote the status message. If it is reliable, then we obtain the checking sequence $sasbsasbsasbs$. If it is unreliable, then we have the sequence $ssasbssasbssasbs$.

We say that machine A has a *reset capability* if there is an initial state s_1 and an input symbol r that takes the machine from any state back to s_1, i.e., $\delta_A(s_i, r) = s_1$ for all states s_i. We say that the reset is *reliable* if it is guaranteed to work properly in the implementation machine B, i.e., $\delta_B(s_i, r) = s_1$ for all s_i; otherwise it is *unreliable.*

For machines with a reliable reset, there is a polynomial time algorithm for constructing a checking sequence [Chow 1978; Chan, Vuong and Ito, 1989; Vasilevskii, 1973]. Let Z_i, $i = 1, ..., n$ be a family of separating sets; as a special case the sets could all be identical (i.e., a characterizing set). We first construct a breadth-first-search tree (or any spanning tree) of the transition diagram of the specification machine A and verify that B is similar to A; we check states according to the breadth-first-search order and tree edges (transitions) leading to the nodes (states). For every state s_i, we have a part of the checking sequence that does the following for every member of Z_i: first it resets the machine to s_1 by input r, then it applies the input sequence (say p_i) corresponding to the path of the tree from the root s_1 to s_i and then applies a separating sequence in Z_i. If the implementation machine B passes this test for all members of Z_i, then we know that it has a state similar to s_i, namely the state that is obtained by applying the input sequence p_i starting from the reset state s_1. If B passes this test for all states s_i, then we know that B is similar to A. This portion of the test also verifies all the transitions

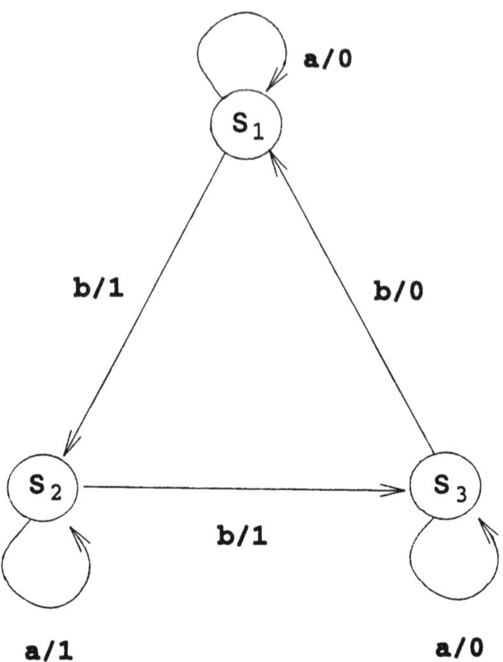

Fig. 1. Transition diagram of a finite state machine

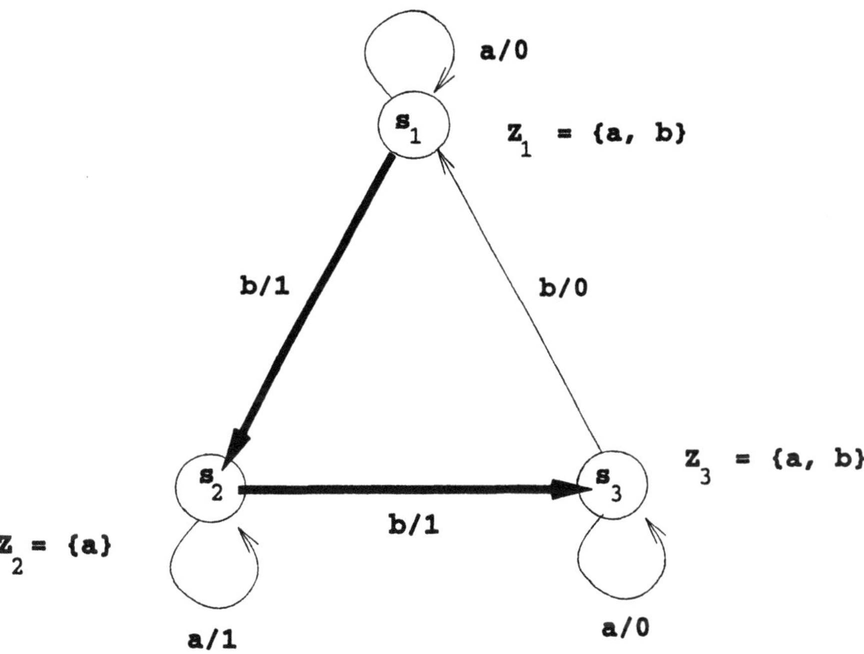

Fig. 2. A Spanning tree of machine in Fig. 1

of the tree. Finally, we check nontree transitions. For every transition, say from state s_i to state s_j on input a, we do the following for every member of Z_j: reset the machine, apply the input sequence p_i taking it to the start node s_i of the transition along tree edges, apply the input a of the transition, and then apply a separating sequence in Z_j. If the implementation machine B passes this test for all members of Z_j then we know that the transition on input a of the state of B that is similar to s_i gives the correct output and goes to the state that is similar to state s_j. If B passes the test for all the transitions, then we can conclude that it is isomorphic to A.

For the machine in Figure 1, a family of separating sets is: $Z_1 = \{a, b\}$, $Z_2 = \{a\}$, and $Z_3 = \{a, b\}$. A spanning tree is shown in Figure 2 with thick tree edges. Sequences ra and rb verify state s_1. Sequence rba verifies state s_2 and transition (s_1, s_2): after resetting, input b verifies the tree edge transition from s_1 to s_2 and separating sequence a of Z_2 verifies the end state s_2. The following two sequences verify state s_3 and the tree edge transition from s_2 to s_3: $rbba$ and $rbbb$ where the prefix rbb resets the machine to s_1 and takes it to state s_3 along verified tree edges, and the two suffixes a and b are the separating sequences of s_3. Finally, we test nontree edges in the same way. For instance, the self-loop at s_2 is checked by the sequence $rbaa$.

With reliable reset the total cost is $O(pn^3)$ to construct a checking sequence of length $O(pn^3)$. This bound on the length of the checking sequence is in general best possible (up to a constant factor); there are specification machines A with reliable reset such that any checking sequence requires length $\Omega(pn^3)$ [Vasilevskii, 1973]. For machines with unreliable reset, only randomized polynomial time algorithms are known [Yannakakis and Lee, 1995; Lee and Yannakakis, 1996a]; we can construct with high probability in randomized polynomial time a checking sequence of length $O(pn^3 + n^4 logn)$.

2.2 Distinguishing sequences

For machines with a *distinguishing* sequence there is a deterministic polynomial time algorithm to construct a checking sequence [Hennie, 1964; Kohavi, 1978] of polynomial length. A distinguishing sequence is similar to an unreliable status message in that it gives a different output for each state, except that it changes the state. For example, for the machine in Figure 1, ab is a distinguishing sequence, since $\lambda(s_1, ab) = 01$, $\lambda(s_2, ab) = 11$, and $\lambda(s_3, ab) = 00$.

Given a distinguishing sequence x_0, first check the similarity of implementation machines by examining the response of each state to the distinguishing sequence, then check each transition by exercising it and verifying the ending state, also using the distinguishing sequence. A *transfer* sequence $\tau(s_i, s_j)$ is a sequence that takes the machine from state s_i to s_j. Such a sequence always exists for any two states since the machine is strongly connected. Obviously, it is not unique and a shortest path [Aho, Hopcroft and Ullman, 1974] from s_i to s_j in the transition diagram is often preferable. Suppose that the machine is in state s_i and that distinguishing sequence x_0 takes the machine from state s_i to t_i, i.e., $t_i = \delta(s_i, x_0)$, $i = 1, \cdots, n$. For the machine in the initial state s_1, the following test sequence takes the machine through each of its states and displays each of the n different responses to the distinguishing sequence:

$$x_0\tau(t_1, s_2)x_0\tau(t_2, s_3)x_0 \cdots x_0\tau(t_n, s_1)x_0 \ . \tag{1}$$

Starting in state s_1, x_0 takes the machine to state t_1 and then $\tau(t_1, s_2)$ transfers it to state s_2 for its response to x_0. At the end the machine responds to $x_0\tau(t_n, s_1)$. If it operates correctly, it will be in sate s_1, and this is verified by its response to the final x_0. During the test we should observe n different responses to the distinguishing sequence x_0 from n different states, and this verifies that the implementation machine B is similar to the specification machine A.

We then establish every state transition. Suppose that we want to check transition from state s_i to s_j with I/O pair a/o when the machine is currently in state t_k. We would first take the machine from t_k to s_i, apply input a, observe output o, and verify the ending state s_j. We cannot simply use $\tau(t_k, s_i)$ to take the machine to state s_i, since faults may alter the ending state. Instead, we apply the following input sequence: $\tau(t_k, s_{i-1})x_0\tau(t_{i-1}, s_i)$. The first transfer sequence is supposed to take the machine to state s_{i-1}, which is verified by its response to x_0, and as has been verified by (1), $x_0\tau(t_{i-1}, s_i)$ definitely takes the machine to state s_i. We then test the transition by input a and verify the ending state by x_0. Therefore, the following sequence tests for a transition from s_i to s_j:

$$\tau(t_k, s_{i-1})x_0\tau(t_{i-1}, s_i)ax_0 \tag{2}$$

After this sequence the machine is in state t_j. We repeat the same process for each state transition and obtain a checking sequence. Observe that the length of the checking sequence is polynomial in the size of the machine A and the length of the distinguishing sequence x_0.

Recall that a distinguishing sequence for the machine in Figure 1 is: $x_0 = ab$. The transfer sequences are, for example, $\tau(s_1, s_2) = b$. The sequence in (1) for checking states is *abababab*. Suppose that the machine is in state s_3. Then the following sequence *babbab* tests for the transition from s_2 to s_3: b takes the machine to state s_1, ab definitely takes the machine to state s_2 if it produces outputs 01, which we have observed during state testing, and, finally, *bab* tests the transition on input b and the end state s_3. Other transitions can be tested similarly.

We can use *adaptive* distinguishing sequences to construct a checking sequence. An adaptive distinguishing sequence is not really a sequence but a decision tree that specifies how to choose inputs adaptively based on observed outputs to identify the initial state. An adaptive distinguishing sequence has length $O(n^2)$, and, consequently, a checking sequence of length $O(pn^3)$ can be constructed in time $O(pn^3)$ [Lee and Yannakakis, 1994].

2.3 Identifying sequences

The previous three methods are based on knowing where we are during the experiment, using status messages, reset, and distinguishing sequences, respectively. However, these sequences may not exist in general. A method was proposed by Hennie that works for general machines, although it may yield exponentially long checking sequences. It is based on certain sequences, called *identifying sequences* in [Kohavi, 1978] (*locating sequences* in [Hennie, 1964]) that identify a state in the *middle* of the execution. Identifying sequences always exist and checking sequences can be derived from them [Hennie, 1964; Kohavi, 1978].

Similar to checking sequences from distinguishing sequences, the main idea is to display the responses of each state to its separating family of sequences instead of one distinguishing sequence. We use an example to explain the display technique. The checking sequence generation procedure is similar to that from the distinguishing sequences and we omit the detail.

Consider machine A in Figure 1. We want to display the responses of state s_1 to separating sequences a and b. Suppose that we first take the machine to s_1 by a transfer sequence, apply the first separating sequence a, and observe output 0. Due to faults, there is no guarantee that the implementation machine was transferred to state s_1 in the first place. Assume instead that we transfer the machine (supposedly) to s_1 and then apply aaa which produces output 000. The transfer sequence takes the machine B to state q_0 and then aaa takes it through states q_1, q_2, and q_3, and produces outputs 000 (if not, then B must be faulty). The four states q_0 to q_3 cannot be distinct since B has at most three states. Note that if two states q_i, q_j are equal, then their respective following states q_{i+1}, q_{j+1} (and so on) are also equal because we apply the same input a. Hence q_3 must be one of the states q_0, q_1, or q_2, and thus we know that it will output 0 on input a; hence we do not need to apply a. Instead we apply input b and must observe output 1. Therefore, we have identified a state of B (namely q_3); that responds to the two separating sequences a and b by producing 0 and 1 respectively, and thus is similar to state s_1 of A.

The length of an identifying sequence in the above construction grows exponentially with the number of separating sequences of a state and the resulting checking sequence is of exponential length in general.

2.4 A Polynomial time randomized algorithm

With status messages, reset, or distinguishing sequences, we can find in polynomial time checking sequences of polynomial length. In the general case without such information, Hennie's algorithm constructs an exponential length checking sequence. The reason of the exponential growth of the length of the test sequence is that it deterministically displays the response of each state to its separating family of sequences. Randomization can avoid this exponential "blow-up"; we now describe a polynomial time randomized algorithm that constructs with high probability a polynomial length checking sequence [Yannakakis and Lee, 1995; Lee and Yannakakis, 1996a]. As is often used in theoretical computer science, "high probability" means that we can make the probability of error arbitrarily small by repeating the test enough times; specifically, the probability that it is not a checking sequence is squared if the length of the testing sequence is doubled. Note that the probabilities are with respect to the random decisions of the algorithm; we do not make any probabilistic assumptions on the specification A or the implementation B. For a test sequence to be considered "good" (a checking sequence), it must be able to uncover *all* faulty machines B.

We break the checking experiment into two tests. The first test ensures with high probability that the implementation machine B is similar to A. The second test ensures with high probability that all the transitions are correct: they give the correct output and go to the correct next state.

SIMILARITY

For $i = 1$ to n **do**

Repeat the following k_i times:

Apply an input sequence that takes A from its current state to state s_i;

Choose a separating sequence from Z_i uniformly at random and apply it.

We assume that for every pair of states we have chosen a fixed transfer sequence from one state to the other. Assume that z_i is the number of separating sequences in Z_i for state s_i. Let x be the random input string formed by running Test 1 with $k_i = O(nz_i \min(p, z_i) logn)$ for each $i = 1, \cdots, n$. It can be shown that, with high probability, every FSM B (with at most n states) that is not similar to A produces a different output than A on input x.

TRANSITIONS

For each transition of the specification FSM A, say $\delta_A(s_i, a) = s_j$, **do**

Repeat the following k_{ij} times:

Take the specification machine A from its current state to state s_i;

Flip a fair coin to decide whether to check the current state
or the transition;

In the first case, choose (uniformly) at random a sequence from Z_i
and apply it;

In the second case, apply input a followed by a randomly selected
sequence from Z_j.

Let x be the random input string formed by running Test 2 with $k_{ij} = O(\max(z_i, z_j) log(pn))$ for all i, j. It can be shown that, with high probability, every FSM B (with at most n states) that is similar but not isomorphic to A produces a different output than A on input x.

Combining the two tests, we obtain a checking sequence with a high probability [Yannakakis and Lee, 1995; Lee and Yannakakis, 1996a]. Specifically, given a specification machine A with n states and input alphabet of size p, the randomized algorithm constructs with high probability a checking sequence for A of length $O(pn^3 + p'n^4 logn)$ where $p' = \min(p, n)$.

2.5 Heuristic procedures and optimization

Checking sequences guarantee a complete fault coverage but sometimes could be too long for practical applications and heuristic procedures are used instead. For example, in circuit testing, test sequences are generated based on fault models that significantly limit the possible faults. Without fault models, covering paths are often used in both circuit testing and protocol testing where a test sequence exercises each transition of the specification machine at least once. A short test sequence is always preferred and a shortest covering path is desirable, resulting in a Postman Tour [Aho, Dahbura, Lee and Uyar, 1991; Naito and Tsunoyamma, 1981; Uyar and Dahbura, 1986].

A covering path is easy to generate yet may not have a high fault coverage. Additional checking is needed to increase the fault coverage. For instance, suppose that each state has a *UIO sequence* [Sabnani and Dahbura, 1988]. A UIO sequence for a state s_i is an input sequence x_i that distinguishes s_i from any other states, i.e., for any state $s_j \neq s_i$, $\lambda(s_i, x_i) \neq \lambda(s_j, x_i)$. To increase the coverage we may test a transition from state s_i to s_j by its I/O behavior and then apply a UIO sequence of s_j to verify that we end up in the right state. Suppose that such a sequence takes the machine to state t_j. Then a test of this transition is represented by a test sequence, which takes the machine from s_i to t_j. Imagine that all the edges of the transition diagram have a white color. For each transition from s_i to s_j, we add a red edge from s_i to t_j due to the additional checking of a UIO sequence of s_j. A test that checks each transition along with a UIO sequence of its end state requires that we find a path that exercises each red edge at least once. It provides a better fault coverage than a simple covering path, although such a path does not necessarily give a checking sequence [Chan, Vuong and Ito, 1989]. We would like to find a shortest path that covers each red edge at least once. This is a *Rural Postman Tour* [Garey and Johnson, 1979], and in general, it is an NP-hard problem. However, practical constraints are investigated and polynomial time algorithms can be obtained for a class of communication protocols [Aho, Dahbura, Lee and Uyar, 1991].

Sometimes, the system is too large to construct and we cannot even afford a covering path. To save space and to avoid repeatedly testing the same portion of the system, a "random walk" could be used for test generation [Lee, Sabnani, Kristol and Paul, 1996; West 1986]. Basically, we only keep track of the current state and determine the next input on-line; for all the possible inputs with the current state, we choose one at random. Note that a pure random walk may not work well in general; as is well known, a random walk can easily get "trapped" in one part of the machine and fail to visit other states if there are "narrow passages". Consequently, it may take exponential time for a test to reach and uncover faulty parts of an implementation machine through a pure random walk. Indeed, this is very likely to happen for machines with low enough connectivity and few faults (single fault, for instance). To avoid such problems, a *guided random walk* was proposed [Lee, Sabnani, Kristol and Paul, 1996] for protocol testing where partial information of a history of the tested portion is being recorded. Instead of a random selection of next input, priorities based on the past history are enforced; on the other hand, we make a random choice within each class of inputs of the same priority. Hence we call it a guided random walk; it may take the machine out of the "traps" and increase the fault coverage.

In the techniques discussed, a test sequence is formed by combining a number of subsequences, and often there is a lot of overlaps in the subsequences. There are several papers in the literature that propose heuristics for taking advantage of overlaps in order to reduce the total length of tests [Sidhu and Leung, 1989; Yang and Ural, 1990].

3 EXTENDED FINITE STATE MACHINES

For testing of data portions of protocol systems finite state machines are not powerful enough to model in a succinct way the physical systems any more. Extended finite state machines, which are finite state machines extended with variables, have emerged from the design and testing of such systems. For instance, IEEE 802.2 LLC [ANSI, 1989] is specified by 14 control states, a number of variables, and a set of transitions (pp. 75-117). A typical transition is (p. 96):

current_state SETUP
input ACK_TIMER_EXPIRED
predicate S_FLAG = 1
outputCONNECT_CONFIRM
action P_FLAG := 0; REMOTE_BUSY := 0
next_state NORMAL

In state SETUP and upon input ACK_TIMER_EXPIRED, if variable S_FLAG has value 1, then the machine outputs CONNECT_CONFIRM, sets variables P_FLAG and REMOTE_BUSY to 0, and moves to state NORMAL.

In our efforts in test generation for Personal HandyPhone Systems (PHS), a 5ESS based, ISDN wireless system [Lee and Yannakakis, 1996b] and for 5ESS Intelligent Network Application Protocols (INAP) [Huang, Lee and Staskauskas, 1996], we use the following model.

For a finite set of variables $\vec{x}$, a predicate on variable values $P(\vec{x})$ returns FALSE or TRUE. Given a function $A(\vec{x})$, an action is an assignment: $\vec{x} := A(\vec{x})$. Informally, an *extended finite state machine* (EFSM) has a finite set of states, inputs, outputs, and transitions between states, which are associated with inputs and outputs. In addition each transition is also associated with a predicate $P(\vec{x})$ and an action $A(\vec{x})$; the transition is executable if the predicate returns TRUE for the current variable values and in this case the variable values are updated by an assignment: $\vec{x} := A(\vec{x})$. Initially, the machine is in an initial state s_1 initial variable values: $\vec{x}_{init}$. Suppose that the machine is at state s with the current variable values $\vec{x}$ and that $< a, P(\vec{x})/o, A(\vec{x}) >$ is an outgoing transition from state s to q. Upon input a, if the predicate $P(\vec{x})$ returns TRUE, then the machine follows the transition, outputs o, changes the current variable values by action $\vec{x} := A(\vec{x})$, and moves to state q. Each combination of a state and variable values is called a *configuration*. Given an EFSM, if each variable has a finite number of values (Boolean variables for instance), then there is a finite number of configurations, and hence there is an equivalent (ordinary) FSM with configurations as states. Therefore, an EFSM with finite variable domains is a compact representation of an FSM.

We now discuss testing of EFSM's, which has becoming an important topic, especially in the network protocol area [Favreau and Linn, 1986; Miller and Paul, 1993; Koh and Liu, 1994; Huang, Lee, and Staskauskas, 1996; Lee and Yannakakis, 1996b]. An EFSM usually has an initial state s_1 and all the variables have an initial value $\vec{x}_{init}$, which consists of the *initial configuration*. A *test sequence* (or a *scenario*) is an input sequence that takes the machine from the initial configuration back to the initial state (possibly with different variable values). We want to construct a set of test sequences of a desirable *fault coverage*, which ensures that the implementation machine under test *conforms* to the specification. The fault coverage is essential. However, it is often defined differently from different models and/or practical needs. For testing FSM's we have discussed checking sequences, which guarantee that the implementation machine is structurally isomorphic to the specification machine. However, even for medium size machines it is too long to be practical [Yannakakis and Lee, 1995; Lee and Yannakakis, 1996a] while for EFSM's hundreds of thousands of states (configurations) are typical and it is in general impossible to construct a checking sequence. A commonly used heuristic procedure in practice is: each transition in the specification EFSM has to be executed at least once. A *complete test set* for an EFSM is a set of test sequences such that each transition is tested at least once.

To find a complete test set, we first construct a *reachability graph* G, which consists of all the configurations and transitions that are reachable from the initial configuration. We obtain a directed graph where the nodes and edges are the reachable configurations and transitions, respectively. Obviously, a control state may have multiple appearances in the nodes (along with different variable values) and each transition may appear many times as edges in the reachability graph. In this reachability graph, any path from the initial node (configuration) corresponds to a feasible path (test sequence) in the EFSM, since there are no predicate or action restrictions anymore. Therefore, a set of such paths in G, which exercises each transition at least once, provides a complete test set for the EFSM. We thus reduce the testing problem to a graph path covering problem.

The construction of the reachability graph is often a formidable task; it has the well-known state explosion problem due to the large number of possible combinations of the control states and variable values. We shall not digress to this topic. From now on we assume that we have a graph G that contains all the transitions of a given EFSM and we want to construct a complete test set of a small size. For clarity, we assume that each path (test sequence) is from the initial node to a *sink* node, which is a configuration also with the initial control state.

Formally, we have a directed graph G with n nodes, m edges, a *source* node s of in-degree 0, and a *sink* node t of out-degree 0. All edges are reachable from the source node and the sink node is reachable from all edges. There is a set C of $k = |C|$ distinct *colors*. Each node and edge is associated with a subset of colors from C. * A path from the source to sink is called a *test*. We are interested in a set of tests that cover all the colors; they are not necessarily the conventional graph covering paths that cover all the edges. The path (test) length makes little difference and we are interested in minimizing the number of paths. We shrink each strongly connected component [Aho, Hopcroft and Ullman, 1974] into a node, which contains all the colors of the nodes and

*Each transition in the EFSM corresponds to a distinct color in C and may have multiple appearances in G. We consider a more general case here; each node and edge have a set of colors from C.

edges in the component. The problem then is reduced to that on a directed acyclic graph (DAG) [Aho, Hopcroft and Ullman, 1974]. From now on, unless otherwise stated, we assume that the graph is a DAG.

We need a complete test set - a set of paths from the initial node to the sink node that cover all the colors C. On the other hand, in the feature testing of communication systems, setting up and running each test is time consuming and each test is costly to experiment. Consequently, we want to minimize the number of tests. Therefore, our goal is: *Find a complete test set of minimum cardinality*. However, the problem is NP-hard. We need to restrict ourselves to approximation algorithms. Similar to the standard approximation algorithm for Set Cover [Garey and Johnson, 1979], we use the following procedure. We first find a path (test) that covers a maximum number of colors and delete the covered colors from C. We then repeat the same process until all the colors have been covered. Thus, we have the following problem: *Find a test that covers the maximum number of colors.* This problem is also NP-hard. In view of the NP-hardness of the problem, we have to content ourselves with approximation algorithms again.

Suppose that an edge (node) has c uncovered colors so far. We assign a weight c to that edge (node), and we have a weighted graph. Find a longest path from the source to sink; it is possible since the graph is a DAG. This may not provide a maximal color test due to the multiple appearances of colors on a path. However, if there are no multiple appearances of colors on the path, then it is indeed a maximal color test.

There are known efficient ways of finding a longest path on a DAG [Aho, Hopcroft and Ullman, 1974]. The time and space needed is $O(m)$ where m is the number of edges. How does this heuristic method compare with the optimal solution? An obvious criterion is the *coverage ratio*: the number of maximal number of colors on a path over the number of colors covered by the algorithm. It can be really bad in the worst case; the coverage ratio can be $\Omega(k)$ where k is the maximal number of uncovered colors on a path.

We now discuss a greedy heuristic procedure. It takes linear time and works well in practice. We topologically sort the nodes and compute a desired path from each node to the sink in a reverse topological order as follows. When we process a node u and consider all the outgoing edges (u, v) where v has a higher topological order and has been processed, we take the union of the colors of node u, edge (u, v), and node v. We compare the resulting color sets from all the outgoing edges from u and keep one with the largest cardinality. This procedure is well defined since G is a DAG. The time and space complexity of this approach is $O(km)$ where k is the number of uncovered colors and m

is the number of edges. Although the second method seems to be better in many cases, its worst case coverage ratio is also $\Omega(k)$.

We now describe briefly an improved procedure. This is similar to the greedy heuristic, except that when we process a node u, we do not consider only its immediate successors but all its descendants. Specifically, for each outgoing edge (u, v) and descendant v' of v (possibly $v = v'$), we take the union of the colors of node u, edge (u, v), and node v'. We compare the resulting color sets from all the outgoing edges from u and descendants v' and keep one with the largest cardinality. The time complexity of this algorithm is $O(knm)$, since we may examine on the order of n descendants when we process a node. The worst case coverage ratio of this method is somewhat better: $O(\sqrt{k})$.

In spite of the negative results in the worst case, the greedy heuristic procedures were applied to real systems [Huang, Lee and Staskauskas, 1996; Lee and Yannakakis, 1996b] and proved to be surprisingly efficient; a few tests cover a large number of colors and, afterwards, each test covers a *very small* number of colors. A typical situation is that the first 20% tests cover more than 70% of the colors. Afterwards, 80% of the tests cover the remaining 30% of the colors, and each test covers 1 to 3 colors. Consequently, the costly part of the test execution is the second part. To reduce the number of tests as much as possible exact procedures for either maximal color paths or minimal complete test sets are needed. The question is, can we obtain more efficient algorithms if we know that there is a bound on the maximum number of colors on any path that is a small constant $c << k$. The problem can be solved in time and space polynomial in the number of colors k and the size of the graph. The detailed algorithm is more involved and we refer the readers to [Lee and Yannakakis, 1996b].

4 PARAMETERIZED EFSM'S

Finally, we consider testing of parameterized EFSM's. As a case study we discuss modeling and test generation of ATM Traffic Management protocol for the ABR (Allowed Bit Rate) services [ATM, 1996]. A formal specification is given in [Lee, Ramakrishnan, Moh and Shankar, 1996], using *Communicating Parameterized Extended Finite State Machines with Timers.* Suppose that two end stations send data and Resource Management (RM) cells to each other via a virtual circuit. Each end station consists of three communicating EFSM's, sending cells to each other. Cells contain *parameters* for traffic monitoring and rate control. Furthermore, there are timers to determine the transmission of data and RM cells. The following is a typical transition:

current_state: S_2
next_state: S_1
event: $Y \geq Crm \& T > Trm$
actions:
$ACR := ACR * (1 - CDF)$;
$CCR_{FRM} := ACR$;
send an FRM cell;
$Y := Y + 1$; $X_1 = 0$;
$t := 0$; $T := 0$

Here CCR_{FRM} is a parameter in the Forward RM (FRM) cell to be sent, ACR, Y and X_1 are variables, T and t are timers, and Crm, Trm and CDF are system parameters, which are constants determined at the connection set up. When the current variable Y and timer T values satisfy the conditions in **event**, the following **actions** are taken and the system moves from state S_2 to S_1: the allowed cell rate ACR is reduced multiplicatively and then copied to CCR_{FRM} parameter in the FRM cell to be sent next, and the involved variable and timer values are updated.

Similar to testing of EFSM's we want to generate tests such that each transition is exercised at least once. Furthermore, we want to exercise the boundary values of the variables and parameters. The timers complicate the test generation; the timer expiration may take a long time and that makes the execution of some test sequences substantially more expensive than others. Furthermore, it takes significantly longer to make some events happen than others. For instance, a large number of data cells have to be sent to enforce the transmission of an RM cell. We need a different optimization criterion than that in the previous section; we want to minimize the test execution time rather than the number of tests. This can be formulated as follows. Each edge and node has a weight - the execution time. We want to generate tests such that each transition is executed at least once, and, furthermore, each boundary value of the variables and parameters is also exercised. On the other hand, we want to minimize the test execution time.

Similar to EFSM testing, we first generate a reachability graph of one end station and then assign to each transition a distinct color, which appears in the corresponding edges in the graph. Furthermore, each boundary value of a variable and parameter is also assigned a distinct color, which appears in the corresponding nodes in the graph. We want to find a shortest tour (test) of the graph such that each color is covered at least once. It can be easily shown that the problem is NP-hard.

We use the following heuristic method. While deciding a tour on the graph from the current node we find a shortest path to a node, which is closest to the current node and contains an uncovered boundary value color. Using this technique, for the ABR protocol, 13 tests cover all the boundary values and transitions in the original specification [Lee, Su, Collica and Golmie, 1997].

The following is a sample test, which is a repeated execution of the sample transition in this section. It verifies that the Implementation Under Test (IUT) reduces its ACR by $ACR*CDF$ but not lower than MCR when the number of outstanding FRM cells is larger than CRM where UT is the Upper Tester and LT is the Lower Tester.

(1) Have UT send Mrm data cells to the IUT.

(2) LT waits for an FRM from the IUT.

The value of the CCR_{FRM} in the received FRM cell must satisfy:

$MCR \leq CCR \leq previous\_CCR * (1 - CDF)$.

(3) Set $previous\_CCR := CCR$.

(4) Repeat (1) to (3) until $CCR_{FRM} = MCR$ twice consecutively.

Obviously, the parameter CCR_{FRM} in the output cell FRM complicates the testing process. However, it adds to the observability of the system behavior; in this case we can read ACR variable values from this parameter.

5 CONCLUSION

We have studied various techniques for conformance testing of protocol systems that can be modeled by finite state machines or their extensions. For finite state machines, we described several test generation methods based on status messages, reliable reset, distinguishing sequences, identifying sequences, characterization sets, transition tours and UIO sequences, and a randomized polynomial time algorithm. For extended finite state machine testing, it can be reduced to a graph path covering problem, and we present several approaches to ensure the fault coverage, to reduce the number of tests and to minimize the execution time.

We have discussed testing of *deterministic* machines. A different notion of testing of *nondeterministic* machines is studied in several papers [Brinksma, 1988] and an elegant theory is developed. In this framework, the tester is allowed to be nondeterministic. A test case is an (in general nondeterministic) machine T, the implementation under test B is composed with the tester machine T, and the definition of B failing the test T is essentially that there exists a run of the composition of B and T that behaves differently than the composition of the specification A and T. It is shown in [Brinksma, 1988] that every specification can be tested in this sense, and there is a "canonical" tester. However, it is not clear how to use this machine T to choose test sequences to apply to an implementation. It has been shown [Alur, Courcoubetis, and Yannakakis, 1995] that testing of nondeterministic machines is in general a hard problem.

Acknowledgement. We are deeply indebted to the insightful and constructive comments from Jerry Linn.

6 REFERENCES

A. V. Aho, A. T. Dahbura, D. Lee, and M. U. Uyar (1991) An optimization technique for protocol conformance test generation based on UIO sequences and rural Chinese postman tours, *IEEE Trans. on Communication*, vol. 39, no. 11, pp. 1604-15.

A.V. Aho, J. E. Hopcroft, and J. D. Ullman (1974) *The Design and Analysis of Computer Algorithms*, Addison-Wesley.

R. Alur, C. Courcoubetis, and M. Yannakakis (1995) Distinguishing tests for nondeterministic and probabilistic machines", *Proc. 27th Ann. ACM Symp. on Theory of Computing*, pp. 363-372.

ANSI (1989) International standard ISO 8802-2, ANSI/IEEE std 802.2.

ATM (1996) ATM Forum Traffic Management Specification Version 4.0, ATM Forum/96, af-tm-0056.000, April.

E. Brinksma (1988) A theory for the derivation of tests, *Proc. IFIP WG6.1 8th Intl. Symp. on Protocol Specification, Testing, and Verification*, North-Holland, S. Aggarwal and K. Sabnani Ed. pp. 63-74.

W. Y. L. Chan, S. T. Vuong, and M. R. Ito (1989) An improved protocol test generation procedure based on UIOs, *Proc. SIGCOM*, pp. 283-294.

S. T. Chanson and J. Zhu (1993) A unified approach to protocol test sequence generation, *Proc. INFOCOM*, pp. 106-14.

M.-S. Chen Y. Choi, and A. Kershenbaum (1990) Approaches utilizing segment overlap to minimize test sequences, *Proc. IFIP WG6.1 10th Intl. Symp. on Protocol Specification, Testing, and Verification,* North-Holland, L. Logrippo, R. L. Probert, and H. Ural Ed. pp. 85-98.

T. S. Chow (1978) Testing software design modeled by finite-state machines," *IEEE Trans. on Software Engineering,* vol. SE-4, no. 3, pp. 178-87.

A. D. Friedman and P. R. Menon (1971) *Fault Detection in Digital Circuits,* Prentice-Hall.

J.-P. Favreau and R. J. Linn, Jr. (1986) Automatic generation of test scenario skeletons from protocol specifications written in Estelle, *Proc. PSTV*, pp. 191-202. North Holland, B. Sarikaya and G. v. Bochmann Ed.

M. R. Garey and D. S. Johnson (1979) *Computers and Intractability: a Guide to the Theory of NP-Completeness,* W. H. Freeman.

F. C. Hennie (1964) Fault detecting experiments for sequential circuits, *Proc. 5th Ann. Symp. Switching Circuit Theory and Logical Design,* pp. 95-110.

G. J. Holzmann (1991) *Design and Validation of Computer Protocols,* Prentice-Hall.

J. E. Hopcroft and J. D. Ullman (1979) *Introduction to Automata Theory, Languages, and Computation,* Addison-Wesley.

S. Huang, D. Lee, and M. Staskauskas (1996) Validation-based Test Sequence Generation for Networks of Extended Finite State Machines, *Proc. FORTE/PSTV*, North Holland, R. Gotzhein Ed.

L.-S. Koh and M. T. Liu (1994) Test path selection based on effective domains, *Proc. of ICNP*, pp. 64-71.

Z. Kohavi (1978) *Switching and Finite Automata Theory,* 2nd Ed., McGraw-Hill.

M.-K. Kuan (1962) Graphic programming using odd or even points, *Chinese Math.*, vol. 1, pp. 273-277.

David Lee, K. K. Ramakrishnan, W. Melody Moh, and A. Udaya Shankar (1996) Protocol Specification Using Parameterized Communicating Extended Finite State Machines - A Case Study of the ATM ABR Rate Control Scheme, em Proc. of ICNP'96, October.

D. Lee, K. K. Sabnani, D. M. Kristol, and S. Paul (1996) Conformance testing of protocols specified as communicating finite state machines - a guided random walk based approach, *IEEE Trans. on Communications,* vol. 44, no. 5, pp. 631-640.

D. Lee, D. Su, L. Collica and N. Golmie (1997) Conformance Test Suite for the ABR Rate Control Scheme in TM v4.0, *ATM Forum/97-0034*, February.

D. Lee and M. Yannakakis (1994) Testing finite state machines: state identification and verification, *IEEE Trans. on Computers*, Vol. 43, No. 3, pp.

306-320.

D. Lee and M. Yannakakis (1996a) Principles and Methods of Testing Finite State Machines - a Survey, *The Proceedings of IEEE*, Vol. 84, No. 8, pp. 1089-1123, August.

D. Lee and M. Yannakakis (1996b) Optimization Problems from Feature Testing of Communication Protocols, *The Proc. of ICNP*, pp. 66-75.

R. E. Miller and S. Paul (1993) On the generation of minimal length test sequences for conformance testing of communication protocols, *IEEE/ACM Trans. on Networking*, Vol. 1, No. 1, pp. 116-129.

E. F. Moore (1956) Gedanken-experiments on sequential machines, *Automata Studies*, Annals of Mathematics Studies, Princeton University Press, no. 34, pp. 129-153.

S. Naito and M. Tsunoyama (1981) Fault detection for sequential machines by transitions tours, *Proc. IEEE Fault Tolerant Comput. Symp.* IEEE Computer Society Press, pp. 238-43.

K. K. Sabnani and A. T. Dahbura (1988) A protocol test generation procedure, *Computer Networks and ISDN Systems*, vol. 15, no. 4, pp. 285-97.

B. Sarikaya and G.v. Bochmann (1984) Synchronization and specification issues in protocol testing, *IEEE Trans. on Commun.*, vol. COM-32, no. 4, pp. 389-395.

D. P. Sidhu and T.-K. Leung (1989) Formal methods for protocol testing: a detailed study, *IEEE Trans. Soft. Eng.*, vol. 15, no. 4, pp. 413-26, April.

M.U. Uyar and A.T. Dahbura (1986) Optimal test sequence generation for protocols: the Chinese postman algorithm applied to Q.931, *Proc. IEEE Global Telecommunications Conference.*

M. P. Vasilevskii (1973) Failure diagnosis of automata, *Kibernetika*, no. 4, pp. 98-108.

C. West (1986) Protocol validation by random state exploration, *Proc. IFIP WG6.1 6th Intl. Symp. on Protocol Specification, Testing, and Verification*, North-Holland, B. Sarikaya and G. Bochmann, Ed.

B. Yang and H. Ural (1990) Protocol conformance test generation using multiple UIO sequences with overlapping, *Proc. SIGCOM*, pp. 118-125.

M. Yannakakis and D. Lee (1995) Testing finite state machines: fault detection, *J. of Computer and System Sciences*, Vol. 50, No. 2, pp. 209-227.

7 BIOGRAPHIES

David Lee is a Distinguished Member of Technical Staff of Bell Laboratories Research and also an adjunct professor at Columbia University and a Visiting Faculty of National Institute of Standards and Technology. His current research interests are communication protocols, complexity theory, and image processing. David Lee is a Fellow of The IEEE, Editor of IEEE/ACM Transactions on Networking and and an associate editor of Journal of Complexity.

David H. Su is the manager of the High Speed Network Technologies group at the National Institute of Standards and Technology. His main research interests are in modeling, testing, and performance measurement of communications protocols. Prior to joining NIST in 1988, he was with GE Information Service Company as the manager of inter-networking software for support of customers' hosts on GEIS's data network. He received his Ph.D. degree in Computer Science from the Ohio State University in 1974.

A pragmatic approach to test generation

E. Pérez, E. Algaba, M. Monedero
Telefónica I+D
Emilio Vargas, 6
28043 - Madrid (Spain)
emilla@tid.es, algaba@tid.es, mmr@tid.es

Abstract
This paper presents a pragmatic approach to the problem of the automatic generation of a test suite for a given system. It introduces the GAP tool, embedded in the HARPO toolkit, which is capable of generating TTCN test suites starting from a SDL specification of the system and test purposes written in MSC notation. In addition to this, GAP computes coverage measures for these tests, which represent an evaluation of their quality.

Keywords
Test generation, SDL, MSC, ISO 9646, TTCN, coverage

1 INTRODUCTION

The ISO conformance testing methodology, ISO 9646 [7], is widely accepted as the main framework in telecommunication systems testing. This methodology includes general concepts on conformance testing and test methods, the test specification language TTCN [8], and the process of specifying, implementing and executing a

Testing of Communicating Systems, M. Kim, S. Kang & K. Hong (Eds)
Published by Chapman & Hall

test campaign. The major point in a test process lies on the availability of a test suite, which must be closely related to the system specification. Unfortunately, the manual design of a test suite is an error prone, time and resource consuming task.

The use of FDTs, especially SDL [1], in system and protocol specifications establishes a suitable environment for the development of automatic test generation tools. These tools, whose main input is the formal specification of the system, help solving the problem of a manual test suite production. Furthermore, the automatic nature of the process ensures the correctness of the generated tests and eases the computation of test quality measures, namely coverage.

GAP, embedded in the HARPO toolkit [6] for the development of testing tools, represents a practical approach to these automatic test generation tools. It focuses on SDL system specifications, test purposes described using MSC[3][4] notation and test cases written in TTCN. In order to generate the test cases, the GAP tool simulates the behaviour of the system under test. The simulation is guided by the MSC test purpose throughout the entire generation process. This approach takes advantage of the increasing number of SDL formal system specifications available to derive test cases without further interpretations of the standards. Moreover, an executable test suite can be implemented with these test cases in an automatic way using the remaining tools within the HARPO toolkit. The whole process of developing a testing tool is thus greatly automatized, being feasible to obtain an executable test suite starting from the formal specification of the system in just one process.

The purpose of this paper is to describe the GAP tool (described in section 3) in its environment. Section 2 describes the methodology chosen in GAP and the tool architecture.

Even though references are made to ISO 9646 methodology in this paper, the GAP automatic test generation tool does not restrict its output to conformance tests.

2 GAP METHODOLOGY

Conventional test generation methodologies can be classified into two main categories:

- Computer Aided Test Generation techniques (CATG): the tests are obtained in a semiautomatic manner, via an interaction of the user with the formal specification of the system. They are usually based on simulation techniques.
- Automatic Test Generation techniques (ATG): the tests are automatically generated by a program that exhaustively explores the behaviour of the system under test, represented by a FSM derived from its formal specification.

Both approaches have their pros and cons.

The test generation methodology chosen in the GAP tool combines both techniques in order to take full advantage of their positive features while minimizing their respective problems. GAP makes use of ATG techniques in the sense of exploring the behaviour of the system so as to generate tests. The difference is that the behaviour of the system is not exhaustively explored, but guided by a test purpose specified by the user. Thus, the ATG state explosion problem is avoided, because the explored behaviour is only the subset of the behaviour of the system that verifies the test purpose.

As we just said, ATG techniques are usually based on a FSM derived from the specification of the system, producing a test suite skeleton (behaviour) which has to be manually completed. GAP uses CATG simulation techniques to explore the behaviour of the system, thus producing complete test suite specifications (behaviour and data). Moreover, coverage measures can be computed over the original system specification, avoiding the need to keep links between the FSM and the original specification. Another advantage of GAP methodology is that the generated test suite is structured in the way it is expected to be, due to the use of user specified test purposes, and the test suite is kept down to a manageable size, unlike the one generated with ATG techniques.

To sum up, the inputs to GAP methodology are the formal specification of the system and a set of test purposes, which enables it to produce a complete test suite specification and coverage measures related to the original system specification.

HARPO is a test tool development environment including a TTCN to C compiler (behaviour and data), PICS and PIXIT proforma editors and a testing tool operation environment. Integrating the GAP tool in the HARPO toolkit provides a high degree of automatization in the test tool development and operation process. Such an environment minimizes the problems of manual test suite production, but its complexity does not completely disappear. It is shifted to the specification of the system and the selection and specification of the test purposes. Nevertheless, this methodology provides great advantages compared to the manual specification of tests: reduced costs and increased quantity and quality of generated tests.

2.1 Methodology elements

The elements of the GAP methodology and the languages used for their specification are presented below.

System under test

The system under test is specified using SDL-92. Nowadays SDL is the most accepted FDT in telecommunication industry, has the largest commercial tool support and is also being used by standardization organizations (ITU, ISO, ETSI). GAP can process either ACT ONE or ASN.1[5] (following Z.105[2] recommendation) data type specifications.

Test purposes

Test purposes in GAP are used to specify behaviour skeletons that drive the test generation, avoiding the simulation of correct test behaviours that are not useful for a given purpose. MSC notation has been extended with annotations to ease the specification of test purposes. A MSC is no longer interpreted as a complete trace of an execution of the system, but as an open pattern that lets the tool derive a subset of the system behaviour that fits it. Thus, a test purpose in GAP is used to generate several test cases.

Test suite specification

The generated test suite, ATS (Abstract Test Suite) according to ISO 9646 terminology, is written in TTCN (MP format). This is a complete test suite including overview, data types, constraints (data values) and behaviour. It can be processed by the TTCN compiler included in HARPO, thus producing an executable version of the tests, ETS (Executable Test Suite) according to ISO 9646 terminology.

Coverage

An important quality measure of a test suite is the coverage degree of the system it provides. The goal is to select the appropriate measures from a practical point of view. GAP computes state, signal and transition coverage measures. Another interesting concept is incremental coverage. The quality of two test suites can be compared with this measurement.

2.2 Work methodology with GAP (HARPO)

Figure 1 depicts the tool development and operation process using GAP within the HARPO toolkit.

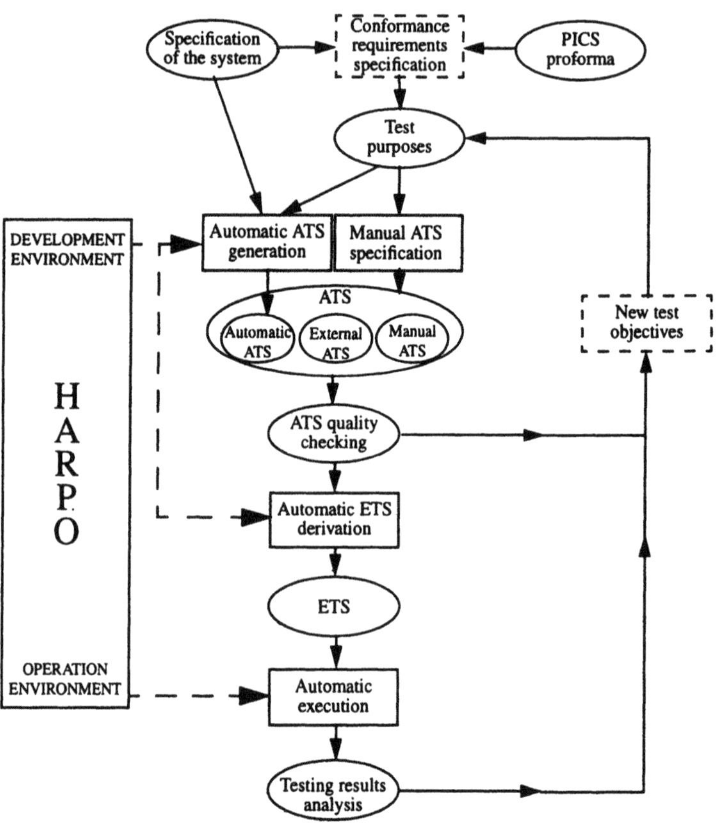

Figure 1 Work methodology with GAP (HARPO).

The process comprises several steps:

1. Identifying conformance requirements. This task requires a previous knowledge of the system under test.

2. Obtaining a SDL formal system specification. It may be obtained from external sources such as ITU, ISO, ETSI, etc. Otherwise, this specification will have to be produced manually.
3. Development of test purposes for the conformance requirements.
4. Execution of the GAP tool taking as inputs the specification of the system and test purposes, to produce a TTCN test suite (ATS).
5. Analysis of the generated test suite to check that it meets the expectations. In case of failure, new test purposes may be defined covering requirements that were not tested. It is possible to manually modify the ATS.
6. Compilation of the test suite with the TTCN compiler included in HARPO, to produce an executable version of the test suite.
7. Execution of the tests against the system and analysis of the results. This step may drive the user to the definition of new testing purposes.

The methodology defined above is in accordance with that defined by the ISO 9646 standard. The main advantage achieved using GAP tool within HARPO is the high degree of automatization in the process of developing a testing tool, starting from a formal description of a system and obtaining an executable test tool.

3 GAP TOOL ARCHITECTURE

Figure 2 depicts the architecture of the GAP tool. There are three main blocks: syntactical front-end (block 1), test generator (block 2) and data type translator and data values validator (block 3).

3.1 Syntactical front-end

This module, block 1 in figure 2, reads input data and stores it in memory in a suitable format to ease navigation for simulation and data type translation purposes. The specification of the system is written in SDL/PR language while test purposes are specified in MSC/PR notation. There are several commercial SDL and MSC graphical editors (SDT, GEODE, etc.) that can be used to dump these specifications from graphical to textual format.

The goal of the module is to store in memory a representation of the specification of the system and the test purposes, which is a well known problem amongst compiler developers. The final result is a memory structure known as abstract syntax tree (AST). Each node of this tree stores a SDL construction and its associated information. An application programming interface (API) is supplied to ease the accesses performed by the other modules. This API provides:

- Easy navigation throughout (the behaviour of) the system to derive test cases, guided by the test purposes.
- Capability to use the AST to store additional information produced while generating tests.

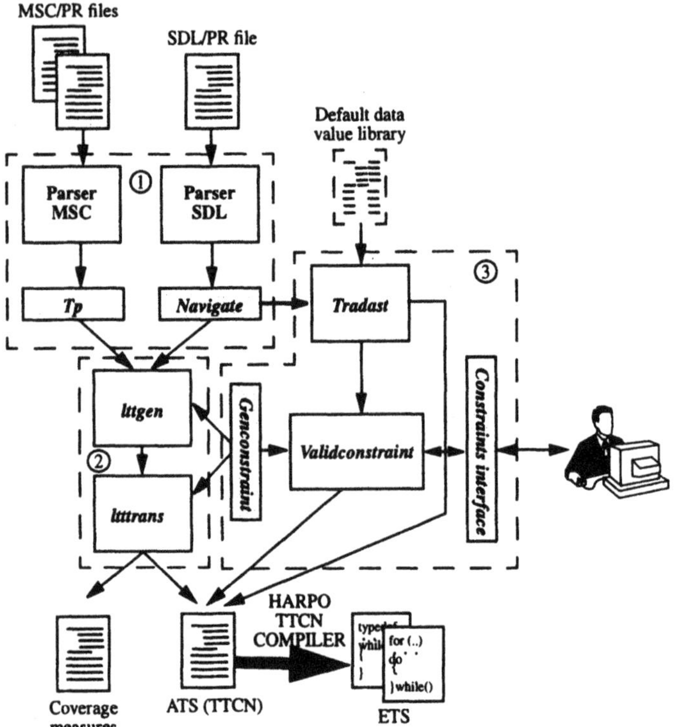

Figure 2 GAP tool architecture.

3.2 Test generator

The test generator, block 2 in figure 2, comprises two modules: lttgen and ltttrans.

The lttgen module simulates the specification of the system as stated in the test purpose and generates an intermediate structure called labelled transition tree (LTT).

The LTTs, the algorithm used by the generator and LTT translator and coverage measures computing module are described below.

Labelled transition trees

A LTT is a data structure that symbolically represents the evolution of a system. The LTT is a behaviour tree composed of nodes and arrows. Starting from a root node, it does not contain backward links (no cycles) nor subtree sharing (it is not a graph). Each node represents a state of the system and each arrow a transition. One LTT is generated for each test purpose, and it comprises the activity of the SDL objects: processes, channels and queues.

The LTT is dynamically built during the simulation, then transformed in its mirror image to reflect the point of view of the tester (see figure 3) and finally dumped in behaviour and data predicates.

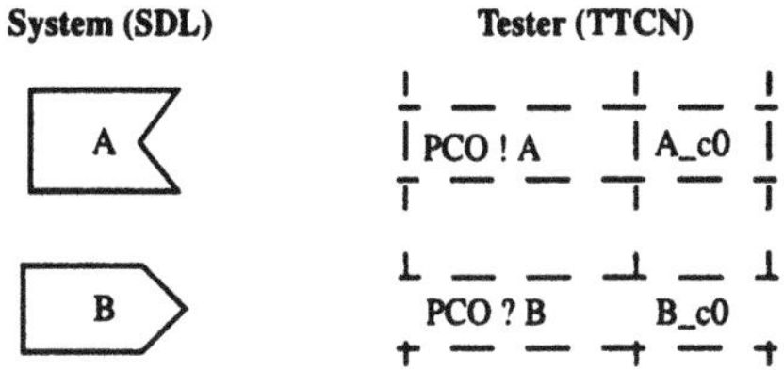

Figure 3 System-tester duality.

Test purposes

The MSC notation is a trace language that models the observational behaviour of a system. Its design shares a common philosophy with SDL.

GAP uses MSCs to describe test purposes. Extensions have been added to the MSC notation to allow for a better description of test purposes. This extensions are called annotations and are inserted in the comments part of MSCs. At a graphical level they only represent part of the documentation of the generated tests. For processing purposes, they are used to limit the exploration of possible behaviours while simulating the system. The goal of annotations is to keep the generated LTT down to a manageable size. Some of the most characteristic annotations are: maximum number of signals in channels, maximum number of process instantiations, maximum depth between two signals, maximum number of occurrences of a signal, signal disallowing, preamble and postamble annotations, etc. Useless loops are avoided via annotations. The user decides which loops are useless, and annotates the test purpose accordingly.

An example of test purpose is depicted in figure 4. It was used to test the call diversion service of the telephone network. This service allows the user to redirect incoming phone calls to another phone number. On the left side of the figure, an example of test purpose for this service is shown. The test purpose is not a complete trace of an execution of the system. Thus the MSC is interpreted as an open behaviour pattern, and several test cases will be generated for it. In fact, every possible test behaviour that fits in the test purpose will be generated. Two of the generated test cases are illustrated on the right side of the figure (MSC notation is used for test cases instead of TTCN in order to simplify it).

Test case 2 contains a loop (Pick, ..., Hang, Pick, ...). The user is responsible of determining if the loop in Test case 2 is useless or not. Such loops may be useful in some situations, i.e. while testing the data transfer phase of a protocol. The generation of these loops is controlled via annotations: signal disallowing, maximum number of occurrences of a signal, etc. For example if the user decides not to generate test cases with Pick loops, the test purpose can be described as in figure 5, where a disallowing signal annotation for Pick has been added. This test purpose will generate Test case 1 but not number 2.

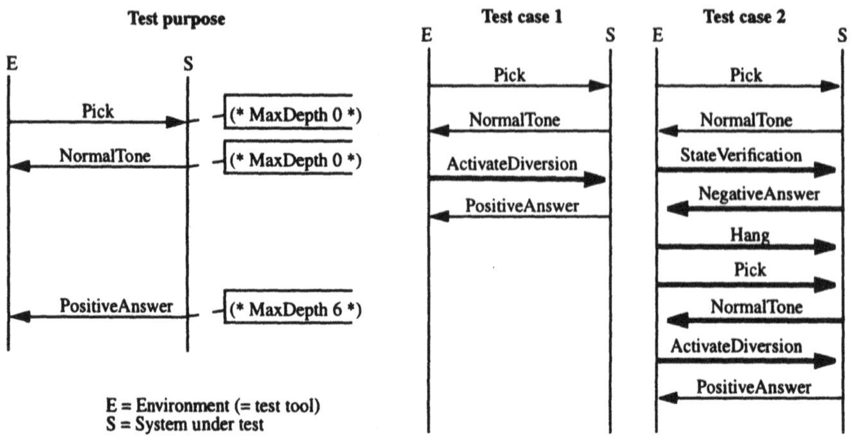

Figure 4 Test purpose and two generated tests for it.

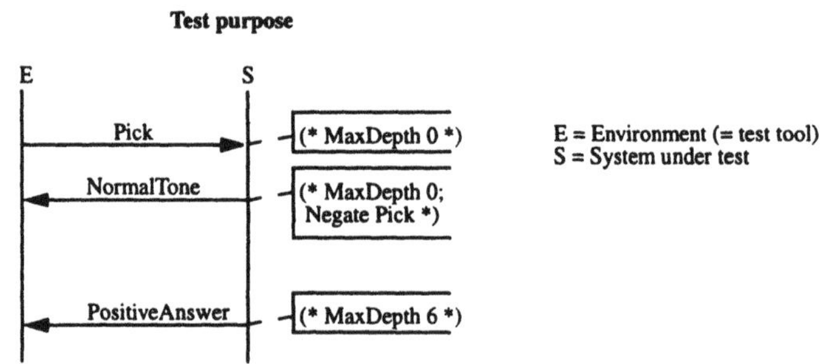

Figure 5 Modified test purpose.

As stated before, the test purpose constitutes a mere behaviour pattern, and not a complete trace of the behaviour of the system.

LTT generation

A symbolic simulator or SDL machine is used to generate the LTTs. Starting from the initial state where processes are at their start point and queues and channels are empty, the SDL machine executes the specification generating the allowed transitions. The state of the system (variable values, processes state, queues and channels state and active timers) is stored in the LTT nodes, whilst those events that produce a state change are stored as transitions. Some of these changes are determined by the appearance of a signal in an input channel from the environment, that is to say, an input transition. On the other hand, the disappearance of a signal in a channel pointing to the environment constitutes an output transition. The remaining transitions are merely internal. The LTT is generated following the rules stated in the test purpose (one LTT is generated for each test purpose). Starting from the initial state, all the possible evolutions of the system are computed and carefully stored in the LTT. Those branches that do not verify the test purpose are pruned during the generation. Thus, the generated LTT represents the behaviour subset of the system that is significant for the test purpose.

Predicates on data are gathered during the generation by means of decision clauses in SDL. In the example in figure 6, if the generation evolves through the TRUE branch, an associated predicate a>=5 will be stored. All these predicates are dumped out to the data validator module, whose purpose is to find the appropriate data values that satisfy them. Subsection 3.3 describes this module.

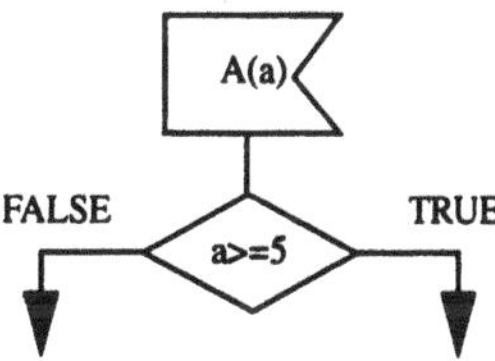

Figure 6 Decision clause.

It is important to state that after the LTT has been generated, only those events that can be observed from the environment are taken into account, namely external signals and timers. Those transitions with no reflection in the environment are pruned, thus simplifying the LTT.

LTT transformation

The LTT reflects a subset of the possible behaviour of the system, but it vaguely resembles a list of TTCN test cases. The final goal is to generate TTCN so several transformations must be applied.

First of all, the direction of send and receive events must be inverted to reflect the point of view of the tester (see figure 3).

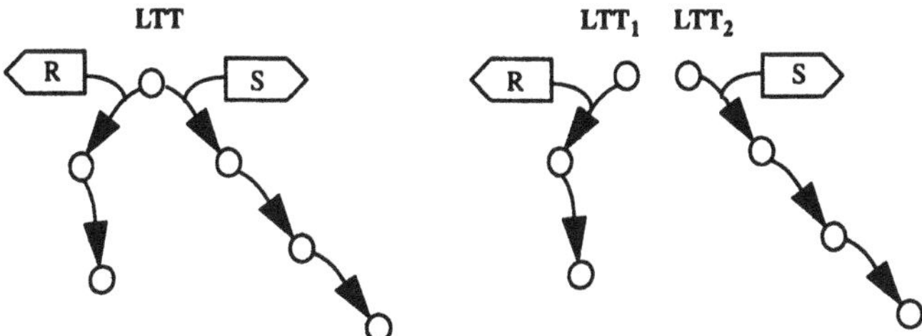

Figure 7 LTT splitting.

Secondly, the LTT must be split into a list of test cases. In the example in figure 7, there are two send events at the same level (from the point of view of the tester). The tester can send either one or the other but never both. The LTT is therefore split in two, LTT_1 and LTT_2, from which two test cases will be derived. Finally, once the initial LTT has been inverted and split into smaller LTTs (test cases), these are dumped in TTCN.

Coverage
The ltttrans module is also in charge of computing coverage measures. It counts the number of states, signals, timers and transitions the LTT covers. A distinction is made between observable signals (black box coverage) and internal ones (white box coverage). The GAP tool calculates state, signal and transition coverage measures.

State coverage is a white box measure. It is a weak measure, because there are usually many different paths to go from one state to another. Therefore the generated tests can cover all the states without completely exercising the system. Thus, the usually accepted value for this measure is 100% of states covered.

Two signal coverage measures are computed: black box and white box. These measures are more demanding than state coverage measures. Nevertheless, 100% coverage is required for them as well. The reason is that they do not take into account the dynamics of the system (the sequence in which the signals are ordered).

Transition coverage, a variant of branch coverage, is a white box measure. It is the stronger measure computed by GAP. The accomplishment of 100% depends on the absence of dead (not reachable) text in the specification and the feasibility of finding values to satisfy all the predicates. Working values near 100% are usually accepted.

Apart from these absolute measures, GAP can also compute an incremental measure of the coverage achieved by the tests generated for one test purpose with respect to those generated for another one. This incremental coverage gives an estimation about what is being tested with a set of tests that has not been tested with another one. Suppose there are two sets of tests for the same system, A and B. If the incremental coverage of B with respect to A is zero, this implies that B is not checking anything that has not already been tested by A. In such a situation, the tests included in B can be discarded. GAP computes state, signal and transition incremental coverage.

3.3 Data type translator and value validator

It corresponds to block 3 on figure 2, and comprises two modules: Tradast and Validconstraint.

The Tradast module translates signals and data types appearing in the specification of the system to a valid notation in TTCN, either ASN.1 or tabular data types.

The Validconstraint module displays the constraints values needed to complete the tests and their associated predicates, if any. An external data value library supplied by the user can be read by the tool in order to ease the process of filling in the constraint values. This library is always dependant on the system under test. Moreover, the values introduced by the user are syntactically and semantically checked, and the corresponding predicates are validated.

Data type translator
This module is responsible for the translation of the signals in the SDL system. It also translates ACT ONE data types to ASN.1 or tabular types, and directly dumps those types already written in ASN.1 (Rec. Z.105). Signals are translated into ASPs and

PDUs, and data types into TTCN types, either in ASN.1 or tabular declarations.

SDL uses ACT ONE notation to define abstract data types. An ACT ONE data type is determined by:

- Literal values
- Operators for data types, defined by their signature
- Axioms defining the semantic of the operators

A data type containing these elements is called a partial type description in SDL.

ACT ONE operators are used as patterns to translate data types. The translation is carried out by applying several heuristics that establish parallelisms between the ACT ONE data type definition and its TTCN type translation. The tool tries to identify these parallelisms and, if it succeeds, executes the translation.

Some predefined data types in SDL can be directly translated, due to their semantic equivalence. GAP needs heuristics to translate the remaining SDL predefined data types.

Data value validator

The function of this module is to supply correct data values (constraints in TTCN terminology) that verify the predicates collected during the simulation, and to dump them into TTCN constraints (ASN.1 or tabular).

Send and receive constraints are computed in a different manner.

A reception constraint in the tester is derived from a send event from the system to the environment (see figure 8). A send event in the system must be completely specified based on constants and variables of the system. Therefore, the GAP tool knows the exact value of the constraint that must be generated. If this reception constraint depends on the value of a variable which cannot be solved, the evolution of the variable is dumped within the TTCN behaviour and passed as a parameter to the constraint (see figure 8).

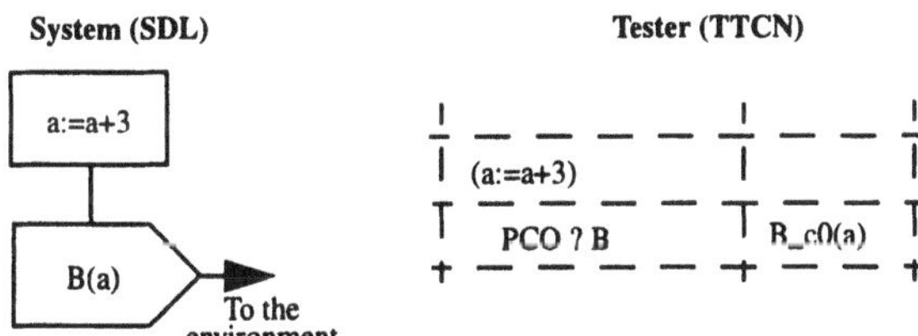

Figure 8 Signal to reception constraint translation.

Send constraints in the tester are derived from reception events in the system (see figure 9). These values are received from the environment, so they are not known at generation time. Therefore, these values must be filled in during the final test value validation phase. Values may have associated predicates: for instance, in figure 9, if the test case evolved through the TRUE branch in the decision clause, the tool would need a constraint A_c0, whose first field (because a is the first parameter of signal A) should be equal or greater than 5.

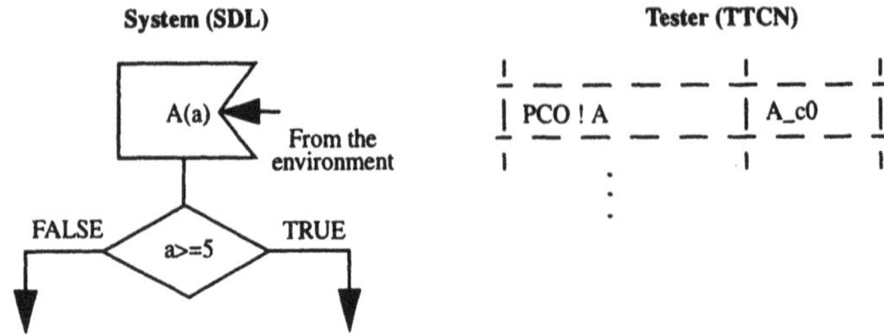

Figure 9 Signal to send constraint translation.

There are two data value generation modes in GAP: semiautomatic and automatic. In the semiautomatic mode, the tool automatically generates all the constraints with no associated predicates, leaving to the user the task of filling in those constraints with associated predicates. The generated values either come from an external default data values library or have been automatically generated choosing any valid data value for each constraint field according to its type definition. The tool also helps the user to fill in the constraints with associated predicates by means of suggesting default values from the external library. Once the constraints have been filled in, the tool checks that the supplied data verify both their type definitions and their associated predicates.

When working in automatic mode, the tool fills in all the needed constraints, either with default values from the library or with generated values. No value checking against associated predicates is performed in this mode. This mode is useful in the first development phases of the test specification, when the user is focused on checking if the dynamic behaviour of the generated tests fits the initial testing goals for the system.

4 CASE STUDY FOR THE TP0 PROTOCOL

In this section a brief example of the generated TTCN, for OSI Transport Protocol class 0 (TP0), is introduced. The test purpose is specified in figure 10. One of the generated test cases and its default are depicted in figure 11: *nconreq*, *nconcnf* and *ndisreq* are the connection and disconnection signals for the network layer, *tcrsignal* is the transport connection request PDU, *tcasignal* the transport connection accept PDU, *tccsignal* the transport connection clear (connection reject) PDU and *tdatindsignal* is the ASP carrying the reassembled transport data PDU up to the session layer.

The first part of the test purpose in figure 10, shows the network connection phase (*nconreq*, *nconcnf*). The open part is carried out between *nconcnf* and *tdatindsignal*, where a maximum of three signals (MaxDepth 3) may be generated. Next, a *tdatindsignal* must be generated. *ndisreq* is the postamble of the tested system, as well as the default. The goal for the postamble and the default is to drive the system under test to the initial state.

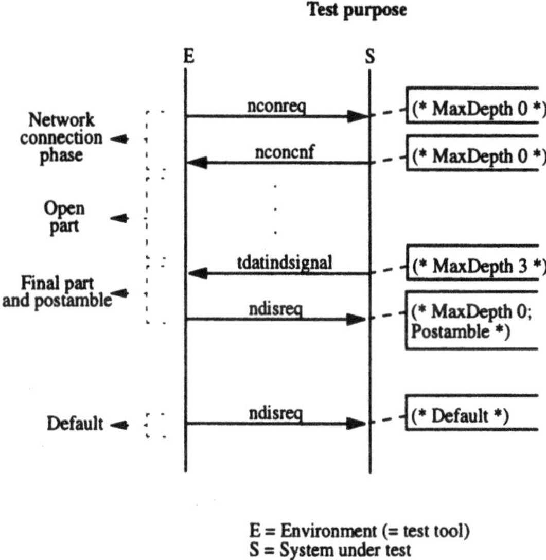

Figure 10 Test purpose for TP0.

Test Case Dynamic Behaviour

Test Case Name: tp0ed_0000
Group: tp0ed_tpexample
Purpose:
Default: tp0ed_default_000
Comments:

Nr	BehaviourDescription	ConstraintsRef	V.
1	START Global		
2	npco ! nconreq	nconreq_c0	
3	npco ? nconcnf	nconcnf_c0	
4	tpco ! tcrsignal (tcr.sref:= tcrsignal.sref)	tcrsignal_c0	
5	(tca.dref:=tcr.sref, auxtca:=tca.dref, tcc.dref:=tcr.sref, auxtcc:=tcc.dref)		
6	tpco ? tcasignal	tcasignal_c0(auxtca)	
7	tpco ! tdtsignal (tdt.em:= tdtsignal.em)	tdtsignal_c0	
8	(eot:=(tdt.em='80'O))		
9	[eot = TRUE]		
10	auxmsg:=Print("spco tdatindsignal")		
11	npco ! ndisreq	ndisreq_c0	P
12	tpco ? tccsignal	tccsignal_c0(auxtcc)	
13	npco ! ndisreq	ndisreq_c0	I

Default Dynamic Behaviour

Default Name: tp0ed_default_000
Group:
Objective:
Comments:

Nr	BehaviourDescription	ConstraintsRef	V.
1	? TIMEOUT Global		
2	npco ! ndisreq	ndisreq_c0	I
3	? TIMEOUT		
4	npco ! ndisreq	ndisreq_c0	F
5	npco ? OTHERWISE		
6	npco ! ndisreq	ndisreq_c0	F
7	tpco ? OTHERWISE		
8	npco ! ndisreq	ndisreq_c0	F
9	spco ? OTHERWISE		
10	npco ! ndisreq	ndisreq_c0	F

Figure 11 Example of a generated test case and its corresponding default for TP0.

The test case depicted in figure 11, is generated for the test purpose of figure 10. Changing the test purpose, for example, MaxDepth 3 by MaxDepth 5 would generate three test cases (including the one depicted in the figure). Lines 2 and 3 match the first two events in the test purpose. Lines 4, 6 and 7 are the generated signals between the second and the third event. They are the transport connection phase plus one transport data PDU. Line 10, represents the *tdatindsignal* produced in the system under test. It is dumped as a *Print* message in TTCN because it is not an observable event in the tester. The operator of the testing tool should verify that it has been received in the system under test. Line 11 is the postamble of the test, carrying the system under test to the initial state, in order to run several test cases in sequence. Line 12 is at the same level of line 6, i.e., the test generation tool does also generate correct alternative behaviour lines. This line means that it is correct behaviour for the system under test to reject the transport connection request issued in line 4. Inconclusive verdicts are assigned to this type of branches, because they state correct behaviour, but they do not fit the test purpose. The default in figure 11 comprises every event not included in the test case, leading the execution of the test to a fail verdict.

Data types are automatically generated from the information included in the specification of the system (and the external data library if needed). Constraints without associated predicates are automatically generated also. In the example, all the constraints, except *tdtsignal_c0*, are automatically generated; *tdtsignal_c0* has an associated predicate (gathered while simulating), i.e., tdtsignal_c0.em = '80'O. The user has to provide a value for the *em* field, that fits the predicate. In this case the system of equations to solve is very simple. Figure 12 illustrates a PDU type definition and a constraint declaration.

PDU Type Definition		
PDU Name: tdtsignal **PCO Type:** SAP **Comments:**		
Field Name	**Field Type**	**Comments**
li	OCTETSTRING[1]	
code	OCTETSTRING[1]	
em	OCTETSTRING[1]	
t_user_data	OCTETSTRING[1..2048]	

PDU Constraint Declaration		
Constraint Name: tdtsignal_c0 **PDU Type:** tdtsignal **Derivation Path:** **Comments:**		
Field Name	**Field Type**	**Comments**
li	'02'O	
code	'02'O	
em	'80'O	
t_user_data	'0123456789ABCDEF'O	

Figure 12 Example of PDU type and constraint.

Coverage measures are automatically computed by the tool. The figures achieved with the test purpose depicted in figure 10 are:

- Signal coverage (black box): 80% (8/10)
- Signal coverage (white box): 65% (11/17)
- State coverage: 86% (6/7)
- Branch coverage: 32% (8/25)

5 CONCLUSIONS

Up to date there is not a final and complete answer for the test automatic generation problem. The GAP tool provides a global and practical solution, easing the test specification, by means of automatizing the process as much as possible in order to obtain optimum test cases reducing costs and time. GAP is embedded in the HARPO toolkit, forming a complete, modular, flexible and upgradeable environment, useful for test suite derivation, validation, execution and maintenance using SDL, MSCs and TTCN.

The defined architecture provides automatic support for test suite generation, which impacts directly on the specification process productivity:

- reducing the testing specification time.
- generating many more test cases than in a manual process.
- the correctness of the test cases is ensured by their automatic nature (derived from the reference system specification).
- better quality test suites (coverage measures).
- being included in HARPO reduces the final testing tool development time.

At the moment of writing this paper, the GAP tool is in its final development phase. Several specifications such as INRES, Transport Protocol TP0 and the Call Diversion Service are being used to test the tool. Since there is no definitive solution to the automatic test generation problem, the intermediate results obtained with a prototype of the tool let us be confident on the fact that GAP is on the right way to achieve its final goal: generate a complete, compilable, TTCN test suite (behaviour, data types and constraints) with realistic (executable against an implementation under test) test cases.

6 REFERENCES

[1] CCITT. Specification and Description Language (SDL). Rec. Z.100. The International Telegraph and Telephone Consultative Committee, 1988. Revised 1993 (SDL-92).

[2] ITU-T. SDL Combined with ASN.1 (SDL/ASN.1). Rec Z.105. The International Telecommunications Union, 1995.

[3] CCITT. Message Sequence Charts (MSC). Rec. Z.120. The International Telegraph and Telephone Consultative Committee, 1992.

[4] ITU-T. Extensions of Basic Concepts for Z.120. The International Telecommunications Union, Draft, December 1994.

[5] ISO. Information Processing Systems - Open Systems Interconnection - Specification of Abstract Syntax Notation One (ASN.1). IS-8824. International Standards Organization, 1987. (See CCITT X.208).

[6] E. Algaba, C. F. Cano, J. I. Sanz, O. Valcárcel. HARPO: Herramientas Automáticas para la Realización de Pruebas OSI. Comunicaciones de

Telefónica I+D, June 1994.

[7] ISO. Information Processing Systems - Open Systems Interconnection - OSI Conformance Testing Methodology and Framework. IS-9646. International Standards Organization, 1989.

[8] ISO. Information Processing Systems - Open Systems Interconnection - OSI Conformance Testing Methodology and Framework - Part 3: The Tree and Tabular Combined Notation. IS-9646-3. International Standards Organization, December 1991.

[9] A. Ek, J. Ellsberger, A. Wiles. Experiences with computer aided test suite generation. Proceedings of IWPTS'93, Pau, September 1993.

[10] J. R. Horgan, S. London, M. R. Lyu. Achieving Software Quality with Testing Coverage Measures. Computer, 27(9):60-69, September 1994.

[11] A. R. Cavalli, B. Chin. Testing Methods for SDLSystems. Proceedings of the Seventh SDL Forum, Oslo, Norway, September 1995.

7 BIOGRAPHY

Esteban Pérez joined Telefónica I+D (R&D Labs.) testing engineering group in 1993 and has been engaged with test generation techniques and in automatizing the development and operation of protocol testing tools.

Enrique Algaba is the project manager of testing engineering group in Telefónica I+D (R&D Labs.). Since joining Telefónica I+D in 1988, he has been engaged in the research and development of protocol testing tools and automatizing the testing process.

Miguel Monedero joined Telefónica I+D (R&D Labs.) testing engineering group in 1995. He has been working since then in the fields of test generation and testing automatization.

Towards abstract test methods for relay system testing

Jun Bi and Jianping Wu
Dept. of Computer Science, Tsinghua Univ.,
Beijing, 100084, P.R.China
bj@csnet1.tsinghua.edu.cn
jinaping@cernet.edu.cn

Abstract

The traditional conformance testing theory and practice have been well used for testing the end system in the network. However, the relay system testing will play an important role in the computer network and distributed system. Abstract test method aims to enable test suite designers to use the most appropriate methods for their circumstances. This paper discusses the abstract test methods for relay system testing. At first, we introduce the model of the R-SUT (Relay System Under Test) and give the conceptual architecture of relay system testing. Then, several abstract test methods are proposed in this paper. At last, we illustrate some practical experience for testing the relay system, such as IP router, SMTP email server, and Packet Assemble/Disassembly (PAD), with the methods we present. These methods could be used for testing ATM switch too.

Keywords

Protocol conformance test, relay system, abstract test method

Testing of Communicating Systems, M. Kim, S. Kang & K. Hong (Eds)
Published by Chapman & Hall

1 INTRODUCTION

With the development of computer netowrks, lots of protocol software and hardware had been implemented by different manufacturers. At the same time, we have to spend more and more time to ensure the correctness of these different implementations. The aim of protocol conformance testing (PCT) is to verify the conformance between protocol implementation and its corresponding standard. Today, it is one of the most active fields on computer network and distributed system.

ISO/IEC 9646 [1] provides the OSI conformance testing methodology and framework. It had been widely used in the test practice for end system ([2][3][4][5][6][7] etc.). Notice that there are two kinds of system in the networks: end system and relay system. The traditional theory and practice of PCT usually focus on the end system testing, and the research for relay system testing is less. Today, relay is an important concept in TCP/IP, switched LAN, and high speed networks. The relay system, such as IP router, LAN switch, and ATM switch, played the important roles in these technologies [7]. Since the peer-to-peer and end-to-end model in ISO/OSI could not fit these relay technologies well, it is very important to study the test theory of relay systems.

Abstract test method aims to enable abstract test suite (ATS) designers to use the most appropriate method for their circumstances [8]. The testers test the behavior of implementation under test (IUT) by protocol data units (PDU) and abstract service primitives (ASP). In ISO/IEC 9646, there are some ripe abstract test methods for end system. These methods are based on ISO/OSI reference model. These test methods could be classified by point of control and observation (PCO), test coordination procedure and the position of tester. Because there is difference between the IUT of end system and the IUT of relay system, it is necessary to study the abstract test methods for relay system. Although there are two relay system test methods, "loop-back" (YL) and "transverse" (YT), defined in ISO 9646, their capabilities are limited. The YL test method is used for testing a relay system from only one subnetwork. Thus the disadvantage of this method is that the behaviour of the relay on only one side is directly observed [1]. The YT method has two PCOs, one on each subnetwork, and uses two test systems external from the IUT. So the procedures for coordinating the control applied to the two testers would be a big problem. To solve these problems and put the relay test into practice, we propose some new relay test methods. We hope it could help the test laboratory make the real test process continuously and high-efficiently.

This paper discusses the characteristics of relay system and presents abstract test methods for relay system testing. The rest of this paper is organized as follows. Section 2 analyzes the R-SUT model. A conceptual architecture of relay system testing is given in section 3. In section 4, several abstract test methods, RL, DL, LT, DT, CT and RT are proposed, then their characteristics are discussed in section 5. After a brief view of the protocol integrated test system (PITS) developed by Tsinghua University in section 6, we will introduce some practical experiences with relay system testing, such as the IP router, the SMTP mail relay, and the Packet Assemble/Disassembly (PAD), using the methods we present in section 7. Finally, we give the conclusion.

2 THE R-SUT MODEL

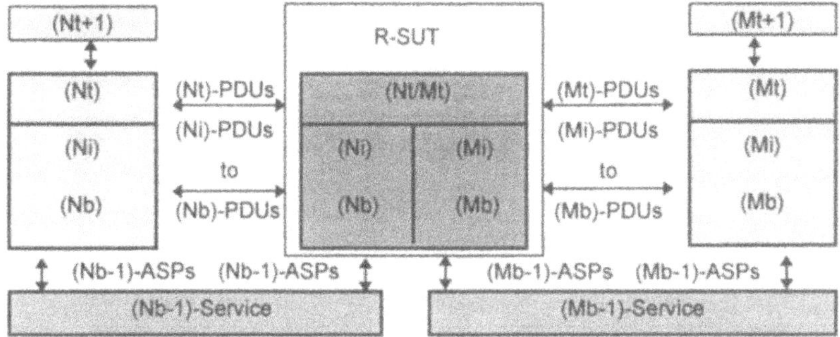

Figure 1 A model of the R-SUT.

There exists a relationship between the test methods and the configurations of the real network system to be tested [1]. There are two main configurations of system in a network:

(1) End system;

(2) Relay system.

Neither the term "relay system" nor "end system" has been defined by ISO nor other standard organizations, even though they are widely used in the field of data communication. The definition given by Cerf and Kirstein [9] is adopted here. It says that the collection of required hardware and software effect the interconnection of two or more data networks, enabling the passage of user data from one to another, is called a "relay system". This infers that a system connected only to one network will not be regarded as a relay system. All system other than relay system could be classified as end systems.

Now, we present a model of relay system under test (R-SUT) and it is shown in figure 1. In this model, there are two protocol suites of subnetworks connected by the relay system. These two suites could be named "N" and "M". If the two subnetworks have the same protocol architectures, N is equal to M. The highest layer in the R-SUT is numbered "Nt" or "Mt" (for "top"), and the lowest is numbered "Nb" and "Mb" (for "bottom"). Notice that Nt is usually equal to Mt, and they realize the function of relay. For single-layer protocol R-

SUTs, Nt (or Mt) is equal to Nb (or Mb). In the following sections, the same notation will be used to refer to layers within the tester. The R-SUT may implement protocols in layers lower than "Nb", but these are not of interest in the test method descriptions. For all test methods, ATSs specify test events at the lower tester PCO in terms of (Nb-1)-ASPs and (Mb-1)-ASPs and/or (Nt) to (Nb)-PDUs and (Mt) to (Mb)-PDUs. There are some features in R-SUTs:

(1) The relay layer is always the highest layer in a relay system. In another word, there is no upper layer above a relay function. So it is not necessary to control and observe its upper boundary by the (Nt+1)-ASPs and (Mt+1)-ASPs.

(2) There are at least two subnetwork under a relay system, so the test events must be control and observed by the two sets of ASPs and PDUs.

3 CONCEPTUAL ARCHITECTURE OF RELAY SYSTEM TESTING

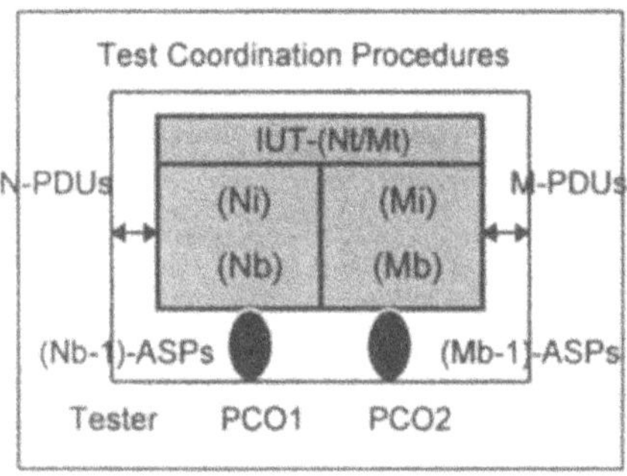

Figure 2 Conceptual architecture of relay system testing.

Abstract test methods are described in terms of what output from the IUT is observed and what inputs to it can be controlled. The starting point of developing abstract test methods is the conceptual testing architecture [1]. The conceptual architecture of the relay system testing is illustrated in figure 2. It is a "black-box" active testing architecture, based on the definition of behavior required by the IUT. The actions in this conceptual tester involve two sets of interactions: one for (N)-protocols and one for (M)-protocols. These can be controlled and observed at PCO1 and PCO2. Because of the ASPs above (Nt) is not specified, the tester is only lower tester (LT). LT would control and observe the (Nb-1)-ASPs including (Nt) to (Nb)-PDUs at PCO1 and (Mb-1)-ASPs including (Mt) to (Mb)-PDUs at PCO2.

4 ABSTRACT TEST METHODS OF RELAY SYSTEM

An abstract test method describes an abstract testing architecture consisting of testers and test coordination procedures, and their relationships to the test system and SUT. Each test method determines the PCOs and test events (i.e., ASPs and PDUs) which shall be used in an abstract test case for that test method. In this section, referring to the concepts and methods provided by ISO/IEC 9646, we propose 6 abstract test methods: RL, CL, LT, DT, CT, and RT. The ATSs should be specified in accordance with these methods.

4.1 Remote loop-back test method (RL)

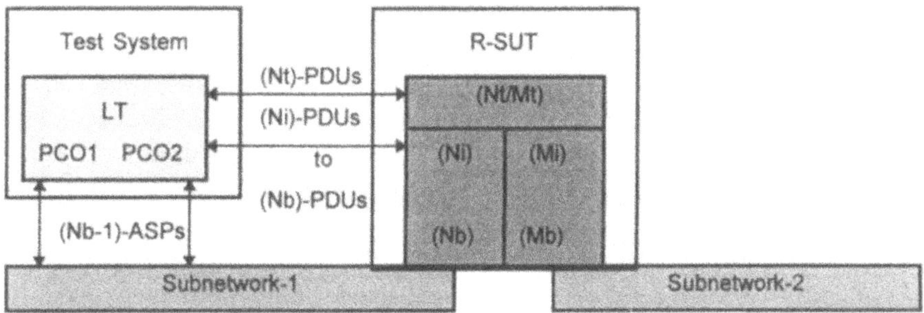

Figure 3 The RL test method.

The remote loop-back test method (RL) is illustrated in figure 3, just like the loopback method presented in ISO 9646. In this test method, there are two PCOs on one subnetwork at SAPs external from the (Nt)-Relay. For connection-oriented protocols, it requires that the two test connections are looped together on the far side of the relay system. This looping could be performed within the relay system or in the second subnetwork. For connectionless protocols, it requires that the PDUs are looped back within the second subnetwork and addressed to return the second PCO. This method enables a relay system to be tested without requiring test systems on two different subnetworks. Because there is only one lower tester (LT), the test coordination procedure of two PCOs would be very simple.

4.2 Distributed loop-back test method (DL)

The distributed loop-back test method (DL) is illustrated in figure 4. It uses a test responder (TR) in the extra destination host on the second subnetwork to send/ receive the PDUs to/from R-SUT. In test system, there are two PCOs for both side functions of R-SUT. When LT sends a PDU from PCO1 to the destination host, it would be relayed by R-SUT. TR located in the second subnetwork then controls and observes the events from R-SUT and returns it to LT through subsidiary test path (STP). This returned message could be obtained by LT from PCO2. The STP is also used for the test coordinating messages. In fact, this method combined the two lower testers (one should be in test system, and another in destination host) into one test system. Because the test suite including two PCOs is executed in one test system, the coordination of PCOs for both sides of R-SUT could be solved. It makes the test process automatically continuously and high-efficiently.

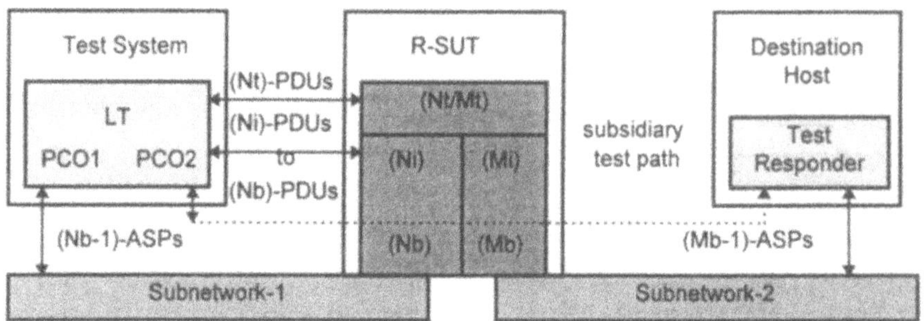

Figure 4 The DL test method.

4.3 Local transverse test method (LT)

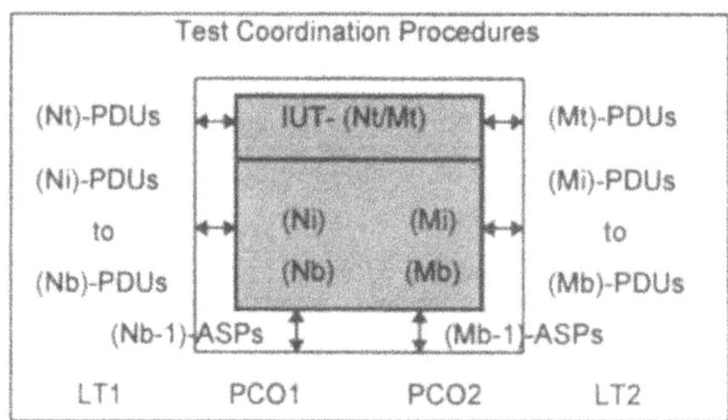

Figure 5 The LT test method.

Considering the transverse method presented in ISO 9646, we give the following four methods, LT, DT, CT and RT. The local transverse test method (LT) is illustrated in figure 5. This method have the following characteristics,

(1) There is no upper tester.

(2) LT1, LT2, and SUT are in one local system, so the events occurred in SUT will be controlled and observed directly. However in many IUT, it seems not easy to find the required API for control and observation.

(3) The test events are specified by (Nb-1)-ASPs/(Nt), (Ni) to (Nb)-PDUs at PCO1 for LT1, and (Mb-1)-ASPs/(Mt), (Mi) to (Mb)-PDUs at PCO2 for LT2.

(4) Test coordination procedure between two lower testers may be realized in this local system by inter-process communication.

4.4 Distributed transverse test method (DT)

The distributed transverse test method (DT) is illustrated in figure 6. This method have the following characteristics,

(1) There is no upper tester.

(2) There are two lower testers, LT1 and LT2, in different test system. The events occurred in R-SUT will be controlled and observed on different subnetwork directly.

(3) The test events are specified by (Nb-1)-ASPs/(Nt), (Ni) to (Nb)-PDUs at PCO1 for LT1, and (Mb-1)-ASPs/(Mt), (Mi) to (Mb)-PDUs at PCO2 for LT2.

(4) Test coordination procedure between LT1 and LT2 would be realized by software or human, so it may be a problem for a real test system.

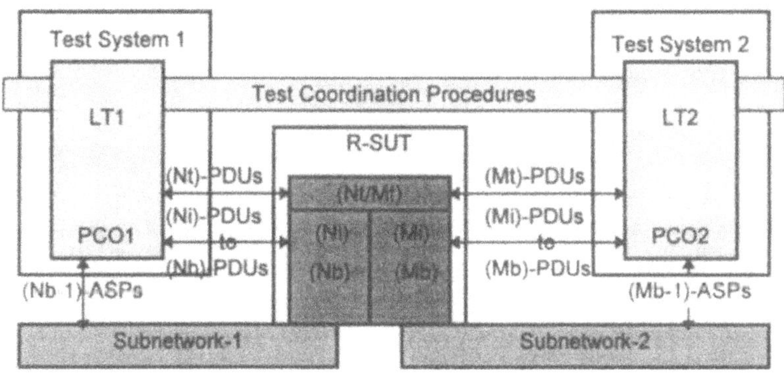

Figure 6 The DT test method.

4.5 Coordinated transverse test method (CT)

The coordination transverse test method (CT) is illustrated in figure 7. This method have the following characteristics,

(1) There is no upper tester.

(2) Test coordinating procedures would be realized as a test management protocol (TMP) between LT1 and LT2, so it may be more difficult in practice.

(3) The test events are specified in terms of (Nb-1)-ASPs/(Nt), (Ni) to (Nb)-PDUs at PCO1 for LT1, and (Mb-1)-ASPs/(Mt), (Mi) to (Mb)-PDUs at PCO2 for LT2.

(4) LT1 and LT2 are in different test system, so the events occurred in R-SUT will be controlled and observed on different subnetwork directly.

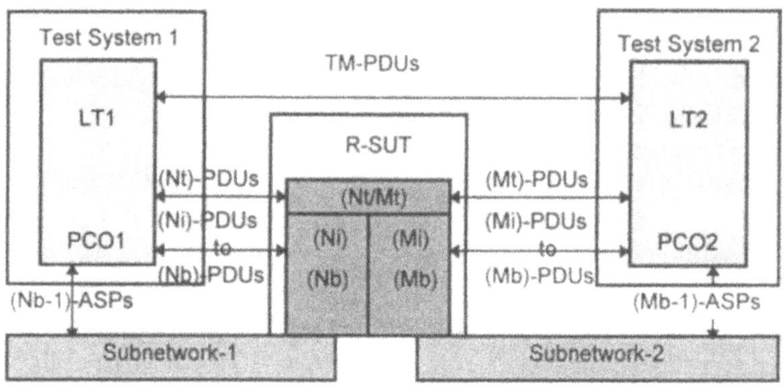

Figure 7 The CT test method.

4.6 Remote transverse test method (RT)

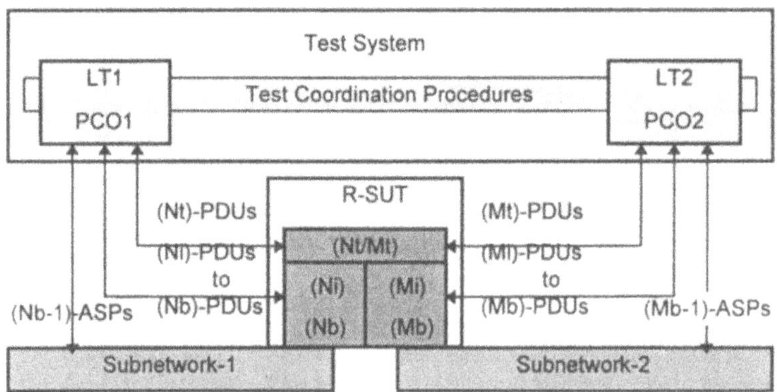

Figure 8 The RT test method.

The remote transverse test method (RT) is illustrated in figure 8. This method have the following characteristics,

(1) There is no upper tester.

(2) LT1 and LT2 are in one test system and the events occurred in R-SUT will be controlled and observed on different subnetwork directly.

(3) The test events are specified in terms of (Nb-1)-ASPs/(Nt), (Ni) to (Nb)-PDUs at PCO1 for LT1, and (Mb-1)-ASPs/(Mt), (Mi) to (Mb)-PDUs at PCO2 for LT2.

(4) Test coordination procedures between LT1 and LT2 would be realized in one test system as inter-process communication, so the test process would be highly efficient.

5 COMPARISON OF THESE ABSTRACT TEST METHODS

The abstract test methods we proposed could be generally divided into two kinds: loop-back method and transverse method.

The loop-back method is used for testing a relay system from one subnetwork. The advantage is that the procedures for coordinating the control applied to the two PCOs can be realized within a single test system. The disadvantage is that the relay behavior on only one side is directly observed. Thus, its behavior on the second subnetwork can not be properly assessed.

The transverse method is used for testing a relay system from two subnetworks. The advantages are:

(1) The behavior on each subnetwork could be controlled and observed.

(2) This method enables the relay system to be tested in the normal mode of operation.

The disadvantage is that the test coordination procedure may be much complex, because there are two LTs for different subnetworks. It is a big problem for the real test system designers.

In the methods of DT and CT, two LTs are located in two different test systems separately. So they could be used in the distributed test environment. In a real test system, the former is more simple but the test coordination procedure is more difficult. If the coordination could not be solved well, the test process would not be automatically and continuously. The later method could solve the coordination successfully. However, because of the implementation of TMP, there will raise more system cost.

In the methods of LT and RT, two LTs are located in one test system. So their test coordination procedures could be solved well. LT method could be used for a IUT which has clear interface. Because the tester and SUT are in the same system, its application would be limited. So, we think the realization of RT method has the following advantages:

(1) The two LTs are in one test system, so the common model and software could be used by these two testers when developing a real test system. It would deduce the system cost.

(2) Test coordination procedures are simple and high-efficient. It could be realized as inter-process communication. It is better than TMP.

(3) The design of abstract test suite is simple. The designer only concerns the test event of two sides of R-SUT and need not pay attention to the coordination of the two sides.

(4) Use "black-box" testing and we need not the upper interface of IUT. So we need not add extra model in R-SUT. It could be used for different IUTs.

Moreover, the characteristics of these methods are shown in table 1.

Table 1 Characteristics of these test methods

	Loop-back		Transverse			
	RL	DL	LT	DT	CT	RT
Observation of two subnetwork	indirectly	directly	directly	directly	directly	directly
Test systems	1	1	1	2	2	1
Test coordination procedures	inter-process, simple, automatic	inter-process, simple, automatic	inter-process, simple, automatic,	by human, and may be single step	using TMP, complex, automatic	inter-process, simple, automatic
Need extra TR	no	yes	no	no	no	no
IUT independence	yes	yes	no	yes	yes	yes

6. PROTOCOL INTEGRATED TESTING SYSTEM (PITS)

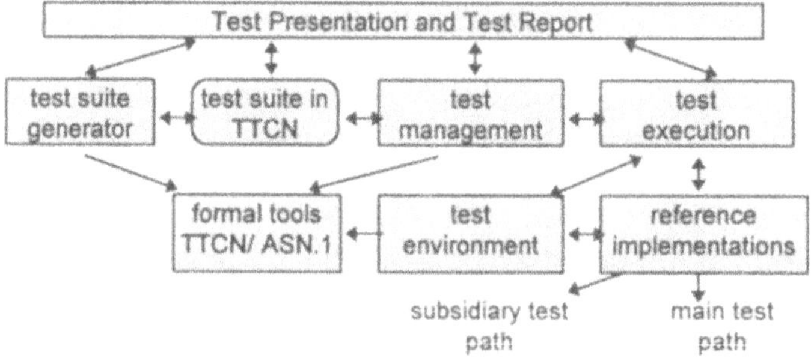

Figure 9 The protocol integrated testing system PITS.

In this section, we will introduce the protocol integrated test system PITS. The

PITS aims to provide a basic platform to test different protocols by different test methods. It could be used for both the conformance testing, the interoperability testing, and the performance testing. It has been used for testing many implementations of end systems and relay systems. The PITS shown in figure 9, is composed of the following main components: test presentation and test report, test management [10], test execution [11], test suite generator, reference implementations, formal support tools and test software environment. The TTCN test suite is generated from EBE specification, which could be translated from LOTOS and Estelle specification [12]. The Reference Implementation (RI) is a very important part in this test system. It is the special protocol implementation and acts as the lower communicating support for controlling and observing the events occurred in test execution (TE).

The following objectives guided our design and implementation effort:

(1) Accordance with ISO protocol testing standards. All the ideas, methods and terminology adopted in our PITS strictly follow ISO 9646 protocol testing standard. In our PITS, all the protocol reference implementations and the services accord with corresponding ISO standards. The test suite (TS) is formally described in TTCN, which is defined in ISO 9646-3 [1].

(2) Flexibility and independence. In our TTCN based TE, test suite is executed according to the operational semantics of TTCN, so this method is flexible and independent on the protocol being tested. It could be regard as a general executing mechanism for any TS in TTCN. So it can test all the protocols whose TS is in TTCN. Then any new protocol can be tested by our PITS, only with fulfillment of its TS in TTCN.

(3) Efficient test execution. The parallel interpreting improves the test executing efficiency. When a test case is being interpreted, the most possible next test case is being interpreted. TTCN based TE interprets a test case just before its execution. It will allow testing operator with the more possibilities to control the testing process, such as single step testing and supervising.

7 PRACTICAL EXPERIENCE WITH RELAY SYSTEM TESTING

In this section, we will introduce some practical testing experiences with the relay test methods using PITS.

7.1 Testing IP router

Today, the IP router is one of the most important relay system in Internet. The function or purpose of IP is to move datagrams through an interconnected set of networks. This is done by passing the datagrams from one Internet module to another until the destination is reached. The IP modules reside in hosts and routers in the Internet. The datagrams are routed from one IP module to another through individual networks basing on the interpretation of the Internet address. Thus, one important mechanism of the Internet protocol is the IP addressing. In the routing from one IP module to another, datagrams may need to traverse a network whose maximum packet size is smaller than the size of the datagram. To overcome this difficulty, a fragmentation mechanism is provided in the IP protocol. Errors detected may be reported via the Internet Control Message Protocol (ICMP).

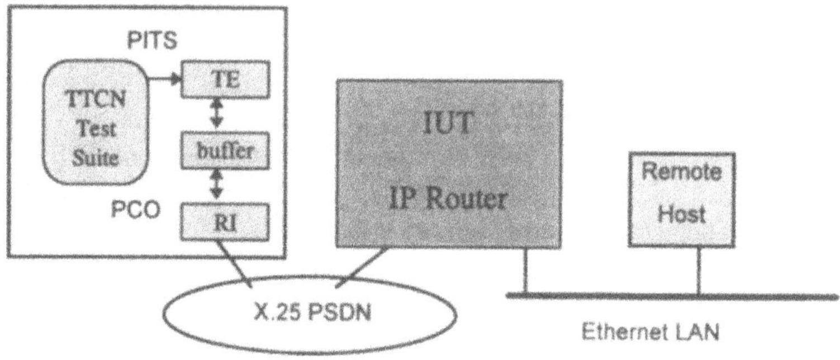

Figure 10 Testing IP router with DL.

We use PITS to test IP router with RL method. It is shown in figure 10. This IP router connects two subnetworks: Ethernet LAN and X.25 public data network. When PITS sends a IP/ICMP datagram (for example ECHO) to the remote host, after routing and addressing, it will be forward by IP router from X.25 PSDN to Ethernet LAN. The response IP datagram could address to PITS and be observed at PCO. We have designed a TTCN based test suite for IP router. This test suite contains 32 test cases and the following is an example. The test purpose is shown in this test case. Now the test suite is only a prototype for verifying the new test architecture. We are developing the complete IP test suite. Then we could test IP from more subnetwork and test more IP options.

Table 2 A test case of IP routing

Test Case Dynamic Behavior					
Test Case Name : Route_01 **Group** : **IP/Route/Route_01** **Purpose** : To verity an ECHO datagram could be relayed by IP router from a host with address of 166.222.1.1 in X.25 PSDN to a host with address 166.111.166.8 in Ethernet LAN **Default** : **Comments** :					
Nr	**Label**	**Behavior Description**	**Constraint Reference**	**Verdict**	**Comments**
1		! Echoreq	Echoreq_01		(1)

2		Start T01			
3 4		? Echorsp	Echorsp_01	PASS	(2)
5		? Otherwise		FAIL	
6 7		? Timeout T01		FAIL	
Detail Comments: (1) SA=166.222.1.1, DA=166.111.166.8 (2) SA=166.111.166.8, DA=166.222.1.1					

7.2 Testing PAD

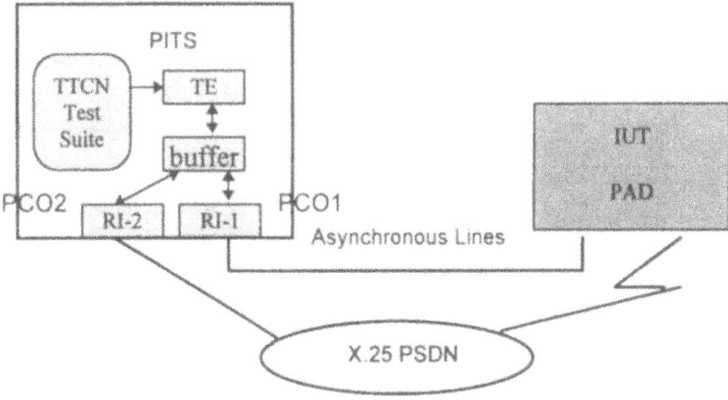

Figure 11 Testing PAD with RT.

CCITT defined three recommendations (X.3/X.28/X.29) about packet assembly/disassembly device (PAD) in public network. Recommendation X.3 defines a set of parameters for PAD. X.29 defines the procedures between a PAD and a packet mode DTE or another PAD, and X.28 defines the DTE/DCE interface for a start-stop mode DTE accessing PAD. The PAD is a special relay system. One side of a PAD is the X.25 public data network for packet mode DTE, and another side is asynchronous lines for terminals.

We use RT method to test PAD. There are two PCOs in the test suite. So we implement two RIs to control and observe the test event in/out the IUT. When TE interpreting and executing the TTCN based test suite, the test events would be send to the corresponding RI from the buffer according to their PCOs. Figure 11 shows using PITS to test the relay function of PAD. Because of RT's advantages, we think this architecture is a good approach to test switch equipment in LAN and WAN.

The TTCN based PAD test suite we designed contains 234 test cases. The following is an example. The parameter 1 allows the start-stop DTE to initiate an escape from the "data transfer" state or the "connection in progress" state in order to send PAD command signals. Value 0 of parameter 1 indicates that recall is impossible; value 1 indicates that recall using a character DLE; value 32 to 126 using graphic character defined by user. In this test case, we verify the function of value 1. Pre_9 is a preamble and CHK_9 is a verification sequence for state 9.

Table 3 A test case of PAD

Test Case Dynamic Behavior

Test Case Name: SP_101
Group : **PAD/SP/**
Purpose : Verity that with parameter 1 being set to 0, PAD recall using a character is impossible.
Default :
Comments :

Nr	Label	Behavior Description	Constraint Reference	Verdict	Comments
1		+ Pre_9			
2		PCO2 ! N_DT_req (MSET)	MSET_20		set 1, 0
3		PCO1 ! DLE	DLE_01		
4		PCO2 ! N_DT_req (DT)	DT_11		'A'
5		Start TM01			
6		PCO1 ? CHAR	CHAR_01	(PASS)	'A'
7		+ CHK_9			
8		PCO1 ? Otherwise		FAIL	
9		PCO1 ? Timeout TM01		FAIL	

Detail Comments:

7.3 Testing SMTP relay server

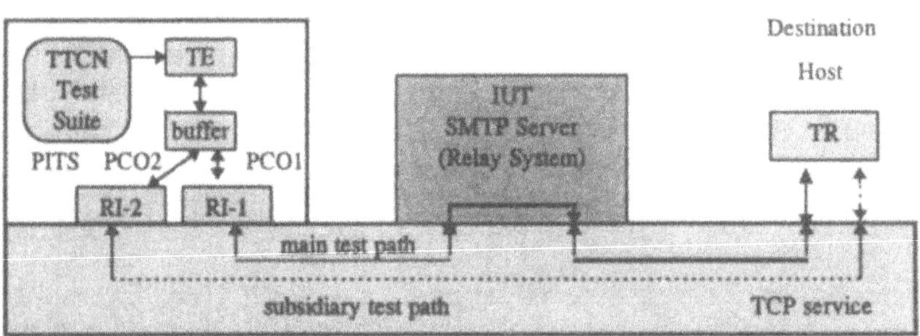

Figure 12 Test architecture of relaying email.

SMTP is designed in the RFC standards. RFC 821 specifies the definition of SMTP and RFC 822 specifies the syntax of test message that sent as email with BNF (Backus-Naur Form). The objective of SMTP is to transfer email reliably and efficiently. SMTP is independent of the particular transmission subsystem and TCP is the most popular transmission subsystem. An important feature of SMTP is its capability to relay mail across transport service environments. A transport service provides an interprocess communication environment (IPCE). Emails can be communicated between processes in different IPCEs by relaying through a process connected to two (or more) IPCEs. More specifically, mail can be relayed between hosts on different transport systems by a server on both transport systems.

We use the DL method to test the relay function of SMTP mail server. Figure 12 shows this testing architecture. There are two PCOs for both side functions of IUT. The TTCN test suite contains 89 test cases in total. There is an example in [13].

8. CONCLUSION

In ISO 9646, there are some standard abstract test methods for the end system and two methods for the relay system. For testing a real relay system, these methods are too simple to direct the test activities well. We have proposed six abstract test methods (RL, DL, LT, DT, CT, and RT) for relay system testing. They are the recommendations for real test system. The characteristics of these test methods had been discussed in section 5. These test methods would be selected according to their characteristic and the situation of SUT. We had implemented three test methods (RL, DL, RT) in PITS using the Sun Sparc workstation and Solaris 2.4. It had been presented to be very successful in the testing of IP router, SMTP mail server, and PAD. Now we are focusing on the other three test methods (LT, DT, CT) in the testing of ATM switch, and the more complex relay system such as the Internet routing protocols. We believe that relay system will be more important in the future, especially for Internet and high speed network. We hope there come more efforts for relay system tesing using the test methods proposed in this paper.

REFERENCES

[1] ISO, *Information Processing System - Open System Interconnection - OSI Conformance Testing Methodology and Framework, ISO/IEC 9646,1991.*

[2] K. Katsuyama and F. Sato, Strategic testing environment with formal description techniques, *IEEE Trans. on Computer*, Vol. 40, No.4, 1991.

[3] S.T.Chanson, et al., The UBC Protocol Testing Environment, in O.Rafiq,

editors, *6th IFIP IWPTS*, 1993.

[4] C.S.Lu, et al., An Implementation of CMIP/CMISE conformance testing system, in T.Mizuno, et. al., editors, *7th IFIP IWPTS*, 1994.

[5] K.Y.Kim, et al., Experiences with the design of B-ISDN integrated test system (BITS), in A.Cavalli, et. al., editors, *8th IFIP IWPTS*, 1995.

[6] W.Geyer, S.Hesse and N.Newrly, TSE-P - a highly flexible tool for testing network management applications using simulation, in B.Baumgarten, et. al., editors, *9th IFIP IWTCS*, 1996.

[7] T.W.Kim, et. al., Field trial and quality test of ATM switching system in Korea, in B.Baumgarten, et. al., editors, *9th IFIP IWTCS*, 1996.

[8] D.Rayner, OSI conformance testing, *Computer network & ISDN systems*, Vol.14, 1987.

[9] V.G.Cerf and R.T.Kirstein, Issues in packet-network interconnection, *proceeding of IEEE*, Vol.66, No.11, 1978.

[10] J. Tian and J. Wu, Test management and TTCN based test sequencing, in A.Cavalli, et. al., editors, *8th IFIP IWPTS*, 1995.

[11] Y. Wang, J. Wu and R. Hao, An approach to TTCN-based test execution, in T.Mizuno, et. al., editors, *7th IFIP IWPTS*, 1994.

[12] J. Wu and S.T.Chanson, Translation from LOTOS and Estelle specifications to extended transition system and its verification, in *2nd IFIP FORTE*, 1989.

[13] J.Bi and J.Wu, et.al., Application of a TTCN based conformance test environment on the Internet email protocol, to appear in *10th IFIP IWTCS*, 1997.

Applying SAMSTAG to the B-ISDN protocol SSCOP

J. Grabowski, R. Scheurer, Z.R. Dai, and D. Hogrefe
Institute for Telematics, University of Lübeck
Ratzeburger Alle 160, D-23538 Lübeck, Germany, Tel.+49 451 500 37 21,
Fax +49 451 500 37 22, {jens,scheurer,dai,hogrefe}@itm.mu-luebeck.de

Abstract
The test generation method SAMSTAG (SDL and MSC based test case generation) has been applied successfully to the B-ISDN ATM Adaption Layer protocol SSCOP (Service Specific Connection Oriented Protocol). For approximately 70% of the identified test purposes complete TTCN test cases have been generated automatically. In this paper we describe the experiment, discuss the results and explain how further improvements of the test generation process can be achieved.

Keywords
Test case generation, protocol validation, SDL, MSC, TTCN, B-ISDN ATM

1 INTRODUCTION

From 1991 to 1993 Swiss PTT promoted a project at the University of Berne which was aimed at suporting the conformance testing process. One objective was the development and implementation of a method for the automatic generation of abstract test cases in TTCN format based on SDL system specifications and MSC test purposes. As a main result of this project we developed the SAMSTAG method and

Testing of Communicating Systems, M. Kim, S. Kang & K. Hong (Eds)
Published by Chapman & Hall © 1997 IFIP

implemented the SAMSTAG tool [Grabowski, 1994; Nahm, 1994]. The applicability of tool and method has been shown by performing a case study based on the ISDN layer 2 protocol CCITT Rec. Q.921. However, that case study was not complete because we generated test cases for some selected test purposes only, but no complete test suite. The reasons for this incompleteness were restrictions imposed on us by lack of time, money and manpower.

In the years 1993–1995 we improved SAMSTAG by providing mechanisms for dealing with complexity, i.e., the state space explosion problem during test generation [Grabowski et al., 1996]. The most important result of these investigations was the development and implementation of partial order simulation methods for SDL specifications [Toggweiler et al., 1995].

Starting in 1995 we performed another case study based on the B-ISDN protocol SSCOP (ITU-T Rec. Q.2110). The choice of SSCOP was influenced by the interest of the ITU-T in a review of the SSCOP SDL specification and by the need for a test suite for SSCOP. This case study has shown that automatic test generation based on SDL specifications and MSC test purposes is feasible. For 69% of the identified MSC test purposes complete TTCN test cases were generated automatically.

In this paper we focus on describing the application of SAMSTAG to SSCOP. We do not compare SAMSTAG with other methods and tools for automatic test generation. For such a comparison the interested reader may have a look at [Doldi et al., 1996]. The paper proceeds as follows: Section 2 describes SAMSTAG whereas Section 3 introduces the SSCOP protocol. Section 4 explains all steps which have to be performed before the SAMSTAG tool can be applied. The results of the test generation process are presented in Section 5. Section 6 describes the expenses of the test suite development using SAMSTAG. Summary and outlook are given in Section 7.

2 SAMSTAG

SAMSTAG supports the generation of conformance test suites according to IS 9646 [ISO/IEC, 1994]. In this methodology test purposes have to be identified which describe the test case objectives. Test purposes are one basis for test case selection and necessary to relate test results to the protocol functions which have been tested.

The test purposes are implemented in form of abstract test cases by using the TTCN notation. The basis for the implementation is the protocol standard. Currently, this implementation is mainly done manually. SAMSTAG automates this implementation step by using formally specified protocol and test purpose descriptions. As indicated by the abbreviation SAMSTAG which stands for *'Sdl And Msc baSed Test cAse Generation'*, it is assumed that the allowed behaviour of the protocol is defined by an SDL specification, and that the purpose of a test case is provided in form of an MSC.* For the understanding of this paper some basic knowledge of MSC [ITU-TS, 1996b], SDL [ITU-TS, 1996a] and TTCN [ISO/IEC, 1994] is required.

*The SAMSTAG method has been generalised in order to cope with protocol specifications and test purposes which are given in other formalisms than SDL and MSC. The SAMSTAG tool implements the SAMSTAG method for SDL and MSC descriptions.

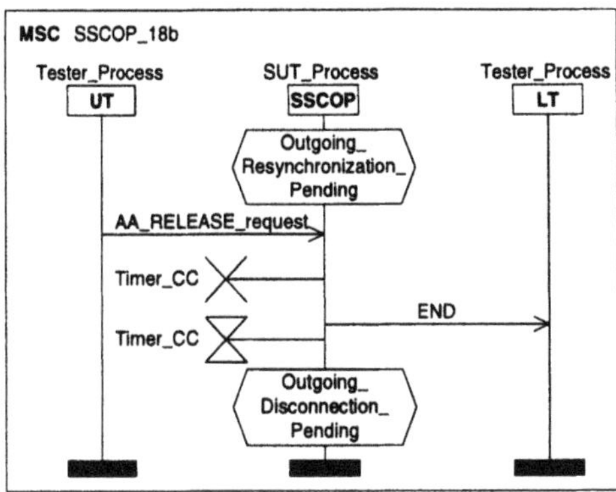

Figure 1 MSC test purpose.

2.1 SAMSTAG input and output

The inputs to the SAMSTAG tool are an SDL specification and an MSC diagram. The test generation process results in a TTCN output.

SDL input
In order to generate the test case SAMSTAG simulates a *closed* SDL system. This means that the SDL specification comprises not only the protocol to be tested, i.e. the *Implementation Under Test* (IUT), but also the tester processes and, optional, other system parts in which the IUT may be embedded or which are required for testing. In the following we use the term *System Under Test* (SUT). An SUT includes no tester processes, but the IUT and all other processes in which the IUT may be embedded or which are necessary for testing.

The specification of a closed system which the IUT only is part of requires additional specification work to be done before test case generation. But, it provides a high degree of flexibility. It allows us to consider different test architectures and to embed the IUT in another system (e.g. [Grabowski et al., 1995]). For simpler cases, tester processes which are able to send and receive all allowed signals at any time can be generated automatically.

MSC input
The SAMSTAG tool accepts test purposes in form of MSCs [ITU-TS, 1996b]. An example is shown in Figure 1. The SAMSTAG tool distinguishes between two types of processes, SUT processes and tester processes. Figure 1 includes one SUT process, SSCOP, and two tester processes, UT and LT. The SSCOP process describes the test purpose from the viewpoint of the SSCOP protocol. In this case, the test purpose is the test of a special state transition. The transition starts in SDL state `Outgoing_Resynchronization_Pending` and ends in state `Outgoing_Disconnection_Pending`. Both states are referred to by means of MSC conditions. During the state transition the SSCOP has to consume an `AA_RELEASE_request` message from UT, cancel timer `Timer_CC`, send `END` to LT and set timer `Timer_CC`

Test Case Dynamic Behaviour					
Test Case Name	: SSCOP_18b				
Group	: CONTROL/RESYNC/RELEASE/				
Purpose	: cf. figures 1 and 6				
Default	: stddefault				
Nr	**Label**	**Behaviour Description**	**Constraints Ref**	**Verdict**	**Comments**
1		PRUT!AA_ESTABLISH_request	AA_ESTABLISH_request_11F_Y		
2		PRLT?BGN	BGN_111_Y		
3		PRLT!BGAK	BGAK_880_N		
4		PRUT?AA_ESTABLISH_confirm	AA_ESTABLISH_confirm_88_E		
5		PRUT!AA_RESYNC_request	AA_RESYNC_request_35_J		
6		PRLT?RS	RS_352_S		
7		PRUT!AA_RELEASE_request	AA_RELEASE_request_23_G		
8		PRLT?END	END_230_Q		
9		PRLT!END	END_230_Q		
10		PRLT?ENDAK	ENDAK_0_M		
11		PRUT?AA_RELEASE_confirm		PASS	
12		PRLT?POLL	POLL_100_S	INCONC	
13		PRLT?END	END_001_D	INCONC	
14		PRLT?BGN	BGN_111_Y	INCONC	
15		PRUT?MAA_ERROR_indication	MAA_ERROR_indication_P_L	INCONC	
16		PRLT?POLL	POLL_100_S	INCONC	
17		PRLT?END	END_001_D	INCONC	
18		PRLT?BGN	BGN_111_Y	INCONC	
19		PRUT?MAA_ERROR_indication	MAA_ERROR_indication_O_K	INCONC	

Figure 2 TTCN dynamic behaviour description.

again. Hence, to drive SSCOP through this state transition the UT has to send a `AA_RELEASE_request` message and the LT has to receive an `END` message.

TTCN output

SAMSTAG produces complete TTCN test case descriptions including the dynamic behaviour tables, message type definitions, and all constraint declarations. Figure 2 presents the TTCN dynamic behaviour description generated for the MSC test purpose shown in Figure 1. The message exchange related to the test purpose can be found in the lines 7 and 8. The lines 1–6 describe all actions of the tester processes in order to drive the IUT into state `Outgoing_Resynchronization_Pending`, i.e., the state from which the test purpose is observable. The lines 9–11 verify that the test purpose has been performed and drive the IUT back into its initial state. The lines 12–19 are related to inconclusive cases.

2.2 SAMSTAG test generation procedure

For a given SDL specification and a given MSC test purpose the SAMSTAG tool generates a TTCN test case by performing the following steps automatically:

1. The SAMSTAG tool simulates the SDL specification and searches for a trace which (a) starts and ends in the initial state of the SDL specification, and (b) includes the MSC test purpose, i.e., during the trace the MSC is performed. The main problem of step 1 is the state space explosion which may occur during the

search. The SAMSTAG tool provides several mechanisms and techniques to cope with this problem. They are described in Section 2.3.

2. For a test case description only observable events are relevant. Observable events describe actions to be performed by tester processes during a test run. In the following, a trace which includes observable events only is called an *observable*. In step 2 SAMSTAG constructs the observable of the trace obtained in step 1, i.e., all events internal to the SUT are removed. As a result we gain a candidate, called *possible pass observable* (PPO), for a test sequence which may lead to a *pass* verdict. It is only a candidate, because the relation between a complete SDL system trace and its observable is not unique. There may exist other traces which have the same observable, but which do not end in the initial state of the SDL specification, i.e., condition (a) of step 1 is violated, or which do not perform the MSC test purpose, i.e., condition (b) is violated.
3. SAMSTAG tries to verify the uniqueness of the PPO. This is done by contradiction, i.e., by the search for at least one trace which has the PPO as observable, but violates condition (a) or (b) of step 1. If such a trace is found, it is shown that the execution of the PPO during a test run does not ensure that the test ends in the expected state, or does not ensure that the test purpose has been fulfilled. According to IS 9646, in both cases no *pass* verdict should be assigned to such a test sequence. If no such trace is found or if all traces found fulfill the conditions (a) and (b), it is verified that the PPO is a test sequence to which SAMSTAG can assign a *pass* verdict. A verified PPO is called *unique pass observable* (UPO).
4. Due to parallelism, the system may behave in a nondeterministic manner. For testing this means that on a stimulus of a tester process the response from the system may be allowed by the specification, but does not follow the intention of the test case. Neither the test purpose can be verified, nor the specification is violated. According to IS 9646, in such a case an *inconclusive* verdict should be assigned. In order to gain a complete test case description, all traces leading to an *inconclusive* verdict have to be considered. Therefore in step 4 SAMSTAG generates *inconclusive observables* for the UPO found in step 3. An inconclusive observable has prefixes which are identical to prefixes of the UPO, but its last event describes a response from the protocol specification which does not follow the UPO, but is allowed by the SDL specification.
5. Finally, the TTCN test case description for the UPO and the corresponding inconclusive observables are generated, i.e. a TTCN dynamic behaviour description which combines all observables is computed. Additionally, TTCN type definitions and constraints declarations are generated for all messages to be send to and received from the system to be tested. The type definitions are based on SDL signal definitions. The constraints follow from the concrete values observed during the computation of UPO and inconclusive observables. In order to cope with *fail* cases in a final way, a TTCN default behaviour description is generated.

All those five steps above are performed automatically by the SAMSTAG tool. The generated TTCN test cases are complete, i.e., no manual completion is needed. Although the sketched five steps procedure works in many cases, it should be noted that due to theoretical reasons it cannot be guaranteed that PPOs and UPOs exist. The problem of finding PPOs and UPOs can be traced back to the halting problem of Turing machines, for which no solution exists [Hopcroft and Ullmann, 1979].

2.3 Dealing with the state space explosion problem

The main problem of test generation is the explosion of the state space which may occur during the search for the required observables. The reasons for this kind of complexity are (1) the increasing power of modern protocol functions leading to complex specifications, (2) characteristics of the chosen specification language, and (3) missing information about the environment in which the IUT should work.

In our case, characteristics referred to by (2) are related to the interleaving semantics of SDL. The problem of (3) is that in general an IUT is modelled as an open system. For an automatic simulation, during which we search for PPOs and UPOs, the behaviour of the environment has to be modelled. The simple assumption that the environment is able to send and receive any valid signal at any time leads to an enormous amount of possible simulation runs.

Complexity due to (1) cannot be avoided. Therefore, we focus on mechanisms that reduce complexity due to (2) and (3). We distinguish between three classes of reduction mechanisms called *heuristics*, *partial order simulation*, and *optimisation strategies*.

Heuristics are based on assumptions about the behaviour of the system to be tested, or of that of its environment. They avoid the elaboration of system traces which are not in accordance with the selected assumptions. *Partial order simulation* methods avoid complexity which is caused by the interleaving semantics of the specification language. They intend to limit the exploration of traces for concurrent executions*. *Optimisation strategies* intend to reduce the possible behaviour of the system environment. This can be done by using external information, e.g., specifications of surrounding services, or by analysing the specification in order to generate optimal input data to be provided by the tester processes.

In SAMSTAG we implemented several heuristics and partial order simulation methods for SDL specifications. Details can be found in [Grabowski et al., 1996; Toggweiler et al., 1995]. Optimisation has been done by hand. We will come back to this point in Section 4.5.

3 SSCOP

The *Service Specific Connection Oriented Protocol* (SSCOP) is used in the *B-ISDN ATM Adaption Layer* (AAL). The purpose of the AAL is to enhance the services

*A concurrent execution can be seen as a partially ordered set of events. All traces which do not violate the partial order describe the interleaved traces of the concurrent execution.

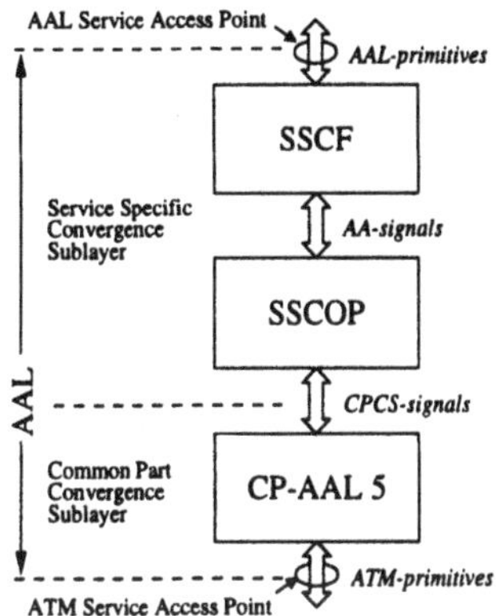

Figure 3 Structure of the ATM Adaption Layer (AAL).

provided by the ATM layer in order to meet the needs of different upper layer applications. One particular AAL type is the *signalling AAL* (SAAL). The SAAL provides communication functions for ATM entities which are responsible for signalling.

As shown in Figure 3, SSCOP can be used within the SAAL. The SAAL is divided into two sublayers, the *Common Part AAL* (CP-AAL) and the *Service Specific Convergence Sublayer* (SSCS). The SSCS comprises an SSCOP entity and a *Service Specific Coordination Function* (SSCF). The objective of SSCF is to map the services provided by the SSCOP protocol to different AAL interfaces. SSCF definitions for *User Network Interface* (UNI) and *Network Node Interface* (NNI) can be found in the ITU-T Recommendations Q.2130 and Q.2140.

3.1 Objective of SSCOP

SSCOP is a connection oriented protocol. Its main purpose is to provide the service of a generic reliable data transfer. In order to implement a reliable data transfer by using the unreliable service of the underlying ATM layer *selective retransmission* is used. This means, all data packets get a sequence number to preserve *sequence integrity*. An SSCOP entity indicates the loss of data packets by sending an `USTAT` PDU. Additionally, SSCOP entities exchange `STAT` PDUs periodically. This is done for keeping track of lost data packets in the special case of lost `USTAT` PDUs. Further characteristics of SSCOP are:

Flow Control. An SSCOP receiver is able to control the rate at which the peer is allowed to send data packets (windowing).

Error Reporting to Layer Management. SSCOP informs the layer management about specific errors such as protocol errors, resynchronization of the connection, or lost data packets.

Keep Connection Alive. SSCOP maintains connections even over periods in which no data transfer is performed. By using a set of timers a connection is partitioned into a *connection control phase*, an *active phase*, a *transient phase*, and an *idle phase*. The status of a connection is communicated between protocol entities by using POLL and STAT PDUs.

Local Data Retrieval. The SSCOP user is able to retrieve data packets which have not yet been released by the transmitting entity. Different access schemes are provided (full, partial, or selective retrieval).

Protocol Error Detection and Recovery. During operation SSCOP detects errors and triggers a recovery mechanism by exchanging ER and ERAK PDUs with the peer entity.

Connection Control. Connection control is related to establishment, release, and resynchronization of an SSCOP connection. A timer is set to protect against PDU loss during the connection control phase.

3.2 SSCOP SDL specification

The SSCOP recommendation Q.2110 [ITU-TS, 1994] includes an SDL specification which has several informal parts. They refer to system parts and data structures which should not be standardised in Q.2110 or which are defined in another manner, e.g., default values of signal parameter are given in tables. In order to get an executable SDL description as SAMSTAG input all informal parts had to be formalised. In the following the main modifications are listed.

Default parameter and field values to AA-signals and PDUs. Default parameter and field values of SSCOP AA-signals and PDUs are provided in form of tables. These values have to be assigned explicitly before sending. In our SDL specification this is done by inserting extra tasks at appropriate places.

Additional queues and buffers. The SSCOP recommendation introduces additional queues and buffers for dealing with SD, MD and UD PDUs. This is done by means of tasks with informal text which refers to and manipulates these queues and buffers. The informal tasks and the corresponding data structures have been formalised. In a first attempt we implemented it by using SDL, but due to complexity, we changed the implementation language to C++. References and manipulations are made by using the /*#code ... */ construct as used in the SDT tool [TeleLogic AB, 1996].

Priority of internal and external signals. The SSCOP specification distinguishes between internal and external signals. Internal signals are sent and received within SSCOP, while external signals are received from the protocol environment. Internal signals are used to trigger the servicing of the queues and buffers. They are handled by the same message queue that handles external signals. The semantics of SDL would imply that internal and external signals have the same priority. But, in contradiction with the SDL semantics the textual SSCOP description prioritises external over internal signals. When confronted with this problem we decided to follow the SDL semantics and to give internal and external signals the same priority.

Modulo arithmetic. For some state variables which are used for storing counter values or sequence numbers the SSCOP specification introduces modulo arithmetics. We did not model these modulo arithmetics, because SAMSTAG always starts in the

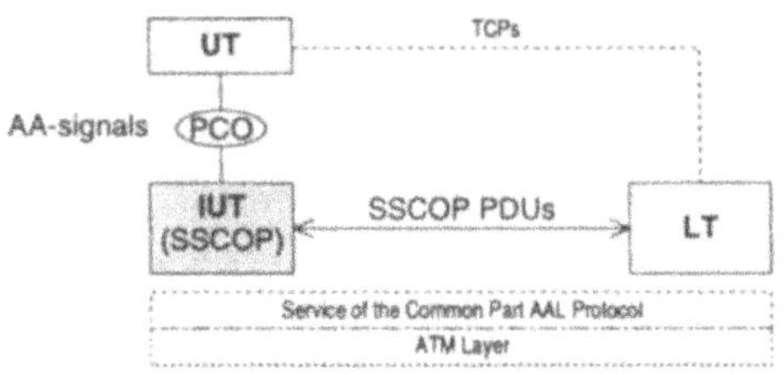

Figure 4 Test method.

initial state of SSCOP and we do not reach the upper bound of affected variables, i.e., modulo arithmetics will never be applied.

3.3 Further modifications

In order to reduce complexity for test generation we implemented some additional modifications and simplifications:

- PDU fields without importance to the function of SSCOP itself have been omitted. All PDUs with variable length have been restricted to a fixed length.
- The handling of PDUs for unassured and management data transfer has been omitted since these features go beyond the scope of conformance testing.
- We abstracted from the CPCS signals by using the PDUs carried by these signals instead.

4 PREPARATORY WORK

Before starting test generation some preparatory work has to be done: (1) a test method and (2) a coverage criterion have to be selected, (3) the structure of the test suite has to be defined, (4) test purposes have to be identified and (5) formalised by means of MSCs, (6) the SSCOP specification has to be adapted to the needs of SAMSTAG , and (7) the tester processes have to be specified. The items (1)-(4) are related to test methodology, the items (5)-(7) are related to SAMSTAG .

4.1 Selection of a test method

IS 9646 [ISO/IEC, 1994] recommends different test methods to be used for protocol conformance testing. These methods mainly differ in the interfaces between tester processes and IUT, and the possibilities to stimulate and observe the IUT during the test. During the definition of our test method we were guided by the distributed test method of IS 9646. Our test method is shown in Figure 4.

There are an upper and a lower interface to the IUT. The upper interface is a point of control and observation (PCO) which is connected to an upper tester (UT). The

UT exchanges AA-signals with the IUT. The lower interface is served by a lower tester (LT). In accordance to IS 9646 we abstracted from the underlaying service, thus the LT exchanges SSCOP PDUs with the IUT. Generally UT and LT coordinate themselves by using *test coordination procedures* (TCPs). We do not model TCPs, because during test case implementation they follow indirectly from the sequence of AA-signals and SSCOP PDUs to be send to and received from the IUT during the test run. Figure 4 does not exactly correspond to the distributed test method as defined in IS 9646. The PCO between IUT and UT is not standardised, i.e., it is not a service access point (SAP). Due to the non-existence of a standardised SAP between IUT and UT it may be more appropriate to use the *remote test method*.* In the remote test method there only exist one PCO at the LT interface. Nevertheless, some responses from the IUT have to be triggered by an upper layer user. In the test case descriptions these stimuli are indicated by using the TTCN construct *implicit send event*. Currently, the SAMSTAG tool is not able to deal with the remote test method. For this IUT events which are triggered by upper layers have to be identified and the corresponding *implicit send events* have to be generated.

4.2 Coverage

A test case checks a particular property of the specification. In order to give some confidence that an IUT conforms to its specification, a test suite should cover as much properties of the specification as possible. We based on the SSCOP SDL specification and looked at all state transitions. For each state transition there exists a number of transition control flow paths leading to a next state. They can be seen as properties or test purposes to be tested. Our intention was to generate a test suite that covers all transition paths.

When starting to implement this coverage criterion for SSCOP we discovered two problems: (1) loops which may lead to an infinite number of transition paths with various lengths and (2) the complex state `Data_Transfer_Ready` which due to loops and a cascade of decisions is a starting point for several hundreds of transition paths.

We tackled (1) by setting the maximum number of loop executions during a test run to 1. The problem of (2) was a little bit more complicated. In order to avoid the combination of different decisions we introduced some internal states before decisions and treated them like SDL states, i.e., we split the state transition graph into smaller and less complex pieces.

4.3 The test suite structure

The structure of the SSCOP test suite is shown in Figure 5. It is a tree structure and reflects the SSCOP functionality. The root of the tree represents the whole test suite.

* A detailed discussion on appropriate test methods for ATM AAL conformance testing can be found in [Yoo et al., 1996].

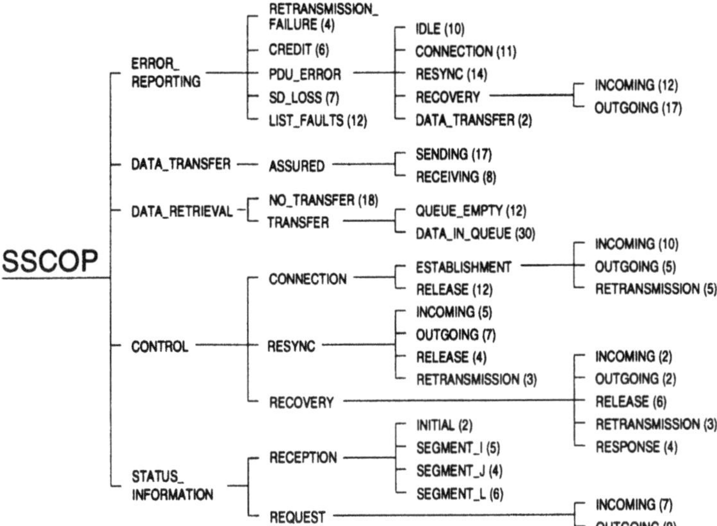

Figure 5 Test suite structure.

Nodes and leafs represent test groups and refer to functions or aspects of SSCOP functions. The test cases in one group should focus on a specific aspect to be tested. The numbers in round brackets following the leaves denote the number of test cases attached to this leaf. The test case SSCOP_18b (Figure 2), for example, is member of the test group called `CONTROL/RESYNC/RELEASE`. The test cases in this group focus on testing the abort of the resynchronization process leading to connection release.

4.4 Identification and specification of test purposes

The identification and specification of the test purposes for the SSCOP test suite follow directly from the coverage criterion (Section 4.2). For each transition path a test purpose is specified. This is done in two steps. In a first step for each test purpose an informal description is produced. In a second step the informal test purposes are formalised by means of MSC diagrams.

An example for an informal description produced for a transition path is shown in Table 1. The informal description is very close to the SDL specification. But, it is aimed to clarify the purpose of a test case and not to specify the entire system behaviour. In case of restrictions imposed by lack of time and money they may be used for the selection of the most important test cases. The formalisation of the test purpose in Table 1 is provided by the MSC in Figure 1.

We identified and specified 281 test purposes. They were assigned to the leaves of the test suite structure according to their functional aspects they focus on. Additionally, in order to do statistical analysis, we arranged the test purposes in 5 groups. This is shown in Table 2. All test purposes of a particular group start in specific states.

Table 1 Informal test purpose description.

Identifier:	SSCOP_18b
Description:	If SSCOP is in state *Outgoing_Resynchronization_Pending* and gets an *AA_RELEASE_request* signal from the SSCOP user, then SSCOP should cancel *Timer_CC*, send an *END* PDU to its peer entity, set *Timer_CC* again, and change into the new state *Outgoing_Disconnection_Pending*.

Table 2 Groups of test purposes.

Group name	*Abbrev.*	*Starting states*	*Number*	*%*
Idle	Idle	Idle	24	9%
Connection Control	ConCo	Incoming_Connection_Pending, Outgoing_Connection_Pending, Outgoing_Disconnection_Pending	51	18%
Resynchronisation	Resyn	Incoming_Resynchronization_Pending, Outgoing_Resynchronization_Pending	38	14%
Recovery	Recov	Incoming_Recovery_Pending, Outgoing_Recovery_Pending, Recovery_Response_Pending	75	27%
Data Transfer	DaTra	Data_Transfer_Ready	93	33%
		total number of test purposes:	281	100 %

4.5 Different models for tester processes

As described in Section 2.1, SAMSTAG needs a closed SDL system as its input. This implies that the tester processes have to be specified as SDL processes.

We started to experiment with general tester processes which are able to send and to receive all allowed signals at any time. But, due to complexity caused by the tester processes we failed even to generate simple test cases. As a result of this experiment we started to use *optimisation strategies* (Section 2.3). This means, we implemented specialised tester processes which stimulate the IUT by using one or more of the following strategies:

- use of special signals to trigger protocol errors;
- use of special sequences of signals as preamble or postamble of the test case in order to reach a particular state quickly;
- focus on a particular signal exchange during which the values of signal parameters are varied;
- specialisation on the role of sender or receiver of data packets.

As a result we gained eight different SSCOP versions which all share the same IUT, i.e., SSCOP process, but differ in the tester processes. Each version focuses on test

Table 3 Number of test purposes covered per version.

			TP Group			*total*	
Version	Idle	ConCo	Resyn	Recov	DaTra	absolute	percental
IDLE	14	31	3	10	0	58	21%
DATA_1	0	0	26	38	16	80	28%
DATA_2	0	0	2	17	4	23	8%
DATA_3	0	0	0	0	3	3	1%
RETRIEVE	10	20	7	10	0	47	17%
RECEIVE	0	0	0	0	17	17	6%
SEND	0	0	0	0	15	15	5%
STAT	0	0	0	0	38	38	14%
total	24	51	38	75	93	281	100%

case generation for one or more selected groups of test purposes. In the following we describe the different SSCOP versions. The Table 3 relates the SSCOP versions to the different groups of test purposes (Section 4.4).

- *IDLE:* This version concentrates on the signal exchange starting in state IDLE and on states dealing with connection control (connecting/disconnecting).
- *DATA_1:* The objective of this version is to cover states dealing with synchronization and recovery of protocol errors.
- *DATA_2 and DATA_3:* These two versions are specialisations of DATA_1. Their objective is to catch the rest of the test purposes that are not directly related to the sending or reception of data packets or the reception of `STAT` PDUs.
- *RETRIEVE:* An important part of the test purposes deal with data retrieval. The data retrieval feature allows the local SSCOP user to retrieve in-sequence data packets which have not yet been released by the SSCOP entity. This is possible in 6 out of the 10 states and requires a preceding phase of data transmission.
- *SEND and RECEIVE:* These two versions concentrate on the sending or reception of data packets.
- *STAT:* This version is very similar to SEND and RECEIVE, but its emphasis is on the transmission of `STAT` PDUs to the IUT. The actual parameter values of the `STAT` PDU are varied.

5 TEST CASE GENERATION

We generated the test suite by using Sun SparcStation 20 and Sun Ultra 2 computers. In this section we describe the result of the test generation procedure and discuss the cases where we failed.

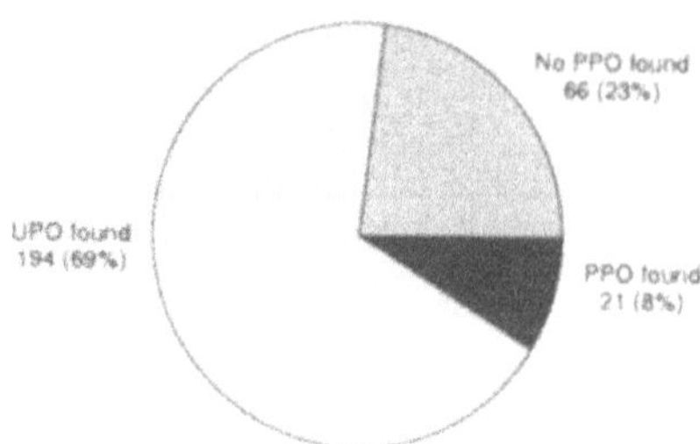

Figure 6 Overall result of the test generation.

5.1 Overall view

For test case generation, SAMSTAG was applied to the different SSCOP models and the 281 test purposes (= 100%). As shown in Figure 6, SAMSTAG generated 194 (= 69%) verified TTCN test cases. For 21 test purposes (= 8%) we found PPOs, but failed to verify their uniqueness (cf. Step 3 in Section 2.2). For 66 test purposes (= 23%) we even did not find PPOs.

In 40% of the cases SAMSTAG generated the verified test cases within 10 minutes, in 53% of the cases it took between 10 minutes and 1 hour. For 7% of the the verified test cases the generation took more than 1 hour.

During the generation process we observed that 3 seconds was the smallest time period to generate a test case and 378 hours was the longest one. For generating the latter one 1 700 560 000 global system states were examined. The longest try where we failed to generate a test case took 837 hours on a Sun Ultra Sparc 2 Workstation. During this attempt more than 2 600 000 000 global system states were investigated.

Our next step was to look at the test generation results for the 5 groups of test purposes. This is done in Figure 7. The groups Idle, ConCo, Resyn, and Recov show approximately the same result. SAMSTAG was able to find UPOs and PPOs for about 80%. The set of PPOs which cannot be verified is relatively small. But, for test purposes related to the DaTra group the result is not that good. The set of found UPOs comprises 46%, the set of PPOs which cannot be verified comprises 17%, and the set of test purposes for which we neither find PPOs nor UPOs comprises 37%. By looking at the SSCOP protocol we see the reason for this result. All test purposes in the DaTra group start in the state `Data_Transfer_Ready` which also is the most complex one. Test generation for test purposes starting in this state is also the most complex part of the entire procedure.

5.2 Failure cases

We identified four reasons for failing to generate test cases:

1. *SSCOP characteristics.* With SSCOP characteristics we refer to the different handling of internal and external signals by the SSCOP protocol. As explained in Section 3.2 the SSCOP standard states that internal and external signals should

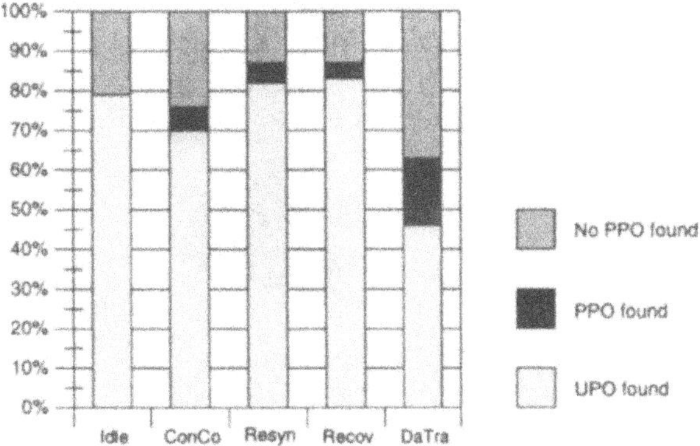

Figure 7 Test generation results related to test purpose groups.

be handled by the same signal queue, but, internal signals should have a lower priority than external ones. This is in contradiction to the SDL semantics. In our SSCOP specification we followed the SDL semantics by assigning the same priority to all signals. As a result we failed to generate test cases for purposes which are somehow related to the different priorities.

2. SAMSTAG *limitations.* The SAMSTAG method is more general than its current implementation. The SAMSTAG tool is a prototype only and includes some restrictions and limitations. In cases where we failed due to SAMSTAG limitations, future SAMSTAG versions may be able to generate test cases.
3. *Complexity.* Due to state space explosion we did not find a trace of the SDL specification which ensures the fulfilment of a given test purpose.
4. *Tester models.* Due to complexity, test case generation using the most general tester process model failed. Furthermore, none of the other models developed turned out to be appropriate to handle these test purposes.

The Figure 8 shows how these reasons are distributed over the failure cases. At least the items 1 and 2, which are responsible for 51 failure cases (= 59%), provide possibilities to improve the result of the test generation process. For item 1 we started discussions with specialists in ITU-T.* In order to align the SSCOP specification to the SDL semantics, the discussions may lead to changes of the SSCOP SDL specification. Failures because of complexity (26 cases, 30%) and inappropriate tester models (10 cases, 11%) need further investigations. At present we cannot say which measures help to avoid these failure cases. Possibly the implementation of further heuristics and optimisation strategies will help.

As shown in Figure 9, the distribution of failures has also been related to the different groups of test purposes. Again the results for the groups Idle, ConCo, Resyn and Recov are comparable. For the test purposes in these groups SAMSTAG mainly fails due to SSCOP characteristics. Complexity and SAMSTAG limitations are only of importance. For the DaTra group we achieved a different result. SAM-

*Within ITU-T, the study group 11 is responsible for the SSCOP recommendation.

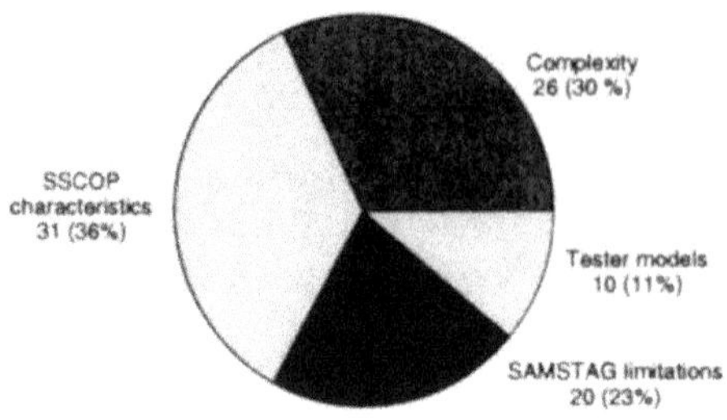

Figure 8 Reasons for failed UPO production.

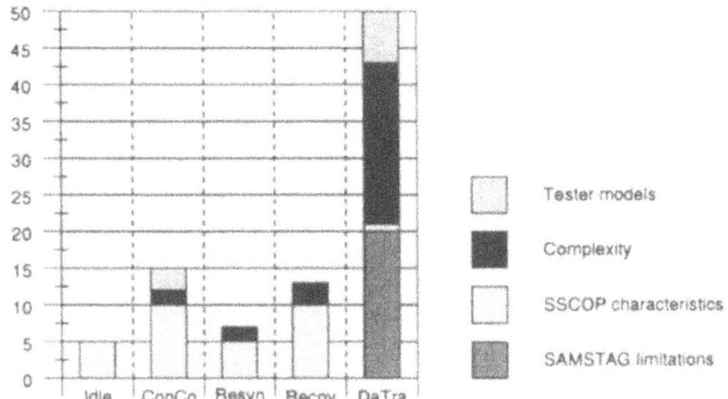

Figure 9 Failure reasons and test purpose groups.

STAG mainly fails due to complexity and SAMSTAG limitations. Our interpretation is, that this result again reflects the difficulty of generating test cases for the SSCOP state `Data_Transfer_Ready`. We believe that lots of failures due to SSCOP characteristics are hidden in the other failure cases.

6 EXPENSES OF TEST SUITE DEVELOPMENT

The goal of SAMSTAG is to improve the conformance testing process in a twofold manner. On the one hand it should save time and money expenses, and on the other hand the application of SAMSTAG should ensure the consistence between specification and test cases. It is obvious that the latter goal has been achieved. For judging time and money savings a comparison with the expenses for the manual development of such a test suite is required.

For SSCOP such a manually specified test suite [ATM Forum, 1996] exists. The test suite has been developed by the ATM Forum in parallel to our work, but without our knowledge. The main differences to the SAMSTAG test suite are that the ATM Forum test suite is based on the *remote test method* (Section 4.1) and that it has a state oriented structure (Section 4.3). The test purposes identified for both test suites are comparable [Grabowski et al., 1997].

Since there was no data on the ATM Forum test suite available at the time of

Table 4 Development expenses.

Phase	*Subphase*	*Expenses*	*to be performed*
Completion of SDL specification		1 month	manually
Preparatory work	Specification of test method and test suite structure	1 month	manually
	Identification and Specification of test purposes	2 months	manually
	Specification of different tester models	2 months	manually
Test suite generation		1 month	automatically

this writing, we are just able to present our expenses. This is done in Table 4. In the table the development process is structured into the main phases *Completion of SDL specification*, *Preparatory work* and *Test suite generation*. The Preparatory work phase is divided into the subphases which have been described in Section 4. The expenses for test case generation not only include the mere generation time, but also the work of relating test purposes and test models and the experimentation on the application of different SAMSTAG heuristics.

The expenses for test case generation not only include the mere generation time, but also the work of relating test purposes and test models and the experimentation on the application of different SAMSTAG heuristics [Grabowski et al., 1996]. In total the expenses for our case study comprises 7 months. We believe that this is a very good result and that due to increasing experience the expenses for the next protocol will decrease. It should also be noted that SAMSTAG generates test cases for only 70 % of the identified test purposes. We did not estimate the expenses for the manual completion of the test suite. Furthermore we did not estimate the expenses for getting familiar with the SSCOP specification and the SAMSTAG method.

7 SUMMARY AND OUTLOOK

In this paper the application of the SAMSTAG tool to the B-ISDN protocol SSCOP has been described. This case study has shown that automatic test generation based on SDL system specifications and MSC test purposes is feasible. Complete TTCN test cases for 68% of the specified test purposes have been generated automatically. The reasons for cases where SAMSTAG fails to generate test cases have been presented and discussed. For this case study, complexity, SAMSTAG limitations and SSCOP characteristics which violate the SDL semantics are the main reasons for failure. The result of the test generation process may be improved by adding functionality to SAMSTAG and by modifications of the SSCOP SDL specification. Our future work will focus on these aspects and on the extension of SAMSTAG in order to cope with the remote test method as well.

Acknowledgements

The work presented in this paper has been supported partially by the F & E project No. 319, funded by Swiss PTT, and the SPP IF project No. 5003-038997, funded by the Swiss National Science Foundation.

The authors would like to thank all people involved in the development of SAMSTAG. We are also grateful to S. Heymer and the anonymous reviewers providing detailed comments and valuable suggestions which have improved contents and presentation of this paper.

8 REFERENCES

ATM Forum [1996]. Conformance Abstract Test Suite for the SSCOP for UNI 3.1, Technical Committee ATM Forum. af-test-0067.000.

Doldi, L., Encontre, V., Fernandez, J.-C., Jéron, T., Le Briquir, S., Texier, N. and Phalippou, M. [1996]. Assesment of Automatic Generation of Conformance Test Suites in an Industrial Context, *in* B. Baumgarten, H.-J. Burkhardt and A. Giessler (eds), *Testing of Communicating Systems*, Chapman & Hall.

Grabowski, J. [1994]. *Test Case Generation and Test Case Specification with Message Sequence Charts*, PhD thesis, University of Berne, Institute for Informatics and Applied Mathematics.

Grabowski, J., Hogrefe, D., Nussbaumer, I. and Spichiger, A. [1995]. Test Case Specification Based on MSCs and ASN.1, *in* A. S. R. Braek (ed.), *SDL'95 - with MSC in CASE*, Proceedings of 7. SDL Forum, North-Holland.

Grabowski, J., Scheurer, R. and Hogrefe, D. [1997]. Comparison of an Automatically Generated and a Manually Specified Abstract Test Suite for the B-ISDN Protocol SSCOP, *Technical Report A-97-07*, University of Lübeck, Germany.

Grabowski, J., Scheurer, R., Toggweiler, D. and Hogrefe, D. [1996]. Dealing with the Complexity of State Space Exploration Algorithms, *Proceedings of the 6th. GI/ITG technical meeting on 'Formal Description Techniques for Distributed Systems', University of Erlangen.*

Hopcroft, J. E. and Ullmann, J. D. [1979]. *Introduction to Automata Theory, Languages, and Computation*, Addison-Wesley Publishing Company.

ISO/IEC [1994]. Information Technology - Open Systems Interconnection - Conformance Testing Methodology and Framework, *International multipart standard 9646*, ISO/IEC.

ITU-TS [1994]. ITU-T Recommendation Q.2110: B-ISDN ATM Adaption Layer - Service Specific Connection Oriented Protocol (SSCOP), ITU Telecommunication Standards Sector, Geneva.

ITU-TS [1996a]. ITU-T Recommendation Z.100: Specification and Description Language (SDL, ITU Telecommunication Standards Sector SG 10, Geneva.

ITU-TS [1996b]. ITU-T Recommendation Z.120: Message Sequence Chart (MSC), ITU Telecommunication Standards Sector SG 10, Geneva.

Nahm, R. [1994]. *Conformance Testing Based on Formal Description Techniques and Message Sequence Charts*, PhD thesis, University of Berne, Institute for Informatics and Applied Mathematics.

TeleLogic AB [1996]. *SDT / ITEX 3.0*, TeleLogic AB, Malmö, Sweden.

Toggweiler, D., Grabowski, J. and Hogrefe, D. [1995]. Partial Order Simulation of SDL Specifications, *in* O. Bræk and A. Sarma (eds), *SDL'95 with MSC in CASE*, North-Holland.

Yoo, S., Collica, L. and Jeong, T. [1996]. Conformance testing of ATM Adaption Layer protocol, *in* B. Baumgarten, H.-J. Burkhardt and A. Giessler (eds), *Testing of Communicating Systems*, Chapman & Hall.

9 BIOGRAPHY

Jens Grabowski studied Computer Science at the University of Hamburg, Germany, where he graduated with a diploma degree in 1990. From 1990 to 1995 he was research scientist at the University of Berne, Switzerland, where he received his Ph.D. degree in 1994. Since October 1995 Jens Grabowski is researcher and lecturer at the Institute for Telematics at the University of Lübeck, Germany. His research activities are directed towards network analysis, formal methods in protocol specification and conformance testing.

Rudolf Scheurer received his Diploma degree in informatics from the University of Berne, Switzerland, in 1994. Since his graduation he has been working as a research assistant at the University of Berne and the University of Lübeck, Germany. His research work is focusing on formal methods in protocol specification and conformance testing.

Zhen Ru Dai is an undergraduate at the Institute for Telematics of the University of Lübeck, Germany. She is working as deputy assistant since 1995. From 1995 to 1997 she was involved in the SSCOP case study.

Dieter Hogrefe is a full professor and director of the Institute for Telematics at the University of Lübeck, Germany. His research activities are directed towards formal description techniques and analysis of specifications. He studied Computer Science and Mathematics at the University of Hannover where he graduated in 1983. After a number of years at Siemens Research and the University of Hamburg he was a full professor at the University of Berne and head of the computer networks and distributed systems group from 1989 to 1995. He is an active member of the ITU-T Study Group 10, in which he is the rapporteur for Question 8/10 on verification and testing based on formal methods and is the chairman of ETSI TC MTS.

Design of Protocol Monitor Emulating Behaviors of TCP/IP Protocols

Toshihiko Kato, Tomohiko Ogishi, Akira Idoue and Kenji Suzuki
KDD R&D Laboratories
2-1-15, Ohara, Kamifukuoka-shi, Saitama 356, Japan
E-mail : {kato, ogishi, idoue, suzuki}@hsc.lab.kdd.co.jp

Abstract

Recently, the TCP/IP protocols are widely used and it is mentioned that, in some cases, throughput is limited due to problems such as network congestion. To solve such problems, the details of communication need to be examined. In order to support these examinations, we are developing an 'intelligent' protocol monitor which can estimate what communication has taken place by emulating the behaviors of the TCP protocol entities in a pair of communicating computers. This paper describes the overview of the monitor and the detailed design of the TCP behavior emulation function for both the state transition based behaviors and the internal procedures for the flow control, such as the slow start algorithm.

Keywords

Interoperability testing, protocol monitor, TCP/IP, TCP behavior emulation

1 INTRODUCTION

Recently, the TCP/IP protocols[1] are widely used in various computer

Testing of Communicating Systems, M. Kim, S. Kang & K. Hong (Eds)
Published by Chapman & Hall

communications. Here, most users of computers use communication functions installed in operating systems or commercial software products as they are, and do not pay attentions to their details. However, some problems occur for TCP/IP communications, in the cases that there are some packet losses due to network congestion and transmission errors, and that the protocol parameters of communicating computers are not matched. Especially, for TCP (Transmission Control Protocol), it is mentioned that the performance may be degraded due to the flow control mechanisms[1], and due to the use of send and receive socket buffers with different sizes[2]. When these problems occur, the details of communications need to be examined to detect the problem sources. For this purpose, it is common to use commercial protocol monitors[3]. However, these monitors have only the functions to capture PDUs (Protocol Data Units) transmitted over networks, to analyze their formats and parameter values, and to display the results. The analysis of packet sequences and the investigation of problem sources need to be performed manually by TCP experts.

In order to support of the analysis of details of computer communications, we have proposed an 'intelligent' protocol monitor which analyze protocol behaviors and detect protocol errors in communicating computers[4]. This monitor captures PDUs over networks, emulates protocol behaviors of communicating computers which send and receive the PDUs, and finds protocols errors if PDU formats and behaviors do not conform to the protocols. We have implemented the intelligent protocol monitor for OSI protocols[4].

By applying the technologies of the intelligent OSI protocol monitor to the TCP/IP protocols, we are currently developing a protocol monitor which emulates the behaviors of TCP/IP protocols. This monitor provides both the PDU monitoring function similar to conventional protocol monitors and the function which emulates the behaviors of a pair of communicating computers according to TCP. One of the largest differences between the TCP/IP protocols and the OSI protocols is that the modern TCP contains some internal procedures which are treated as "local matter" in the OSI protocols. These procedures include the slow start and congestion avoidance algorithms by which a sender controls the rate of injecting data segments into a network. As the results, the protocol emulation becomes much more complicated for TCP than for the corresponding protocol in the OSI protocol stack, i.e. OSI Transport Protocol class 4.

This paper describes the design of our protocol monitor which emulates TCP/IP protocol behaviors. The next section and section 3 describes the requirements and the overview of our TCP/IP protocol monitor, respectively. Section 4 describes the detailed design of the emulation function of TCP protocol. Section 5 gives some discusses on our monitor and section 6 makes some conclusions.

2 REQUIREMENTS FOR TCP/IP PROTOCOL MONITOR

We suppose the following requirements for our TCP/IP protocol monitor.

(1) Since the TCP/IP protocols are used over high speed networks such as LANs, it is required that all PDUs transmitted over the network can be captured in an on-line operation, and that the monitoring results are examined by an operator in an off-line operation.

(2) As described above, the monitored TCP/IP communication needs to be examined in the following two points of view.

- Examining what PDUs are transmitted over the network which the monitor is attached to. This examination handles all of the PDUs captured by the monitor.
- Examining what communication has taken place between a specific pair of computers. Among the TCP/IP protocol suits, TCP has the most complicated protocol behaviors, and therefore this examination mainly focuses on TCP.

(3) In the latter case of (2), the behaviors of a computer for the captured TCP segment sequences are examined. In this examination, the state of TCP protocol entity in the computer is identified and its behaviors are emulated according to the TCP specification. We call this examination the **TCP behavior emulation**.

(4) As described above, modern TCP includes some internal procedures which are not specified in the state transition of the original TCP[5]. Furthermore, these procedures may not be implemented for some TCP/IP software products. The TCP behavior emulation needs to support these internal procedures and take account of the possibility that they are not used in the computers being examined.

(5) In some network configurations, LANs are interconnected via a WAN (wide area network) such as ISDN. In this case, there may be some time difference between the sending of PDUs by computers in the remote LANs and the capturing of them by the monitor. The TCP behavior emulation needs to take account of these time differences.

3 OVERVIEW OF TCP/IP PROTOCOL MONITOR

In order to satisfy the requirements described above, we have designed the following functions in our TCP/IP protocol monitor.

(1) As depicted in Fig. 1, the TCP/IP protocol monitor is attached to a LAN and observes PDUs over the LAN. According to the requirements in Section 2, the monitor provides both the **PDU monitoring function,** which analyzes all TCP/IP PDUs transmitted over the LAN, and the **TCP behavior emulation function** focusing on the TCP protocol behavior of a specific pair of computers.

(2) The PDU monitoring function is similar to the function of conventional protocol monitors. It captures PDUs transmitted over the LAN, and analyzes PDU formats and parameter values according to TCP/IP protocols including

ARP (Address Resolution Protocol), IP, ICMP (Internet Control Message Protocol), UDP (User Datagram Protocol) and TCP.

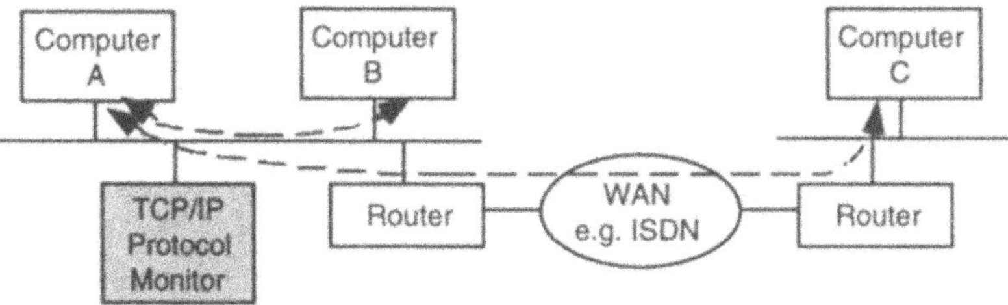

Figure 1 Example of network configuration with TCP/IP protocol monitor.

(3) The TCP behavior emulation function is realized by the following two steps. First, the event sequence is estimated for an individual computer by taking account of the time differences between the PDU capturing by the monitor and the PDU handling in the computers being examined. These time differences may be negligible for the case that the computers are located in the same LAN (computers A and B in Fig. 1), but not negligible for the case that they are located in the remote LANs (computers A and C in Fig. 1).
Next, the behavior of the TCP protocol entity of an individual computer is emulated according to the estimated event sequence.

(4) The monitor function is implemented as software running in UNIX workstations. Figure 2 depicts the software structure of the monitor.

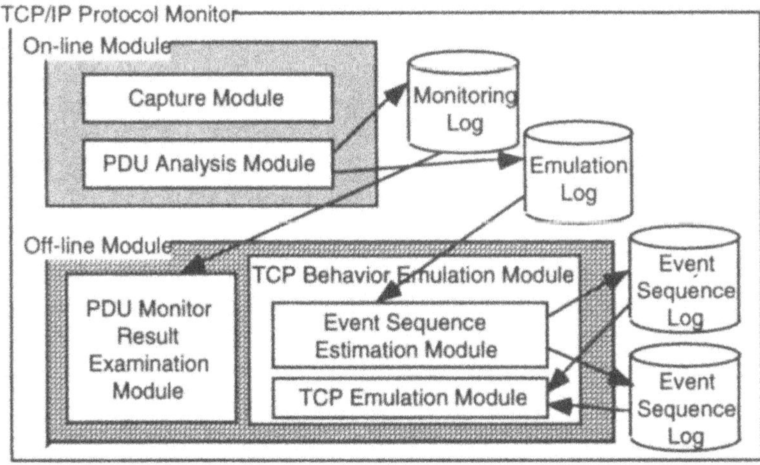

Figure 2 Software structure of monitor function.

It consists of an on-line and off-line module. The on-line module includes the **capture module** which captures PDUs transmitted over the LAN and the **PDU analysis module** which analyzes their format and parameter values according to the TCP/IP protocols. The PDU analysis module outputs the analysis results of the captured PDUs to the display of the workstation. This output is also saved in the **monitoring log** for the purpose of the off-line examination. It also saves the information used by the TCP behavior emulation in the **emulation log**.

(5) The off-line module consists of the **PDU monitor result examination module** and the **TCP behavior emulation module**. The PDU monitor result examination module allows an operator to examine the monitoring log by the help of editor functions such as cursor move and string search. The TCP behavior emulation module includes the **event sequence estimation module** and the **TCP emulation module**, which generate the event sequence for an individual computer and emulate TCP according to the estimated event sequence, respectively.

(6) The event sequence contains TCP protocol events, each of which is a "sent TCP segment" or a "received TCP segment", together with the estimated time of the event. The TCP emulation module maintains the state transition specification for TCP and processes each event according to the following procedure.

- When the event is a received TCP segment, it emulates the TCP behavior when the TCP protocol entity receives the segment. It looks up the corresponding state transitions and performs it. If it sends out a segment, the module checks a sent TCP segment in the event sequence and emulates the received and sent segments.
- When the event is a sent TCP segment, the protocol emulation module searches for the input which generates the segment, and it considers that the input is applied to the TCP protocol entity. If there are no input to generate the segment, it decides that the TCP protocol entity has some protocol errors.

4 DETAILED DESIGN OF TCP BEHAVIOR EMULATION

As described in the previous section, the TCP behavior emulation is realized by the emulation log generation by the on-line module, and the event sequence estimation and the TCP emulation by the off-line module. This section describes the details of these procedures.

4.1 Generation of Emulation Log

The PDU analysis module in the on-line module saves in the emulation log a record containing the following information for every captured TCP segment. This information is necessary to emulate the behavior the TCP protocol entity in computers.

- The time when the beginning of a TCP segment is detected and the time when the end of a TCP segment is detected.
- The source and destination IP addresses.
- The parameters in TCP header except TCP checksum.
- The length of TCP segment including TCP header and TCP data.
- Whether TCP checksum is correct or not.

As for the time described above, the monitor software detects a PDU when the monitor has captured the whole data of the PDU. That is, the monitor software knows only the time when the end of a PDU is detected. The time for the beginning of a PDU is calculated from the length of the PDU and the transmission speed of the network.

In saving the above information, the procedure of IP, especially the reassembling of the fragmented IP datagrams is performed in the PDU analysis module. If a TCP segment is fragmented by IP, the following procedures are used.

- The length of TCP segment and whether TCP checksum is corrected or not are calculated after the reassembling.
- The time for the beginning of the TCP segment corresponds to that for the beginning of the first IP datagram containing the TCP segment. The time for the end of the TCP segment corresponds to that for the end of the last IP datagram containing the TCP segment.

4.2 Estimation of Event Sequence

The TCP behavior emulation module is invoked with a pair of IP addresses, which indicate the computers focused on. In the beginning of the TCP behavior emulation, the event sequence estimation module prepares an event sequence log for each computer from the emulation log. This is performed in the following way for computers A and B.

(1) The module selects a record corresponding to a TCP segment without TCP checksum error exchanged between the specified computers.

(2) When a TCP segment is transferred from computer A to computer B, the module considers that an event of sent TCP segment takes place in computer A and an event of received TCP segment in B.

(3) The module estimates the processing time of the event in computers A and B. For a sent TCP segment, the processing time is estimated as the time for the beginning of TCP segment minus the transmission delay between the monitor and the computer. For a received TCP segment, it is estimated as the time for the end of TCP segment plus the transmission delay. The transmission delay is estimated by the propagation delay and the transmission time for a segment.

(4) The module reorders the records according to the estimated processing time. This reordering is performed independently for A and B.

(5) By applying the procedures (1) through (4) to all records saved in the emulation log, the event sequence logs for A and B are generated.

Figure 3 shows an example of the event sequence estimation for two computers, A and B, attached to LANs interconnected through ISDN with 64 Kbps transmission speed and 100 msec propagation delay. The TCP/IP protocol monitor captures some segments between computers A and B, and generates the emulation log in the figure. As for computer A, the delay with the monitor is negligible, the processing time in the event sequence log for A is the same as either the beginning or end time in the emulation log.

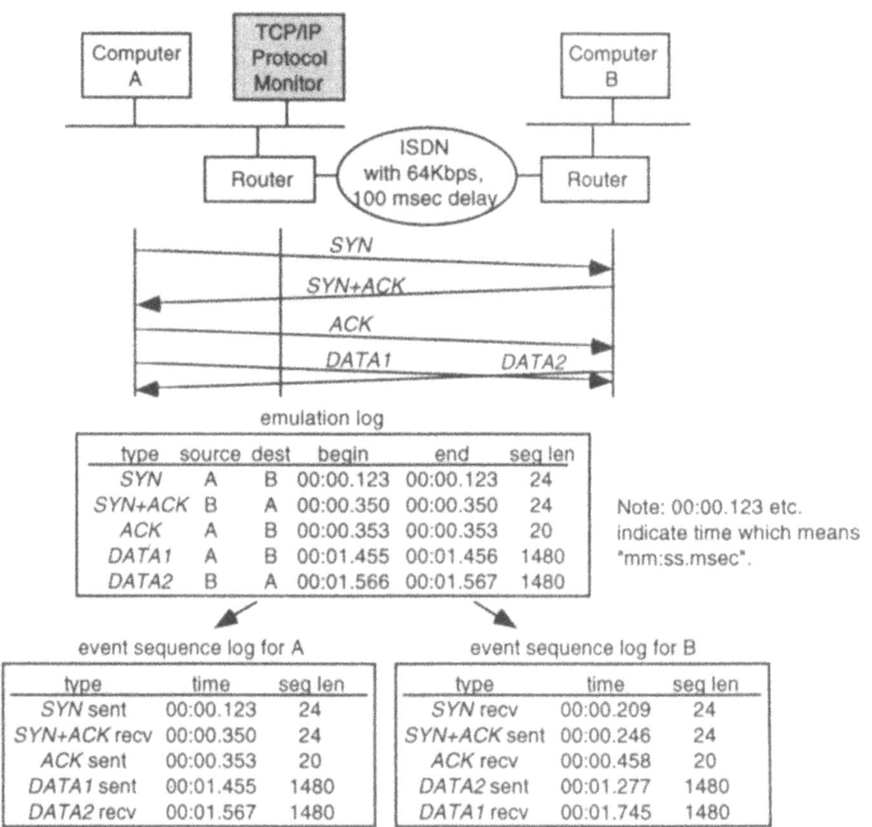

emulation log

type	source	dest	begin	end	seg len
SYN	A	B	00:00.123	00:00.123	24
SYN+ACK	B	A	00:00.350	00:00.350	24
ACK	A	B	00:00.353	00:00.353	20
DATA1	A	B	00:01.455	00:01.456	1480
DATA2	B	A	00:01.566	00:01.567	1480

event sequence log for A

type	time	seg len
SYN sent	00:00.123	24
SYN+ACK recv	00:00.350	24
ACK sent	00:00.353	20
DATA1 sent	00:01.455	1480
DATA2 recv	00:01.567	1480

event sequence log for B

type	time	seg len
SYN recv	00:00.209	24
SYN+ACK sent	00:00.246	24
ACK recv	00:00.458	20
DATA2 sent	00:01.277	1480
DATA1 recv	00:01.745	1480

Figure 3 Example of event sequence estimation.

On the other hand, the processing time in the event sequence log for B is estimated from the propagation delay, 100 msec, and the transmission time of individual segments. For example, the processing time of *DATA2* in computer B is estimated in the following way.

- The length of *DATA2* is 1500 byte including IP header, and it takes 1500 * 8 / 64000 = 188 msec to transmit *DATA2* through ISDN. It takes 1 msec to transmit it through the remote Ethernet.
- Therefore, the estimated processing time of *DATA2* in computer B is given by the equation
  00:01.566 - 0.100 (propagation delay) - 0.188 - 0.001 = 00:01.277.

4.3 Details of TCP Emulation

By use of the event sequence logs, the TCP emulation module emulates the behavior of the TCP protocol entity of each computer. This module maintains the protocol behavior of TCP and traces how the entity behaves on an event by event basis. The behaviors are categorized into state transition based behavior and internal procedures of modern TCP. The rest of this section describes how these two kinds of behaviors are emulated by the TCP emulation module.

Specification of State Transitions

The TCP emulation module maintains the state and the internal variables to specify the state transition based behaviors for each TCP connection. The state takes the following values:

CLOSED, SYN_SENT, SYN_RCVD, ESTABLISHED,
FIN_WAIT_1 (state after the first *FIN* is sent),
FIN_WAIT_2 (state waiting for the second *FIN*),
CLOSING (state in simultaneous close),
CLOSE_WAIT (state waiting for *close* from the application after receiving *FIN*), and
LAST_ACK (state waiting for *ACK* for *FIN*).

The internal variables maintained in the TCP emulation module include the followings:

send sequence variables such as
SND.NXT : the send sequence number to be sent next,
SND.UNA : the least send sequence number which is unacknowledged,
SND.WND : the maximum send sequence number which can be sent by the advertised window,
ISS : the initial send sequence number, and
MSS : the maximum segment size, and

receive sequence variables such as
RCV.NXT : the receive sequence number to be received next,
RCV.WND : the maximum receive sequence number which can be received by the advertised window, and
IRS : the initial receive sequence number.

Figure 4 shows the state transitions of TCP. A state transition for one input and for one state is associated with one or more possibilities, each of which is specified by the condition, the output and the requirements for its parameters, the next state, and the variable update. The followings specify details for some state transitions in the figure.

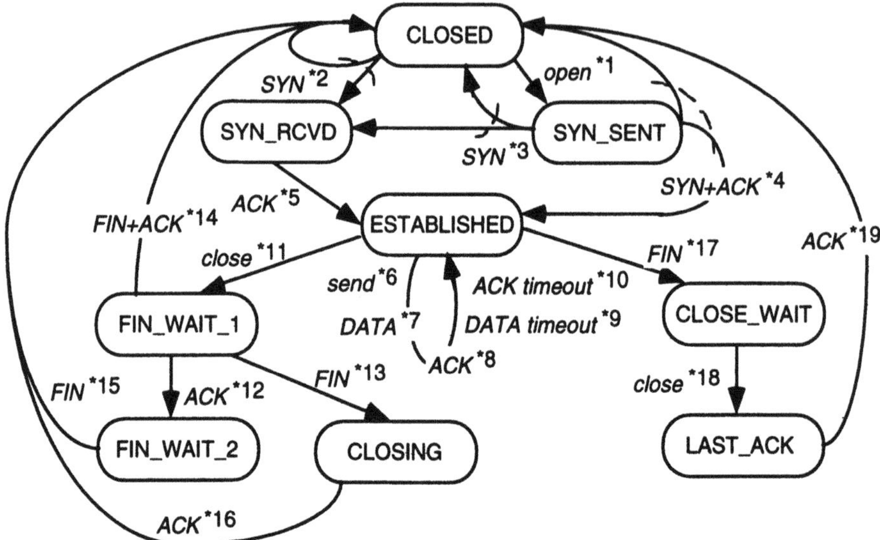

Figure 4 TCP state transition.

State Transition 1 : *open* in CLOSED
output : *SYN*; next state: SYN_SENT;
variable update:
ISS = sequence number (SEQ) in *SYN*; SND.UNA = ISS;
SND.NXT = ISS+1; SND.WND = window size (WND) in *SYN*;
MSS = maximum window size (MSS) in *SYN*;

State Transition 2 : *SYN* in CLOSED
1) output: *SYN+ACK* with acknowledgment number (ACK)
= SEQ in *SYN*+1; next state: SYN_RCVD;
variable update:
ISS = SEQ in *SYN*; IRS = SEQ in *SYN+ACK*; SND.UNA = ISS;
SND.NXT = ISS+1; SND.WND = WND in *SYN*; RCV.NXT = IRS+1;
RCV.WND = WND in *SYN+ACK*;
MSS = min (MSS in *SYN*, MSS in *SYN+ACK*); or
2) output: *RST+ACK*
with SEQ = 0 and ACK = SEQ in *SYN* + TCP data length (LEN) of *SYN*;
next state: CLOSED;

State Transition 4 : *SYN+ACK* in SYN_SENT
1) output: *ACK* with SEQ = SND.NXT and ACK = SEQ in *SYN+ACK* +1;
next state: ESTABLISHED;
variable update:
IRS = SEQ in *SYN+ACK*; SND.UNA = ACK in *SYN+ACK*;
SND.WND = WND in *SYN+ACK*; RCV.NXT = IRS+1;
RCV.WND = WND in *ACK*;
MSS = min (current MSS, MSS in *SYN+ACK*); or
2) output: *RST* with SEQ = ACK in *SYN+ACK*; next state: CLOSED;

Emulation Based on State Transition Specification

By use of the state transitions defined in the previous section, the TCP emulation module traces the behaviors of TCP protocol entity. The algorithm is depicted in Fig. 5, and can be summarized as follows.

(1) The events saved in the event sequence log are traced one by one.

(2) If an event is a sent TCP segment (we call a **sent event**), the TCP emulation module searches for a transition for the current state which sends out the TCP segment. This is performed by looking up all of the transitions for the current state.
If such a transition exist, the TCP emulation module emulates the transition, including changing the state to the next state and updating the internal variables.
If such a transition does not exist, the TCP emulation module considers that there may be some protocol errors.

(3) If an event is a received TCP segment (we call a **received event**), the TCP emulation module looks up the state transition for the current state and the received segment. If the transition does not send out any outputs, then the transition is emulated.
If the transition sends out some outputs, the TCP emulation module looks for the next sent event in the event sequence log. If the next sent event is the correct output for the transition being traced, then the TCP emulation module reads out the sent event, and emulates the current received event and the sent event.
If the correct output is not found, the TCP emulation module supposes that there are no outputs for the event. If there is a possibility with no output for the transition, then the possibility is emulated. If there are no possibilities which do not send out any output for the transition, then the TCP emulation module considers that the current received event did not take place due to loss of the segment.

According to this algorithm, the behaviors of computer A in Fig. 3 are emulated in the following way.

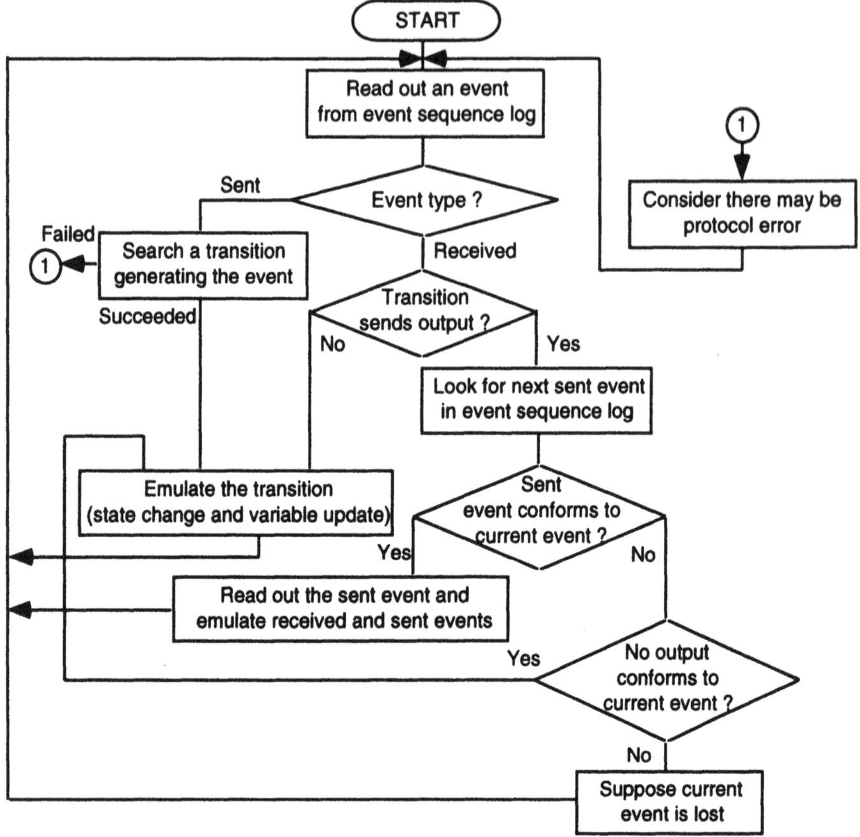

Figure 5 Algorithm for emulating state transitions.

- First, the TCP emulation module reads out an event '*SYN* sent' and selects State Transition 1 for this event. The new state is SYN_SENT.
- Next, the module reads out an event '*SYN+ACK* recv' and looks up State Transition 4. Since this transition includes some outputs, then the module looks for the next sent event in the event sequence log, and finds an event '*ACK* sent'. This output conforms to possibility 1) in State Transition 4, '*SYN+ACK* recv' and '*ACK* sent' are emulated. The new state is ESTBLISHED.

Internal Procedures of Modern TCP

Most of Modern TCP software products contain four flow control algorithms; slow start, congestion avoidance, fast retransmit, and fast recovery[1]. They are considered procedures defined internally by TCP protocol entities.

(1) Slow Start and Congestion Avoidance

In Old TCP procedures, when a connection established, the sender injects multiple segments into the network, up to the window size advertised by the receiver. This may cause a problem that some segments are lost in routers if there is a slower link between the sender and receiver. Similarly, when any TCP segments are lost in the middle of communication, it is considered that there may be congestion some where in the network between the sender and receiver.

In order to solve this problem, the slow start and congestion algorithms are introduced. They use a congestion window, called **cwnd**, which control the sending rate internally in the sender, and a slow start threshold size, called **ssthresh**.

(1) Initialization for a given connection sets cwnd to one segment and ssthresh to 65535 byte.
(2) The sender never sends more than the minimum of cwnd and the advertised window from the receiver.
(3) When congestion occurs (indicated by a timeout or the reception of duplicate ACKs), one-half of the current window size (the minimum of cwnd and the advertised window) is saved in ssthresh. Additionally, if the congestion is indicated by a timeout, cwnd is set to one segment.
(4) When new data is acknowledged by the receiver, cwnd is increased in the following way. If cwnd is less than or equal to ssthresh, TCP is in slow start and cwnd is incremented by one segment every time an ACK is received. This opens the window exponentially. If cwnd is greater than ssthresh, congestion avoidance is being performed and the growth of cwnd is linear.

(2) Fast Retransmit and Fast Recovery

This algorithm allows TCP to retransmit a segment which is considered to be lost and, after that, to invoke congestion avoidance, not slow start. This algorithm may improve the throughput under moderate congestion, especially for large windows.

(1) When the sender receives three duplicate ACKs, it considers that these ACKs indicate that a segment has been lost. It sets ssthresh to one-half of the current cwnd and retransmits the missing segment.
(2) When an ACK arrives that acknowledges new data, the sender sets cwnd to ssthresh and starts the congestion avoidance.

Emulation of Internal Procedures

In order to emulate the internal procedures of TCP, we have adopted the following method.

(1) The TCP emulation module estimates the invocation of the internal procedures by monitoring the TCP segments. For example, it estimates that

the slow start algorithm is invoked when a connection is established or when it detects a DATA segment retransmission caused by timeout (not duplicated ACKs). It also estimates the start of the fast retransmit when it detects a DATA segment retransmission invoked by three duplicated ACKs.

(2) The TCP emulation module maintains the following internal variables associated with the slow start and congestion avoidance algorithms and the fast retransmit and fast recovery algorithms:

variables associated with the slow start and congestion avoidance

CWND : the estimated congestion window,

SSTHRESH : the estimated slow start threshold, and

STATUS : indicates what algorithm is being emulated and takes NORMAL, SS (slow start), CA (congestion avoidance), FR (fast retransmit), and

variables associated with the fast retransmit and fast recovery

D_ACK : number of the received duplicated ACK.

(3) The TCP emulation module estimates the internal procedures by using these variables during the state transition emulation. The way of estimation is specified in the state transitions for *send*, *DATA*, *ACK*, and *DATA timeout* in state ESTABLISHED. The followings show examples.

State Transition 9 : *DATA timeout* in ESTABLISHED

output: *DATA* with SEQ in *DATA* != SND.NXT;

next state: ESTABLISHED;

variable update:

SSTHRESH = max (2*MSS, 1/2 * min (CWND, SND.WND);

CWND = MSS; STATUS = SS;

In this state transition, the invocation of the slow start is detected and the interval variable SSTHRESH and CWND are estimated.

State Transition 6 : *send* in ESTABLISHED

output: *DATA* with SEQ in *DATA* >= SND.NXT;

next state: ESTABLISHED;

variable update:

SND.NXT = SEQ in *DATA*+LEN of *DATA*;

RCV.NXT = ACK in *DATA*; RCV.WND = WND in *DATA*;

if (SND.NXT - SND.UNA > CWND)

the internal procedures may not be used;

That is, whether the internal procedures are used or not is checked every time a DATA segment is transmitted.

State Transition 8 : *ACK* in ESTABLISHED

1) when D_ACK == 2 and ACK in *ACK* == SND.UNA

output: *DATA* with SEQ in *DATA* == SND.UNA;

next state: ESTABLISHED;

```
variable update:
   SSTHRESH = max (2*MSS, 1/2 * min (CWND, SND.WND );
   CWND = SSTHRESH+3*MSS; D_ACK = 0; STATUS = FR;
/* This corresponds to the fast retransmit. */
2) when D_ACK < 2 and ACK in ACK== SND.UNA
next state: ESTABLISHED;
variable update:
   if ( STATUS != FR ) D_ACK = D_ACK +1;
   else CWND = CWND+MSS;
3) when SND.UNA < ACK in ACK< SND.NXT
next state: ESTABLISHED;
variable update:
   SND.UNA = ACK in ACK; SND.WND = WND in ACK; D_ACK = 0;
   if ( STATUS == FR ) CWND = SSTHRESH; STATUS = CA;
   /* This is the start of congestion avoidance following the fast retransmit.
   */
   if ( STATUS == SS )
      if ( CWND <= SSTHRESH ) CWND = CWND+MSS;
      else CWND = CWND + MSS * MSS / CWND; STATUS = CA;
   if ( STATUS == CA )
      CWND = CWND + MSS * MSS / CWND;
      if ( CWND > 65535 ) STATUS = NORMAL; CWND = 65535;
   /* These two are the slow start and congestion avoidance. */
```

This transition specifies both the slow start and congestion avoidance algorithms and the fast retransmit and fast recovery algorithms.

(4) Since computers being examined may not support such internal procedures, the TCP emulation module will stop the emulation of those procedures when the monitored event sequences do not conform to the algorithms.

5 DISCUSSIONS

(1) It is considered that our TCP/IP protocol monitor is used effectively for the detailed analysis of TCP/IP communications. Especially, it is helpful to analyze the behavior of TCP internal procedures for the flow control. For example, our monitor can estimate the numbers of invocation of the slow start and the congestion avoidance algorithms. It can also estimate the number of DATA segment retransmission. These may help the detection of the problem sources of throughput degradation.

(2) We have the following comparisons with the development our intelligent protocol monitor for OSI protocols.

- The design of the protocol emulation function becomes simpler than that of our OSI protocol monitor, for the points that our monitor focuses only on the TCP emulation and that it performs the emulation in an off-line operation.

- The design becomes more complicated for the points that our TCP/IP protocol monitor supports TCP internal procedures of the flow control. For this purpose, the TCP emulation module has introduced two state information, the state corresponding to the state transition based behavior, and internal variable STATUS which maintains what internal procedure is being emulated.

(3) Since TCP/IP is rather a mature protocol, it is considered that available protocol software does not include so many protocol errors. Our TCP/IP protocol monitor is used to analyze the details of communication, especially the details of TCP behaviors, such as the number of invocations of slow start.

(4) It is possible that our TCP/IP protocol monitor performs a wrong estimation of event sequence and a wrong TCP emulation. For example, the estimation of event does not take account of the buffering delay in routers. When such buffering delay is larger than propagation delay and transmission time, estimated event sequence by our monitor may be wrong in the actual processing order. In order to cope with such cases, it is required to reorder event sequence when any protocol error is detected during the emulation. It is possible to apply the rule based programming to implement such reordering based on heuristic algorithm[6].

6 CONCLUSIONS

In this paper, we have described the design of our TCP/IP protocol monitor which supports the detailed analysis of computer communications according to TCP/IP protocol. It provides the PDU monitoring function similar to conventional protocol monitors and, besides that, the function which can estimate what communication has taken place by emulating the behaviors of the TCP protocol entity in a pair of communicating computers. Since modern TCP includes some internal procedures for the flow control, such as the slow start algorithm, our monitor can emulate these procedures as well as the state transition based behaviors. This emulating functions are effective in analyzing TCP/IP communication, including counting how many times the slow start algorithm is invoked and how many DATA segments are retransmitted by the timeout retransmission and the fast retransmit algorithm.

7 ACKNOWLEDGMENT

The authors wish to thank Dr. H. Murakami, Director of KDD R&D Laboratories, for his continuous encouragement to this study.

8 REFERENCES

[1] Stevens, W.R. (1994) TCP/IP Illustrated, Vol. 1 : The Protocols. Addison

Wesley.
[2] Moldeklev, L. and Gunningberg, P. (1995) How a Large ATM MTU Causes Deadlock in TCP Data Transfer. *IEEE Trans. On Networking*, **3**.
[3] Tekelec (1992) Chameleon User's Manual.
[4] Ogishi, T., Idoue, A., Kato, T. and Suzuki, K. (1995) Design of Intelligent OSI Protocol Monitor. in *Proc. of the International Workshop on Protocol Test Systems*, 351-66.
[5] DARPA Internet Program Protocol Specification (1981) Transmission Control Protocol. *RFC 793*.
[6] Kato, T., Ogishi, T., Idoue, A. and Suzuki, K. (1996) Design of Protocol Interoperability Testing System based on Rule Base Programming. in *IPSJ SIG Notes*, **DPS 77-5**. (in Japanese)

9 BIOGRAPHY

Toshihiko Kato is the senior manager of High Speed Communications Lab. in KDD R&D Labs. Since joining KDD in 1983, he has been working in the field of OSI, formal specification and conformance testing, distributed processing, ATM and high speed protocols. He received the B.S., M.E. and Dr. Eng. Degree of electrical engineering from the University of Tokyo, in 1978, 1980 and 1983 respectively. Since 1993, he has been a Guest Associate Professor of Graduate School of Information Systems, in the University of Electro-Communications.

Tomohiko Ogishi is a member of High Speed Communications Lab. in KDD R&D Labs. Since joining KDD in 1992, he worked in the field of computer communication. His current research interests include the protocol testing on TCP/IP communication. He received the B.S. Degree of electrical engineering from the University of Tokyo in 1992.

Akira Idoue is a research engineer of High Speed Communications Lab. In KDD R&D Labs. Since joining KDD in 1986, he worked in the field of computer communication. His current research interests include implementation of high performance communication protocols and communication systems. He received the B.S. and M.E. Degree of electrical engineering from Kobe University, Kobe, Japan, in 1984 and 1986 respectively.

Kenji Suzuki is the senior manager of R&D Planning Group in KDD R&D Labs. Since joining KDD in 1976, he worked in the field of computer communication. He received the B.S., M.E. and Dr. Eng. Degree of electrical engineering from Waseda University, Tokyo, Japan, in 1969, 1972 and 1976 respectively. He received Maejima Award from Communications Association of Japan in 1988, Achievement Award from the Institute of Electronics, Information and Communication Engineers in 1993, and Commendation by the Minister of State for Science and Technology (Persons of scientific and technological research merit) in 1995. Since 1993, he has been a Guest Professor of Graduate School of Information Systems, in the University of Electro-Communications.

A Two-Level Approach to Automated Conformance Testing of VHDL Designs

Jean Moonen[5], *Judi Romijn*[3a], *Olaf Sies*[4], *Jan Springintveld*[2b], *Loe Feijs*[1,5], *Ron Koymans*[5].

[1] *Computing Science Department, Eindhoven University of Technology.*
E-mail: feijs@win.tue.nl
[2] *Computing Science Institute, University of Nijmegen.*
E-mail: jans@cs.kun.nl
[3] *CWI, Amsterdam, The Netherlands.*
E-mail: judi@cwi.nl
[4] *Electrical Engineering Department, Eindhoven University of Technology.*
[5] *Philips Research Laboratories Eindhoven.*
E-mail: {jmoonen,feijs,koymans}@natlab.research.philips.com

Abstract

For manufacturers of consumer electronics, conformance testing of embedded software is a vital issue. To improve performance, parts of this software are implemented in hardware, often designed in the Hardware Description Language VHDL. Conformance testing is a time consuming and error-prone process. Thus automating (parts of) this process is essential.

There are many tools for test generation and for VHDL simulation. However, most test generation tools operate on a high level of abstraction and applying the generated tests to a VHDL design is a complicated task. For each specific case one can build a layer of dedicated circuitry and/or software that performs this task. It appears that the

[a] Research carried out as part of the project "Specification, Testing and Verification of Software for Technical Applications" at the Stichting Mathematisch Centrum for Philips Research Laboratories under Contract RWC-061-PS-950006-ps.

[b] Research supported by the Netherlands Organization for Scientific Research (NWO) under contract SION 612-33-006.

Testing of Communicating Systems, M. Kim, S. Kang & K. Hong (Eds)
Published by Chapman & Hall © 1997 IFIP

ad-hoc nature of this layer forms the bottleneck of the testing process. We propose a *generic* solution for bridging this gap: a generic layer of software dedicated to interface with VHDL implementations. It consists of a number of Von Neumann-like components that can be instantiated for each specific VHDL design.

This paper reports on the construction of and some initial experiences with a concrete tool environment based on these principles.

1 INTRODUCTION

As is well-known, the software embedded in consumer electronics is becoming increasingly voluminous and complex. Accordingly, testing the software takes up an increasing part of the product development process – and hence of the costs of products. Therefore, Philips considers automating (parts of) the test process a vital issue.

More and more, manufacturers of consumer electronics do not completely develop the software themselves but import parts from other manufacturers. To guarantee well-functioning and interoperability of these parts, it is essential that they are tested for functional conformance w.r.t. internationally agreed standards. Therefore, testing efforts in this area concentrate on functional conformance testing (see (ISO 1991, Holzmann 1991, Knightson 1993) for testing terminology and methodology).

To optimise performance (in terms of speed or bandwidth), the lower layers of protocol stacks are often implemented directly in hardware. Testing these layers would imply hardware testing. However, Philips is interested in detecting design errors *before* implementation in silicon, which would mean testing hardware *designs* rather than their implementations.

Nowadays, hardware is designed using internationally standardised Hardware Description Languages. Testing a design then is testing a program in the description language at hand. Among the Hardware Description Languages, VHDL (IEEE 1993) is prominent.

There are many tools for test generation on the one hand and VHDL simulation, analysis and synthesis on the other hand. Moreover a lot of effort is put into extending and refining these tools. Ideally, therefore, the testing process could be automated by generating tests with a test generation tool, and then executing these tests using a simulation tool. However, most test generation tools expect behaviour to be modelled in clean-cut events with a high level of abstraction. Applying such tests to a VHDL design whose interface behaviour consists of complex patterns of signals on ports is by no means a trivial task.

Now, it is always possible to solve this problem by adding a layer of dedicated circuitry and/or software to bridge the gap between low-level events and high-level events, but it appears that the ad-hoc nature of this dedicated circuitry and software forms the bottleneck of the testing process.

We propose a *generic* solution for bridging the gap between generating tests on the abstract level and executing tests on the simulation level. This makes it possible for each of the two different tasks (test generation and test execution) to be performed at the appropriate level within one test trajectory, with a higher degree of automation. The idea is to build a generic layer of software (written in VHDL), dedicated to interface with VHDL implementations. We call this layer the *test bench*. It consists of a number of components that fulfill various tasks: to offer inputs to interfaces of the implementation, to observe outputs at these interfaces and to supervise the test process. The components are Von Neumann-like in the sense that for each *specific* VHDL design they are loaded with sets of instructions. These sets are compiled from user-supplied mappings between high level and low level events and abstract test cases derived from the specification. In order to be maximally generic, the test bench should accept tests described in a standardised test language. In this way, any tool that complies with this test description language can be used for test generation.

Of course, this test bench will not solve all the problems involved in interpreting abstract tests. But by performing many of the routine (and repetitive) tasks, it enables the tester to concentrate on the specific properties of the interface behaviour of the protocol under test.

This paper reports on the construction of and some initial experiences with a concrete tool environment based on these principles. This prototype tool environment is called *Phact* and has been developed at Philips Research Laboratories Eindhoven, in cooperation with CWI Amsterdam and the universities of Eindhoven and Nijmegen. It consists of a test generation part and a test execution part. The intermediate language between the two parts is the standardised test description language TTCN (*Tree and Tabular Combined Notation* (ISO 1991, Part 3)). In the test execution part we find the test bench written in VHDL, with a front-end that accepts TTCN test suites.

In the current version of our tool environment, test generation is done by the *Conformance Kit* (van de Burgt *et al.* 1990, Kwast *et al.* 1991) of Dutch PTT. This tool takes as input a specification in the form of an Extended Finite State Machine (EFSM) and generates a TTCN test suite for the specification. The *Leapfrog* tool from (Cadence 1996) is used for VHDL simulation.

This paper is organised as follows. In Section 2, we globally describe the tool environment and the testing process it supports. Section 3 highlights each important step in the test process. In Section 4, we describe our experiences with the use of the environment and discuss its current limits. Finally, in Section 5, we compare our approach with other approaches for analysis of VHDL designs.

Acknowledgements We would like to thank Nicolien Drost for developing an initial version of the observer compiler, Rudi Bloks for his help in understanding the finer details of VHDL and the Leapfrog tool, and the anonymous referees for their useful comments.

2 GLOBAL DESCRIPTION OF TEST ENVIRONMENT AND TEST PROCESS

In this section, we give an overview of the tool environment and the testing process it supports. The next section treats some interesting aspects in more detail. We begin with a short digression on *functional conformance testing*.

Conformance testing aims to check that an implementation conforms to a specification. *Functional* conformance testing only considers the external (input/output) behaviour of the implementation. Often the implementation is given as a *black box* with which one can only interact by offering inputs and observing outputs.

In the theory of functional conformance testing many notions of conformance have been proposed. The differences between these notions arise from (at least) two issues. The first issue is the language in which the specification is described (and the (black box) implementation is assumed to be described). Specifications can be described, e.g., by means of automata, labelled transition systems, or by temporal logic formulas. Secondly, the differences arise from the precise relation between implementation and specification that is required. Typically the different conformance notions differ in the extent to which the external behaviour of the implementation should match the specification.

Thus conformance testing always assumes a specific notion of conformance. However, for most conformance relations, exhaustive testing is infeasible in realistically sized cases: some kind of selection on the total test space is inevitable. So it is generally not possible to fully establish that an implementation conforms to the specification; the selected tests rather aim to show that the implementation approximately conforms to the specification. Conformance then simply means: the resulting test method has detected no errors. An appropriate mixture of theoretical considerations and practical experience should then justify this approach. This holds in particular for the test process supported by our tool environment.

Following ISO methodology (ISO 1991, Knightson 1993), the conformance test process can be divided in the sequence of steps given in Figure 1.

Our prototype tool environment automates the test generation and test execution phases and to a lesser extent the test realisation phase. It expects two inputs: the VHDL code for the Implementation Under Test (henceforth called IUT) and the (abstract, formal) functional specification, in the form of a deterministic Extended Finite State Machine (EFSM). From the EFSM specification abstract test cases are derived. These test cases are translated to the VHDL level and executed on the IUT. The history of the test execution is written to a log file and the analysis phase just consists of inspecting this file and the verdicts it contains.

Note that the EFSM is required to be deterministic. We believe that the restriction to deterministic machines is not a real restriction since we are mostly interested in testing a single deterministic VHDL implementation.

The tool environment consists of two parts, taking care of test generation and test execution, respectively. Each one contains an already existing tool. Test generation is done by the *Conformance Kit*, developed by Dutch PTT Research (van de Burgt *et*

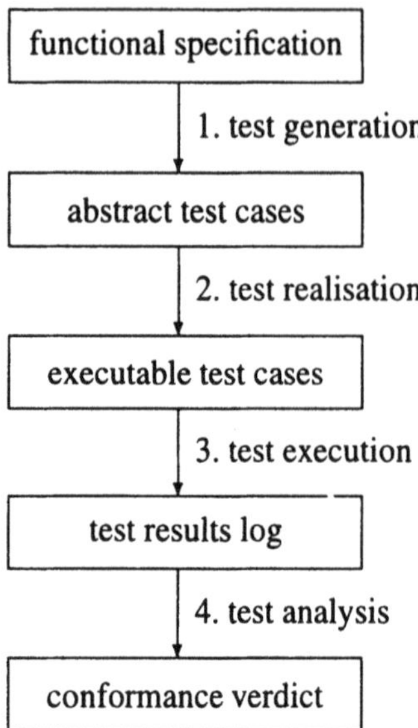

Figure 1 Global conformance testing process

al. 1990, Kwast *et al.* 1991). When given an EFSM as input, this tool returns a test suite for this EFSM in TTCN notation. The user can to a certain extent determine the parts of the EFSM that are tested and the particular test generation method used. We elaborate on this in Section 3.1.

The test cases in the test suite are applied to the IUT by a *test bench*, which is, like the IUT, written in VHDL. The *Leapfrog* tool from (Cadence 1996) simulates the application of the test suite to the IUT using the test bench. Thus testing an IUT here means: simulating it together with the test bench.

The test bench, which is described in more detail in Section 3.3 and in (Sies 1996), consists of several components connected by a *bus*: *stimulators*, *observers*, and a *supervisor*. Stimulators apply input vectors to the IUT. Observers observe the output of the IUT and feed this information back to the supervisor. The stimulators and observers are diligent but ignorant slaves to the supervisor, which operates on the basis of the test suite and feedback from the observers. The test bench has been designed generically and only needs to be instantiated for each particular IUT.

Compilers connect the test generation part, the output of which is in TTCN notation, to the test execution part, the input of which must be readable for VHDL programs. There are three compilers, one for each type of component of the test bench. The compiler for the supervisor translates the TTCN test suite to an executable format. The compilers for the stimulators and observers map abstract events from the EFSM to patterns of bit vectors at the VHDL level. They require user-supplied translations (comparable to PIXITs in ISO terminology). Section 3.2 discusses this in more detail.

Given an IUT written in VHDL and a specification or standard to test against, the

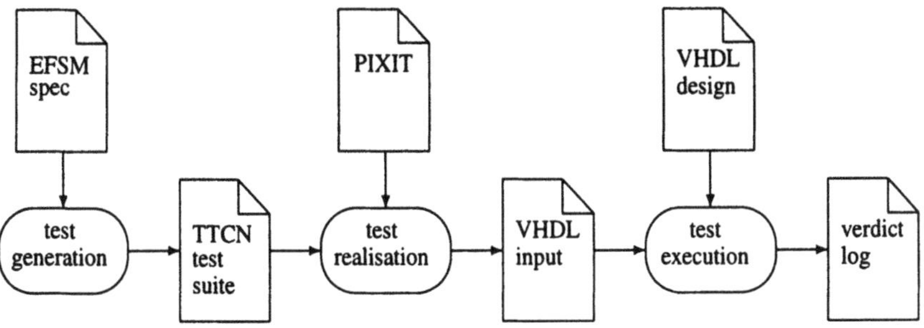

Figure 2 Overview of the test trajectory using *Phact*

global test set-up from Figure 1 leads in our setting to the following sequence of steps, also depicted in Figure 2:

0. (Manual) Write an abstract specification EFSM of the IUT.
1. (Automatic) Use the Conformance Kit to derive a test suite for this EFSM, specifying which parts of the EFSM must be tested and what test generation method must be used.
2. (a) (Automatic) Compile the test suite to the executable format for the supervisor.
   (b) (Manual) Define translations between abstract events and patterns of bit vectors (in Figure 2 called PIXITs).
   (c) (Automatic) Compile the translations to input files for the stimulator and observer, respectively.
   (d) (Manual) Instantiate the test bench as appropriate for the IUT. That is: enter the number of stimulator/observer pairs, the precise name and location of the compiled translation files, etc.
3. (Automatic) Run the Leapfrog tool on the instantiated test bench together with the IUT.
4. (Manual) Inspect the resulting conformance log file.

We end this section by remarking that the Leapfrog tool also allows the use of the Hardware Description Language *Verilog* (IEEE 1995a). In particular, the Leapfrog can simulate combinations of VHDL and Verilog programs, which makes it possible to plug a Verilog program as IUT into the VHDL test bench.

3 STEPWISE THROUGH THE TESTING PROCESS

The following sections explain the consecutive steps in the testing process more thoroughly.

3.1 Generating tests with the Conformance Kit

The Conformance Kit consists of a collection of tools for test generation.

The Extended Finite State Machine model supported by the Kit is a slight extension of the traditional Mealy-style FSM model. Transitions are labelled with input/output pairs, where input and output are treated as simultaneous events (inputs without outputs are allowed). In addition to states and transitions, an EFSM may contain a finite set of variables that range over the booleans or over finite, convex subsets of the integers. Transitions may modify the values of the variables and may be guarded by simple formulas over the variables. There is also the option to mark transitions. For instance, it often happens that certain transitions are added to the EFSM only to make it complete. These transitions are artificial and should not be tested. This is achieved by marking them with a certain marker and excluding all transitions marked thus from the test generation. Finally, it is possible to specify Points of Control and Observation (PCOs) where inputs and outputs occur. They correspond to interfaces of the IUT.

To allow for test generation, the EFSM should be deterministic. Given a deterministic EFSM, one of the tools in the tool set builds a deterministic, trace-equivalent, and minimal FSM (i.e., the FSM exhibits the same external behaviour as the EFSM and contains no pair of distinct but trace-equivalent states). Test generation tools proper take this FSM as input and return a TTCN test suite.

We highlight two of the test generation methods (for more information on test generation methods in general we refer to (Fujiwara *et al.* 1991, Holzmann 1991)).

The *Transition Tour* method. This method yields a finite test sequence (i.e., a sequence of input/output pairs) that performs every transition of the FSM at least once. Thus it checks whether there are no input/output errors.

The *Partition Tour* method. In addition to the previous method this method also checks for each transition whether the target state is correct. It is similar to the UIO-method (Sabnani & Dahbura 1988, Aho *et al.* 1991) which in its turn is a variant of the classical W-method (Chow 1978). Unlike the Transition Tour method, this method yields a number of finite test sequences, one for each transition of the FSM. Each one is a concatenation of the following kinds of sequences:

- A *synchronising sequence*, that transfers the FSM to its (unique) start state. Theoretically, such a sequence need not always exist. In practice however, most machines have a reset option and hence a synchronising sequence.
- A *transferring sequence*, that transfers the FSM from the start state to the initial state of the transition to be tested.
- The input/output pair of the transition.
- A *Unique Input/Output sequence* (UIO) which verifies that the target state is correct (that is, all other states will show different output behaviour when given the input sequence corresponding to the UIO). If this sequence does not exist it is omitted.

Although theoretically the fault-coverage of this method is not total, not even when one correctly estimates the number of states of the implementation (Chan *et al.* 1989), the counter-examples are academic and we expect that the fault coverage in practice is quite satisfactory.

3.2 From abstract tests to executable tests

In the EFSM specification the input and output events of the IUT are described at a very abstract level. For instance, a complicated pattern of input vectors, taking several clock cycles, may have been abbreviated to a single event `Input_Datum_1`. The abstraction is needed to get a manageable set of meaningful tests. But when one wants to use the TTCN test suite derived from the EFSM to execute tests on the IUT, one has to go back from the abstract level of the EFSM to the concrete level of the VHDL implementation. This translation must be such that the VHDL test bench knows for each abstract event exactly what input should be fed to the IUT or what output from the IUT should be observed. For stimulators, the abstract input events have to be translated to patterns of input bit vectors. For the observers we have to write parser-code to recognise a pattern of output bit vectors as constituting a single abstract output event.

These user-supplied translations may be quite involved and hence sensitive to subtle errors. We expect that in the approach outlined in this paper, this is the part that consumes most of the user's effort.

The translation is constructed in four steps:

1. All abstract events used in the EFSM are grouped per PCO in input and output event groups.
2. All ports of the IUT are grouped into the input or output port group of one interface. Each interface should be associated with exactly one PCO.
3. Each event of an input (output) event group at one PCO is translated to sequences of values of the ports in the input (output) port group at the associated IUT interface. This is done for each interface.
4. All event translations are fed to the compilers that generate code which is understood by the test bench during simulation.

We will give a very simple example of a user-supplied translation that is input for the observer compiler.

The IUT for which the example file is intended is a protocol that transfers data from a Sender to a Receiver and, when successful, sends an acknowledgement back to the Sender. For synchronisation purposes, the acknowledgement is an alternating bit. The IUT has two interfaces (PCOs): *Sender* and *Receiver*. We consider the observer at the *Sender* interface, which should observe acknowledgement events. This situation is depicted in Figure 3.

The *Sender* interface has two output ports (which are connected to the input ports

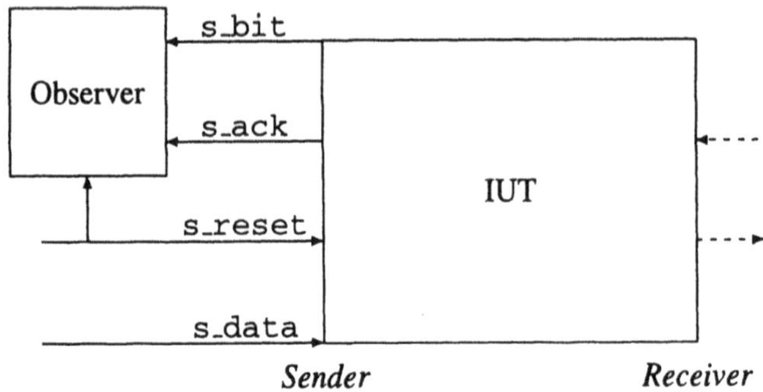

Figure 3 An example IUT

of the observer): `s_bit`, through which the alternating bit is delivered, and `s_ack`, through which arrival and presence of an acknowledgement is indicated. Furthermore, the interface has two input ports: `s_data`, a 4 bit wide port through which the *Sender* communicates data to the IUT, and `s_reset`, which has the value 1 whenever the *Sender* resets the IUT.

An acknowledgement event consists of an announcement that an acknowledgement is coming, followed by the acknowledgement itself. The announcement is indicated by the signal at `s_ack` having the value 1; the value at the `s_bit` port is not yet relevant. Subsequently, the acknowledgement is delivered: port `s_ack` still carries 1, and port `s_bit` has the value 0 or 1 for the alternating bit.

Now we have all information needed to construct the translation that is input for the observer compiler. The translation code is given in Figure 4. Note that the lines preceded with // are comments.

First, the translation contains two so-called *qualifiers*, conditions that determine when the parsing of the output of the IUT at this interface should be started or aborted. Parsing should start when an acknowledgement is coming, so the start qualifier uses the value of the `s_ack` port. Parsing should be aborted whenever the IUT is reset, so the abort qualifier uses the value of the `s_reset` port.

Next, the event translation proper is given. Bit masks are defined to recognise individual output bit vectors. In this case the vectors represent two one-bit ports with `s_bit` at the first position and `s_ack` at the second. So mask `ack_coming` has 1 for `s_ack`, and x for `s_bit`, indicating that both 11 and 01 match here. Mask `ack_0` only matches when `s_bit` is 0 and `s_ack` is 1. Output events are defined as regular expressions over the (names for the) bit masks. Here, the arrival of an acknowledgement is recognised by consecutive matching of the two relevant bit masks. This two-phase definition of events reflects the way the observer parses the output from the IUT during execution.

```
// Observer bit patterns for the PCO at the Sender side

// Observed ports, with number of bits:
// s_bit(1) s_ack(1)

PCO Sender

QUALIFIERS

// Start parsing output when this qualifier is true
  [(:s_ack = '1')]

// Abort parsing when this qualifier is true
  [(:s_reset = '1')]

MASKS

  ack_coming = 'x1'
  ack_0      = '01'
  ack_1      = '11'

EVENTS

  ACK_OUT_0 = ack_coming ack_0;
  ACK_OUT_1 = ack_coming ack_1;
```

Figure 4 Example user-supplied translation for observer

3.3 Executing tests at the VHDL level

In order to test the VHDL implementation with the generated tests, we need to execute the VHDL implementation. Executing VHDL code means hardware simulation, for which we use the Cadence Leapfrog tool.

When simulating a VHDL program which models a reactive system, the program should be surrounded by an environment which behaves – from the program's point of view – exactly like the environment in which the program eventually must operate. This environment should also be able to observe whether the program is operating correctly, and to hand out verdicts reflecting these observations. Finally, since the execution is done by VHDL simulation, the environment itself should be programmed in VHDL too.

Creating the proper environment in VHDL is hard work. However, many tasks remain the same when testing different IUTs. We have therefore created a *generic* VHDL environment, which can easily be instantiated to suit any IUT. The environment we created to perform these tasks is referred to as the *test bench*.

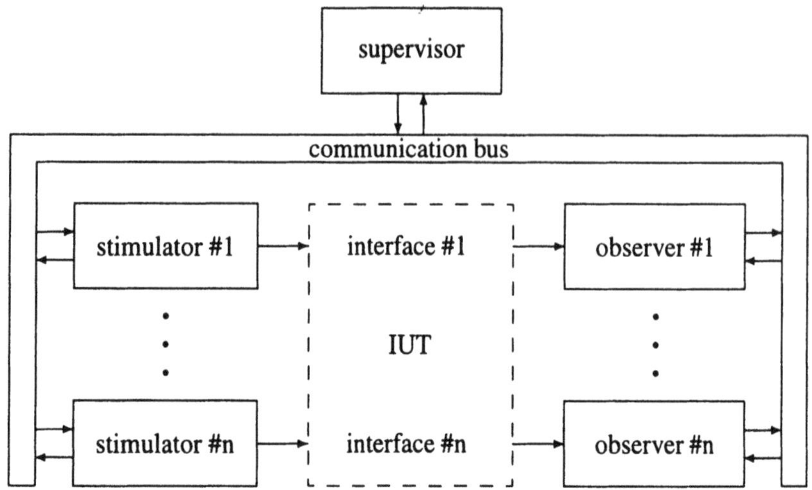

Figure 5 Structure of the VHDL test bench

The test bench consists of three kinds of components: a supervisor, some stimulators and some observers. The components communicate with each other by means of a bus. Figure 5 shows the structure of the test bench.

Each component type is dedicated to perform its particular task for any IUT. To achieve this, each component type has its own instruction set. When plugging an IUT into the test bench, each component is loaded with a sequence of instructions which are specific to the IUT in question. Thus the components can be viewed as small Von Neumann machines.

In the following paragraphs we explain the task of each component type in detail. Thereafter, we describe how the generic test bench is instantiated for testing a certain IUT.

The *supervisor* component has control over the whole test bench. It takes the generated TTCN test suite as input, works its way through each test case and outputs a log file with the verdict and some simulation history. While traversing a test case, it steers the stimulator and observer components and uses a number of timers. Each test case is executed in the following way.

When the current TTCN test case states that input should be provided to the IUT, the supervisor notifies the stimulator at the designated interface. After the stimulator indicates that it has completed this task, the supervisor goes on with the remainder of the test case.

When the TTCN test case states that output should be generated by the IUT, the supervisor checks with the observer at the designated interface to see if this output has been observed. If the output has been observed, the supervisor goes on with the remainder of the test case. If nothing was observed, the supervisor will wait for the observer's notification of new output from the IUT. If output other than the desired output is observed, the TTCN code indicates what action should be taken.

The TTCN generated by the Conformance Kit typically hands out the verdict *fail* in such a situation.

When the TTCN test case states that a verdict should be handed out, the supervisor logs this verdict to the output file, and quits the current test case.

The other TTCN commands handled by the supervisor are timer commands. TTCN offers the possibility to use timers for testing timing aspects of the behaviour of a system. These timers may be started, stopped and checked for a time-out. At the start of the TTCN test suite, all timers with their respective duration are declared. The supervisor handles these timer instructions in the obvious way. It can instantiate any number of timers with different durations and use them in the prescribed way.

The TTCN produced by the Conformance Kit, however, employs the timer construction in only two ways. It uses one timer for the maximum time a test case should take. This ensures that the test bench will not get stuck in the simulation. A second timer is used to test transitions from the EFSM that have an input event but no output. Since no output event is specified, the IUT should not generate one. This is tested by letting a timer run for some time, during which the IUT should not generate output. Any output observed before the timer expires is considered erroneous and leads to the verdict *fail*. The precise value to which the no-output timer should be set is gleaned from the specification.

The *stimulator* component provides input to the IUT. It waits until the supervisor commands it to start providing a certain abstract event, then drives the input ports of the IUT with the appropriate signals. It has access to the user-defined translation of abstract input events to VHDL input signals.

The *observer* component observes output from the IUT and notifies the supervisor of the abstract events it has observed. Like the stimulator component, it has access to the user-defined translation of VHDL output signals to abstract output events.

Observing the ports of a VHDL component and recognising certain predescribed events is no trivial task. The observer must parse the output of the IUT such that the patterns provided by the user are recognised. Parsing is done with the help of a parser automaton, constructed with the UNIX tool Lex (and the user-defined translation). The observer uses this automaton to decide which event matches the current output. When the IUT outputs a sequence of values that does not fit into any of the patterns, the supervisor is notified of an error using a special error event.

The supervisor and stimulators communicate directly in a synchronous way – the supervisor always waits for the stimulators to end their activity before resuming its own task – while the supervisor and observers communicate in an asynchronous way via FIFO queues.

In order to plug an arbitrary VHDL implementation into the test bench as the current IUT, some *instantiating* has to take place. The test bench must have as many instantiations of the observer and the stimulator component as the IUT has interfaces. These instantiations must each be connected to the proper interface of the IUT. The IUT may need some external clock inputs, these have to be provided with the correct speed. The supervisor must have the desired number of timers at its disposal, as specified in the TTCN test suite. Each observer (stimulator) must be given access to the compiled version of the user-defined translation. Likewise, the supervisor must be given access to the compiled version of the TTCN test suite.

When these instantiating actions have been performed, the test bench is ready for simulation.

4 EXPERIENCES

We experimented with our tool environment by running it on a small protocol example. The protocol was derived from the Alternating Bit Protocol (Bartlett *et al.* 1969), with some modifications to test crucial features of the test bench. The features tested mostly concerned the synchronising mechanisms in the test bench.

During the test runs, the VHDL implementation we constructed for the example protocol proved not to conform to its abstract specification. Among other things, the toggling of the alternating bit was not implemented correctly. Already in this small protocol, multiple errors were detected that were subtle enough to escape a manual inspection of the VHDL code.

After conformance was shown for the corrected implementation, we modified the abstract specification EFSM to have discrepancies the other way around. All of these were detected.

Following this small protocol, we considered a fair-sized, more complex and industrially relevant design. For this we selected a part of the 1394 Serial Bus Protocol, which has been standardised in (IEEE 1995b). The 1394 protocol implements a high speed, low cost bus that can handle communication between video and audio equipment, computers, etc. It supports multi-media applications, allows for "plug-and-play", and provides data transfer rates ranging from 100 Mbit/s to 400 Mbit/s.

The experiments have not yet been carried to completion but we can already report some of our findings. We started off with a natural and abstract specification EFSM suggested by the standard document. However, when constructing the translation from abstract events to low-level events, we found that the interface behaviour of the implementation had a very high degree of interleaving of input and output events at different interfaces. In fact, the low-level representation of one abstract event often turned out to be a complete protocol in itself, involving low-level synchronization schemas and corresponding handshake mechanisms. To enable the test bench to deal with this behaviour, these protocols should be encoded into the stimulator and observer components. Given the simple, generic set-up of the stimulator and observer components, this appeared to be virtually impossible. This problem was worsened by the fact that the documentation of the protocol and the PIXIT information both lacked the degree of precision required to construct the translation.

It remains to be investigated whether the problems encountered with the complicated interface behaviour are specific to the 1394 protocol or occur more frequently and require a refinement or extension of the test bench.

The remainder of this section is devoted to the limits of the test generation method currently supported.

The EFSM specification format imposes certain restrictions. It has difficulties in modelling, e.g., output events without an input, events occurring simultaneously at multiple interfaces, data parameters of events, and timers. Solutions here require more research in the theory of testing.

Regarding the Conformance Kit itself, it would be convenient if the test generation process could be steered more directly by the user. For instance, one may want to transfer the implementation to a certain interesting state, and perform certain experiments in that state, whereas the Kit moves in a completely autonomous way through the state space.

5 RELATED WORK

Our tool environment has a modular structure and integrates two well-known techniques: one for automatic generation of TTCN test suites based on finite state machines and the other for the simulation of VHDL hardware designs.

A number of papers that employ similar techniques for analysing VHDL designs have appeared. Only (Geist *et al.* 1996) seems to follow a similar approach to conformance testing. When keeping the phased trajectory from Figure 1 in mind, the focus in (Geist *et al.* 1996) is on the test generation phase, the other phases are not described in detail. The method used for test generation is quite different from the classical graph-algorithmic approach such as applied by the Conformance Kit. Model checking techniques are used to derive the tests automatically from an FSM model of either the implementation or the specification. To test a certain transition, a model checking tool is fed with the FSM and a query asserting the non-existence of this transition. The tool derives a counterexample containing the path to the transition. This path is then used as a test sequence. More general temporal formulas can be used to direct the counterexample to check certain situations. Selection of interesting transitions is based on a ranking of state variables, as opposed to the transition marking supported by the Kit (see Section 3.1). Although coverage is obtained w.r.t. the 'interesting' state variables, there is no measure for coverage w.r.t. exhaustive testing. It seems that theoretic support for dealing with the state explosion problem is as much an issue for this approach, as it is for ours.

In (Ho *et al.* 1995) a tool is described for exhaustive state exploration and simulation of VHDL designs. The VHDL design is transformed into an FSM for which a transition tour is generated (see Section 3.1). This tour induces a finite set of finite sequences of bit vectors which together exercise every transition of the VHDL design. As this tool only concerns simulation, there is no notion of conformance w.r.t. a specification, or a mechanism for automatic error detection.

In (Walsh & Hoffman 1996) a tool environment is described for the automatic execution of test scripts on VHDL components. There is no support for the automation of test script generation itself.

Finally, there exist many tools for the *verification* of VHDL designs (e.g., Beer *et al.* 1996, Bickford *et al.* 1996, Borrione *et al.* 1996). Each of them maps VHDL code to some semantical domain, on which the verification algorithms operate. It may be worthwhile to see whether our approach can benefit from techniques used in these tools.

REFERENCES

A.V. Aho, A.T. Dahbura, D. Lee, and M.U. Üyar. An optimization technique for protocol conformance test generation based on UIO sequences and Rural Chinese Postman Tours. In: IEEE *Transactions on Communications*, Vol. 39, no. 11, pp. 1604–1615, 1991. Also appeared in: Proceedings of *Protocol Specification, Testing and Verification* VIII, pp. 75–86, North-Holland, 1988.

K.A. Bartlett, R.A. Scantlebury, and P.T. Wilkinson. A note on reliable full–duplex transmission over half–duplex links. *Communications of the ACM*, 12:260–261, 1969.

I. Beer, Sh. Ben–David, C. Eisner, and A. Landver. RuleBase: an industry-oriented formal verification tool. In: Proceedings of the 33rd ACM *Design Automation Conference*, Las Vegas, NV, USA, 1996.

M. Bickford and D. Jamsek. Formal specification and verification of VHDL. In M. Srivas and A. Camilleri, editors, *FMCAD'96,* Palo Alto, CA, USA, November 1996, volume 1166 of *Lecture Notes in Computer Science*, pages 310–326. Springer-Verlag, 1996.

D. Borrione, H. Bouamama, D. Deharbe, C. Le Faou, and A. Wahba. HDL-based integration of formal methods and CAD tools in the PREVAIL environment. In M. Srivas and A. Camilleri, editors, *FMCAD'96,* Palo Alto, CA, USA, November 1996, volume 1166 of *Lecture Notes in Computer Science*, pages 143–158. Springer-Verlag, 1996.

S. van de Burgt, J. Kroon, E. Kwast, and H. Wilts. The RNL Conformance Kit. In: Proceedings 2nd *International Workshop on Protocol Test Systems*, pp. 279–94, North-Holland, 1990.

Cadence. Leapfrog VHDL Simulator. Product information at `http://www.cadence.com/software/qx-leap.html`.

W.Y.L. Chan, S.T. Vuong, and M.R. Ito. An improved protocol test generation procedure based on UIOs. In: Proceedings of *SIGCOMM '89*, pp. 283–294, ACM, 1989.

T.S. Chow. Testing software design modeled by finite-state machines. In: IEEE *Transactions on Software Engineering*, Vol. 4, no. 3, pp. 178–188, 1978.

D. Geist, M. Farkas, A. Landver, Y. Lichtenstein, S. Ur, and Y. Wolfsthal. Coverage-directed test generation using symbolic techniques. In M. Srivas and A. Camilleri, editors, *FMCAD'96,* Palo Alto, CA, USA, November 1996, volume 1166 of *Lecture Notes in Computer Science*, pages 143–158. Springer-Verlag, 1996.

S. Fujiwara, G. von Bochmann, F. Khendek, M. Amalou, and A. Ghedamsi. Test selection based on finite state models. In: IEEE *Transactions on Software Engineering*, Vol. 17, no. 6, pp. 591–603, June 1991.

R. Ho, C.H. Yang, M.A. Horowitz, and D. Dill. Architecture validation for processors. In: Proceedings of the *International Symposium on Computer Architecture*, Santa Margerita Ligure, Italy, June 1995.

G.J. Holzmann. *Design and validation of computer protocols*. Prentice Hall, Englewood Cliffs, 1991.

IEEE. Standard description language based on the Verilog(TM) hardware description language. International standard 1364, 1995.

IEEE. Standard for a high performance serial bus. International standard 1394, 1995. For more information see `http://www.1394ta.org`.

IEEE. Standard VHDL language reference manual (ANSI). International standard 1076, 1993.

ISO. Information technology, open systems interconnection, conformance testing methodology and framework. International standard IS–9646, 1991.

K.G. Knightson. *OSI Protocol Conformance Testing: IS 9646 Explained*. McGraw-Hill, 1993.

E. Kwast, H. Wilts, H. Kloosterman, and J. Kroon. User manual of the Conformance Kit. Dutch PTT Research, Leidschendam, October 1991.

K.K. Sabnani and A.T. Dahbura. A protocol testing procedure. In: *Computer Networks and ISDN Systems*, Vol. 15, no. 4, pp. 285-297, 1988.

O. Sies. Automatic techniques for protocol conformance testing. Master's Thesis, Department of Electrical Engineering, Technological University Eindhoven, March 1996.

P. Walsh and D. Hoffman. Automated Behavioral Testing of VHDL Components. In: 1996 *Canadian Conference on Electrical and Computer Engineering*, Calgary, Alberta, Canada, May 1996.

INDEX OF CONTRIBUTORS

Aboulhamid, E. 75
Adilinis, E. 57
Alexandridis, A.A. 57
Algaba, E. 318, 365
Alonistioti, A. 57

Bi, J. 324, 381
Bockmann, G. 167
Boullier, L. 330
Bourhfir, C. 75
Brinksma, E. 143

Cavalli, A.R. 288
Chanson, S.T. 220
Charles, O. 109
Cousin, P. 311
Curgus, J.A. 200

Dai, Z.R. 397
Dangakis, K. 37
Desécures, E. 330
Dssouli, R. 75

Feijs, L. 432
Fukada, A. 239

Gaitanis, N. 57
Grabowski, J. 37, 397
Groz, R. 109

Heerink, L. 143
Heiner, M. 125
Henniger, O. 255
Higashino, T. 239
Hogrefe, D. 397
Hu, S. 65
Hwang, I. 91

Idoue, A. 416

Jang, M. 91

Kaji, T. 239
Kang, D. 183
Kato, T. 416
Katrivanos, P. 57
Kim, M. 91, 183
Kim, T. 91
König, H. 125
Kostarakis, P. 57
Koymans, R. 432

Lee, D. 339
Lee, J. 91
Lima Jr., L.P. 288

Monedero, M. 318, 365
Moonen, J. 432

Mori, M. 239

Ogishi, T. 416
Oh, H. 91

Paschalis, A. 57
Péquignot, B. 330
Pérez, E. 318, 365
Petrenko, A. 167, 272

Rayner, D. 3
Rennoch, A. 21
Rico, N. 75
Romijn, J. 432

Schieferdecker, I. 21
Scheurer, R. 397
Sies, O. 432
Springintveld, J. 432
Stepien, B. 21
Su, D. 339
Suzuki, K. 416

Tan, Q.M. 143
Taniguchi, K. 239
Tretmans, J. 143

Ulrich, A. 125

Valcárcel, O. 318
Vlahakis, A. 57
Vuong, S.T. 200

Walter, T. 37
Wang, N. 65
Wattelet, J.-C. 65
Wu, J. 324, 381

Xiroutsikos, S. 57

Yevtushenko, N. 272
Yoo, S. 183

Zhang, S. 220
Zhu, J. 200

KEYWORD INDEX

Abstract test method 381
ASN.1 330
ATS 65, 318
Automata composition 288

B-ISDN 397
Bottleneck 220

CATG 311
Checking sequence 340
Communicating FSMs 272
Communication protocols 167
Compiler 311
Complete test set 339
Component testing 288
Conformance 3
  testing 37, 57, 75, 91, 109, 167, 272, 324, 330, 339
Control flow testing 75
Cost-effective testing 3
Coverage 109, 365
Covering path 339
Cycle analysis 75

Data flow testing 75
Data part testing 91
DCS 65
DECT 57
Directed graph 339
Distributed architectures 288

EFSM 75
Embedded
  systems 288
  testing 272
ETS 318
ETSI 65
European Commission 311
Executability 75
Extended finite state machine 339

Fault models 272
Finite state 220
  machine 339
Formal description techniques 167

Generation of test suites 330
Generic 311
GSM 65

Integration 57
Interoperability 3
  testing 416
Intool 311
ISO IS-9646 318, 365

Labeled transition systems 167

Mobile 65
Modelling 57
MSC 318, 365, 397

Network protocols 125, 339

Open architecture 311

Performance 220
  testing 21
PICS 318
PIXIT 318
Protocol
  conformance test 381
  monitor 416
  testing 3, 183, 200, 239
    system 324
  validation 397

Quality of service 21

Real-time testing 37
Relay system 381
Requirements / specifications 125

SDL 318, 330, 365, 397
Signalling test 65
Single-module 91
SMG 65
SMTP 324
Specification 220Standard 65
Symbolic evaluation 75

TCP behavior emulation 416
TCP/IP 416
Test
  cases 37
  coverage 200, 239
  derivation 272
  generation 75, 91, 167, 288, 365, 397
  hypothesis 109
  path selection 183
  procedures 57
  selection 200, 239
  sequence tree 183
  suite 21
Testing 125
Traceability 330
TTCN 21, 37, 65, 318, 324, 330, 365
    397

Validation 65
Verification 239

GPSR Compliance
The European Union's (EU) General Product Safety Regulation (GPSR) is a set of rules that requires consumer products to be safe and our obligations to ensure this.

If you have any concerns about our products, you can contact us on

ProductSafety@springernature.com

In case Publisher is established outside the EU, the EU authorized representative is:

Springer Nature Customer Service Center GmbH
Europaplatz 3
69115 Heidelberg, Germany

www.ingramcontent.com/pod-product-compliance
Ingram Content Group UK Ltd.
Pitfield, Milton Keynes, MK11 3LW, UK
UKHW012145240726
13966UKWH00001B/169

9781475767001